IN SILICO LEAD DISCOVERY

By

Maria A. Miteva
MTI, INSERM U973 – University Paris Diderot
Paris, France

CONTENT

FOREWORD

The development of quantitative mathematical models of real phenomena which are consequently evolved on computational devices has become a common practice in science and engineering in the last 50 years. The major advantage of such simulation technologies is the possibility to process large amounts of data in relatively short times and thus to accelerate the discovery rate and to improve the quality of innovation. The application of such technologies in the pharmaceutical industry holds the promise to remediate to the continuously decreasing number of new chemical entities approved by the regulatory authorities in the last decade.

The present book, edited by a leading scientist in the field of computational molecular science, Dr. Maria Miteva from INSERM, France, compiles several state-of-the art reviews of the major techniques of early stage *in silico* lead discovery as they are currently applied in life science industries. Chapter 1 entitled "Chemical libraries for virtual screening" shows to the reader that the drug-likeness of the constituents of virtual chemical libraries is a property which determines the hit rate of virtual screening experiments as well as the potential of the identified leads to reach late stages of development. In an effort to make a bridge between chemistry and biology, the editor invited contributions which review both bioinformatic and cheminformatic structure-based and ligand-based computational methodologies. Chapter 2 is highlighting the role of molecular docking in "Structure-based virtual screening", while Chapter 3 describes "3D similarity search for lead compound identification" as the principal technique applied in ligand-based approaches to *in silico* lead discovery. Chapter 4 brings an account of how the emerging "Fragment-based methods for lead discovery" improved significantly the performance of ligand-based approaches. Further on Chapter 5 and Chapter 6 review two major challenges of structure-based approaches. The first consists in "Analyzing and predicting of protein binding pockets" as these remain unknown for a number of newly solved protein structures, while the second is related to the "Receptor flexibility in ligand docking and virtual screening" caused by atomic thermal fluctuations. Chapter 7 and Chapter 8 describe, respectively, the specific challenges of "Protein-Protein Interaction Inhibition (2P2I)" and of "Application of *in silico* methods to study ABC transporters involved in multidrug resistance" and in the uptake and excretion of drugs.

I would like to finally cite Chapter 9 dedicated to the "Successful applications of *in silico* approaches for lead/drug discovery". The Chapter documents the impact of computational molecular science technologies on the research and development workflow, with real illustrative examples taken from the pharmaceutical industry. Chapter 10 reminds to the reader how computerized molecular graphics and visualization techniques brought to computational molecular science an unprecedented impact by allowing for direct "Visualisation and efficient communication in structure-based lead discovery". These two last Chapters allow me to conclude that computational molecular science has now come of age. When deployed knowledgably the related technologies are capable to generate visual and intelligent scientific hypotheses about the molecular mechanisms of action of drug molecules. These models were proven to accelerate the rate of discovery, to lower attrition, to shorten the time-to-market cycle and ultimately to reduce the costs of research and development.

The documented successes of computational molecular science are however episodic and therefore in the future all our efforts should be directed towards the development of standard reproducible protocols. I believe that with the richness of the material exposed in the book, the reader will find the necessary knowledge and inspiration to bring computational molecular science to the next frontier and to transform it from an esoteric science into a reliable discovery tool at the service of every chemist and biologist.

Dr. Martin G. Grigorov

Head of Bioinformatics
Nestlé Research Center
Vers-chez-les-Blanc
CH-1000 Lausanne 26
Switzerland

PREFACE

The number of promising protein targets that could enter drug discovery programs has significantly increased because of the recent advances in human genomics, proteomics and because new technologies able to address difficult targets are being developed. Today drug discovery campaigns are still time consuming and expensive and in this respect *in silico* methods are attractive since they facilitate efficient handling of large compound collections, mining of the data and improving "hit-rates", thereby contributing to each step of the pre-clinical work. In fact, *in silico* methods, and particularly *in silico* screening, have been used for lead identification and can complement high-throughput and NMR-based screenings. Numerous successful studies employing *in silico* screening suggest that such methods can be very helpful to find novel molecules. While *in silico* approaches have been used for many years to assist hit identification, recent studies also report the discovery of several compounds that have either reached the market or entered clinical trials as an evidence of the contribution of these technologies to pharmaceutical research.

This eBook presents important aspects of *in silico* methodology for discovering new lead compounds. It contains ten individual chapters addressing different topics used in modern *in silico* lead discovery. An overview of the most widely employed methods and the latest advances are provided. The eBook outlines the recent progress made in docking-scoring and ligand-based techniques, as well as fragment-based screening. Freely available tools to assist *in silico* lead discovery are also described. Many examples of successful applications of these methods for lead/drug discovery are given. Leading experts on bioinformatics and drug design address in the eBook challenging issues, such as chemical libraries design, druggable pocket prediction, protein receptor flexibility, targeting complex systems such as protein-protein complexes or ABC transport proteins, as well mixed methodologies for acceleration of lead discovery. I hope that this eBook contains much valuable information and will help to enhance the understanding of these technologies for scientists engaged in drug discovery projects.

The successful completion of this eBook was made possible by the assistance of many people to whom I am very grateful. I express my gratitude to all individual chapter contributors and the reviewers for their suggestions. I would like to thank Dr. Bruno O. Villoutreix (Inserm, Paris, France), Dr. Emil Alexov (Clemson University, USA), Jivko Mitev (Paris, France) and Dr. Martin Grigorov (Nestle Research Center, Lausanne, Switzerland) for comments about this eBook. I also want to thank the publisher and in particular Asma Ahmed (Bentham Science Publishers Ltd.) for the support.

Maria A. Miteva
MTI, INSERM U973
University Paris Diderot
Paris, France

CONTRIBUTORS

Ruben ABAGYAN
University of California, San Diego, Skaggs School of Pharmacy and Pharmaceutical Sciences, 9500 Gilman Drive, MC 0657, La Jolla, Ca 92093-0657, USA

Emil ALEXOV
Computational Biophysics and Bioinformatics, Department of Physics, Clemson University, SC 29634, USA

Andrea BORTOLATO
Syngenta Crop Protection Research, Schaffhauserstrasse, CH- 4332 Stein, Switzerland

Frédéric GUYON
MTi, University Paris Diderot - Inserm UMR-S 973, 35 rue Helene Brion,
75013 Paris, France

Alicia P. HIGUERUELO
Department of Biochemistry, University of Cambridge, 80 Tennist Court Road, Cambridge CB2 1GA, UK

David LAGORCE
MTi, University Paris Diderot - Inserm UMR-S 973, 35 rue Helene Brion,
75013 Paris, France

Wen Hwa LEE
Structural Genomics Consortium, University of Oxford, Old Road Campus Research Building, Roosevelt Drive, Headington, OX3 7DQ, Oxford, UK

Jean-Didier MARECHAL
Departament de Química, Universitat Autònoma de Barcelona, 08193Bellaterra, Spain

Brian D. MARSDEN
Structural Genomics Consortium, University of Oxford, Old Road Campus Research Building, Roosevelt Drive, Headington, OX3 7DQ, Oxford, UK

Maria A. MITEVA
MTi, University Paris Diderot - Inserm UMR-S 973, 35 rue Helene Brion,
75013 Paris, France

Rooplekha C. MITRA
Computational Biophysics and Bioinformatics, Department of Physics, Clemson University, SC 29634, USA

Xavier MORELLI
CNRS & Aix-Marseille Universités, Institut de Microbiologie de la Méditerranée, Laboratoire Interactions et Modulateurs de Réponses (FRE3083),
31 Chemin Joseph Aiguier, 13402 Marseille Cedex 20, France

Stefano MORO
Molecular Modeling Section, Dipartimento di Scienze Farmaceutiche, Università di Padova, Via Marzolo 5, 35131 Padova, Italy

Ilza PAJEVA
Centre of Biomedical Engineering, Bulgarian Academy of Sciences,
Acad. G. Bonchev Str. Bl. 105, BG-1113 Sofia, Bulgaria

David PERAHIA
Institut de Biochimie et Biophysique Moléculaire et Cellulaire, Université Paris-Sud, Bat 430, 91405 Orsay, France

Francesca PERRUCCIO
Syngenta Crop Protection Research, Schaffhauserstrasse, CH- 4332 Stein, Switzerland

Will R. PITT
UCB Celltech, branch of UCB Pharma S.A., Slough, UK; Department of Biochemistry, University of Cambridge, 80 Tennist Court Road, Cambridge CB2 1GA, UK

Charles H. ROBERT
CNRS Laboratoire de Biochimie Théorique, IBPC, 13 rue Pierre et Marie Curie, 75005 Paris, France

Philippe ROCHE
CNRS & Aix-Marseille Universités, Institut de Microbiologie de la Méditerranée, Laboratoire Interactions et Modulateurs de Réponses (FRE3083),
31 Chemin Joseph Aiguier, 13402 Marseille Cedex 20, France

Olivier SPERANDIO
MTi, University Paris Diderot - Inserm UMR-S 973, CDithem Platform,
35 rue Helene Brion, 75013 Paris, France

Nikolay P. TODOROV
Molscape Therapeutics, 1 Woodman Way, Cambridge, CB4 6DS, UK

Pierre TUFFERY
MTi, University Paris Diderot - Inserm UMR-S 973, RPBS, 35 rue Helene Brion,
75013 Paris, France

Bruno O. VILLOUTREIX
MTi, University Paris Diderot - Inserm UMR-S 973, 35 rue Helene Brion,
75013 Paris, France

Michael WIESE
Pharmaceutical Chemistry II, Pharmaceutical Institute, University of Bonn,
An der Immenburg 4, 53121 Bonn, Germany

In Silico Lead Discovery, 2011, 01-19

CHAPTER 1

Chemical Libraries for Virtual Screening

David Lagorce, Olivier Sperandio, Maria A. Miteva and Bruno O. Villoutreix*

MTI, Inserm UMR-S 973, University Paris Diderot, 35 Rue Helene Brion 75013 Paris, France; Fax: +331 57 27 83 72, E-mail: bruno.viloutreix@univ-paris-diderot.fr

Abstract: The number of new drug approvals per year has been decreasing over the past decade for numerous reasons including an increase in regulatory requirements and lack of sufficient knowledge on the patho-physiological processes being targeted. Also, many compounds fail in development because they lack efficacy and safety. One strategy, among many, to circumvent this high attrition rate is through improving the quality of the compound collections as libraries have usually grown in size with little or inadequate attention about the quality. There are many different ways to prepare a compound collection, the process can involve increasing diversity but it can also imply the creation of focused collection dedicated to a specific disease-type and/or target. In parallel, ADMET (Absorption, Distribution, Metabolism, Excretion and Toxicity) properties have to be considered and the parameters assessed tuned according to the project, stage of the project and disease type. In this chapter, *in silico* methods facilitating the creation of a generic target-independent compound collections are explored and several *in silico* ADMET prediction tools are discussed. Key concepts are described to build a compound collection appropriate for hit finding. Overall, this procedure involves several steps: database cleaning, compound filtering step using drug-likeness or lead-likeness criteria, removal of undesirables chemical structures, and structural and chemical quality control. Because for *in silico* screening studies the collection has to be in 3D, a paragraph exposes some recently developed methods for 3D structure generations, and a list of commercial and free standalone packages as well as online tools is provided.

INTRODUCTION

Drug discovery is time consuming, taking on average 10-15 years and costly, in the order of $1 billion to bring a new drug to market [1-4]. This complex process requires a multitude of technologies-skills-steps from bioinformatics and/or chemoinformatics, synthetic/combinatorial chemistry, high-throughput screening (HTS), pharmacology-toxicology to clinical testing.

The drug discovery pipeline (Fig. **1**) usually starts from the characterization of a disease, followed by target identification and validation, hit-compound identification by HTS screening and/or *in silico - in vitro* screening, hit-to-lead and lead optimizations, clinical trials up to the approved drug [5, 6]. Because it is difficult to speed-up the clinical trials, intensive efforts have been devoted to the optimization of the preclinical steps including preparation of the libraries, new assays for compounds screening and new methods for lead finding [7]. A crucial step after target identification is to find small molecule agonists or antagonists that modulate the activity of the desired target. HTS has become a dominant strategy for hit/lead identification within the pharmaceutical industry these last 15 years, while other technologies are also applied including virtual screening (*i.e.*, *in silico* screening) and many other computer-assisted and biophysical approaches.

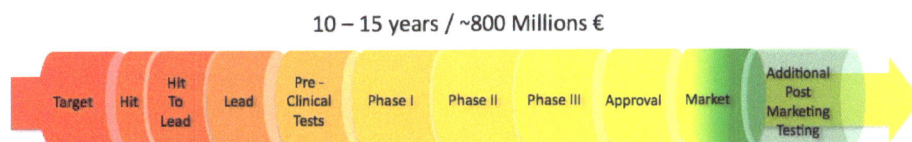

Figure 1: Drug discovery pipeline.

In spite of all these tools and skills, many compounds fail in development due to lack of efficacy and safety: from inappropriate absorption, distribution, metabolism, and excretion to toxicity issues [8]. This high attrition can be attributed to the lack of sufficient knowledge on the patho-physiological processes, etc., and, to the quality of the compound collection. Preparation of a compound collection is thus of paramount importance in drug discovery and only increasing the size of a library is not considered in itself to be an efficient strategy. It is nowadays generally accepted that smaller, high-quality libraries, diverse and/or focused are more desirable than very large collections, since in general, higher hit rates are obtained while preventing the escalating costs of screening millions of compounds.

A possible way to guide the design of a compound collection is to select molecules that are said to possess a lead-like or drug-like profile (Fig. **2**). Another option is to generate a compound collection that focuses on the type of targets. In most cases, it is also of importance to consider the disease type (*e.g.*, cancer or coagulation disorders, acute condition or chronic, route of administration… etc) and the stage of the project (*e.g.*, discovery of new hits or optimization phases). Other practical issues with compound collections are that each year, about 1% of the molecules in a library are no longer available. Further and beside scientific considerations, designing a collection also depends on the available budget.

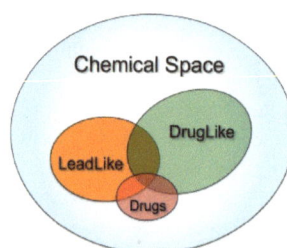

Figure 2: The chemical space conceptualized through analysis of lead-like, drug-like and marketed drugs

Preparation of a compound collection is not trivial and often research groups learn from trial and error. For example, it has been shown that drugs developed some 30 years ago do not occupy the same regions of the chemical space as compared to modern compounds. Many studies underline that the more the drug-candidates progress through Phase I, II and III stages, the more the surviving molecules show a reduction in their molecular weight, suggesting that screening/optimization tends to select relative large molecules that are then not optimal for administration in humans (*i.e.*, they are rejected before or during the clinical trials). Along the same line, the more the compounds tend to be lipophilic the more the chances to be withdrawn increase [9-12]. All these observations led to the development of rules and methods that attempt to select compound candidates that have a higher chance to advance through the drug discovery process. Many concepts and *in silico* tools have been developed over the years to assist the process of database generation, several will be commented in the following paragraphs. Further, compounds are often distributed by chemical vendors in 2D (2-dimension) while knowledge about the three-dimensional structure (3D) can be critical to several *in silico* screening approaches such as docking/scoring or for some similarity search engines. Thus, we will briefly describe some key methods that can generate 3D structures for small compounds. There are also many difficulties evaluating the structural and chemical quality of the small molecules (*e.g.*, predicting pKa values), some methods exploring these problems will be briefly commented.

SMALL MOLECULE DATABASES

Databases of small molecules are basically needed for all virtual screening computations either ligand-based or structure-based and are also essential for experimental screening. A chemical database is a library of compounds (thousands to millions, today over 10 millions molecules can be purchased [13]) in a special electronic format, such as the so-called Structure Data File format (.sdf). Scientists can find many commercial databases or can have access to numerous free chemical collections [13, 14] (Table **1**). For example, *DrugBank* [15] contains nearly 4800 drug entries including around 1350 FDA (United States Food and Drug Administration, which regulates drugs, biologics, medical devices, food, animal feed and drugs, cosmetics and radiation emitting products, in the aim to ensure that all these substances are safe according to the regulatory guideline) approved small molecule drugs, more than 3200 experimental drugs, as well as thousands of proteins or drug target sequences which are linked to these drug entries. *PubChem*, which is the largest freely available public molecular information database, contains to date more than 18 million unique chemical structures and more than 38 million substances. This is a compound repository, maintained by the National Center for Biotechnological Information (NCBI), created to facilitate information exchange and data sharing among the NIH Roadmap-funded Molecular Library Screening Center Network (MLSCN) [16] and the scientific community. *PubChem* is organized in three dynamically growing primary databases: *PubChem Compound*, *PubChem Substance* and *PubChem BioAssay*. Other types of database are for instance *DUD, Directory of Useful Decoys* [17]. This collection contains a set of active compounds against a total of 40 targets, together with molecules expected to be decoys. This collection is specifically designed for benchmarking virtual screening methods. The *WOMBAT Database* [18] is yet another example, covering over 70,000 chemical entries gathered from the scientific literature (*e.g.*, data published in the Journal of Medicinal Chemistry or in the European Journal of Medicinal Chemistry) over a period of 20 years. The database contains also *in vitro* activities of these molecules against over 600 targets.

Table 1: Examples of free and commercial chemical compounds databases and related tools.

Database	URL	Ref.
Free Databases		
ACToR	http://actor.epa.gov/actor/faces/ACToRHome.jsp	[19]
Animal Toxin Database	http://protchem.hunnu.edu.cn/toxin/index.jsp	[20]
Brenda	http://www.brenda-enzymes.info/	[21]
Biometa	http://biometa.cmbi.ru.nl/	[22]
ChEBI	http://www.ebi.ac.uk/chebi/	[23]
ChemBank	http://chembank.broad.harvard.edu/	[24]
ChemSpider	http://www.chemspider.com	[14]
Clan Tox	http://www.clantox.cs.huji.ac.il/	[25]
Ditop	http://bioinf.xmu.edu.cn/databases/ADR/index.html	[26]
DSSTox	http://www.epa.gov/ncct/dsstox/index.html	[27]
DrugBank	http://www.drugbank.ca/	[28]
DUD	http://dud.docking.org/	[29]
eMolecules	http://www.emolecules.com/	[30]
EMBL Databases	http://www.ebi.ac.uk/FTP/	[31]
GDB-13	http://www.gdb.unibe.ch	[32]
HaptenDB	http://www.imtech.res.in/raghava/haptendb/	[33]
Human Metabolome Database	http://www.hmdb.ca/	[34]
KEGG Drug	http://www.genome.jp/kegg/drug/	[35]
KEGG Ligand	http://www.genome.jp/kegg/ligand.html	[35]
Ligand Expo	http://ligand-expo.rcsb.org/	[36]
LigandInfo	http://www.ligand.info/	[37]
MDPI	http://www.mdpi.org/database.htm	[38]
Metabolic Site Predictor	http://www-ucc.ch.cam.ac.uk/msp/htdocs/	[39]
U.S. NCI Database	http://cactus.nci.nih.gov/ncidb2/download.html	[40]
PubChem	http://pubchem.ncbi.nlm.nih.gov/	[41]
Relibase	http://www.ccdc.cam.ac.uk/free_services/free_downloads/	[42]
SuperNatural Database	http://bioinformatics.charite.de/supernatural	[43]
SuperDrug	http://bioinf.charite.de/superdrug/	[44]
SuperLigands	http://bioinformatics.charite.de/superligands/	[45]
SuperToxic	http://bioinformatics.charite.de/supertoxic	[46]
TOXNET	http://toxnet.nlm.nih.gov	[47]

Table 1: cont....

VITIC	http://www.lhasalimited.org/index.php?cat=2&sub_cat=72	[48]
WOMBAT	http://dud.docking.org/wombat/	[18]
ZINC	http://zinc.docking.org/	[49]
Commercial Databases		
4SC	http://www.4sc.de	
ACB blocks	http://www.acbblocks.com	
Asinex	http://www.asinex.com	
CEREP	http://www.cerep.fr/Cerep/Users/index.asp	
ChemBridge	http://www.chembridge.com/	
Chemical Diversity	http://www.chemdiv.com/	
Comprehensive Medicinal Chemistry	http://www.symyx.com/products/databases/bioactivity/cmc/index.jsp	
ChemStar	http://www.chemstar.ru	
ComGenex	http://www.rdchemicals.com/targeted-compound-libraries/comgenex.html	
EMC	http://www.microcollections.de/	
Enamine	http://www.enamine.relc.com	
InterBioScreen	http://www.ibscreen.com	
Leadscope	http://www.leadscope.com/	
Maybridge	http://www.maybridge.com	

FILTERING AND DESIGNING THE COMPOUND COLLECTION

Several suggestions have been made to facilitate the design of a database suitable for screening experiments [50, 51] but the debate is still open. There are indeed many different ways to prepare a chemical collection for a drug discovery project. For example, one solution consists in focusing on the biological property space, *i.e.*, a specific therapeutic drug target. Alternatively, one may want to create a highly diverse library, thereby possibly increasing the likelihood of detecting some novel "hits". This strategy is commonly named diversity-based design of compounds' library [52]. Generating a focused-library or target-oriented library can be accomplished using entry-point criteria such as scaffold, pharmacophore, physicochemical properties and/or the 3D structure of the target [53]. Obviously, generating a diverse library is plagued with difficulties as the definition of diversity is elusive although it is intuitively understood as the need to remove some closely related analogs while trying to cover as much as possible different areas of the vast, almost infinite, chemical space. Also, many compounds are not really amenable to optimization because of solubility, stability and many other problems such as permeability, distribution in the organism, metabolism, toxicity or efficacy [54]. In fact, in most cases, the ideal drug molecule is a molecule that can be administrated orally, and thus has to go through all the different biological barriers and finally reach the desired target. Loosely defined windows of certain physico-chemical values are usually considered to flag a compound as suitable or not to accomplish such a task. In this context, a popular filtering technique, based on the Lipinski's rule of 5 (RO5), is routinely employed for the design of a compound collection [55]. Indeed, investigating a collection of drugs and drug candidates, Lipinski *et al.* [55], realized that only a minor percentage of these compounds had a molecular weight (MW) > 500, a lipophilicity (expressed as log P) >5), more than five hydrogen bond donors (as the sum of OH and NH groups), and more than ten hydrogen-bond acceptors (expressed as the sum of O and N atoms). From this observation the now famous Lipinski RO5 was defined and states that low permeability of a molecule can be expected if more than one of these rules is violated. These rules define drug-like properties with respect to bioavailability, *i.e.*, molecules with poor pharmacokinetic properties for oral administration. This increased awareness about compound quality paved the way to many studies and very rapidly led to the realization that it is important to not only select molecules that have the desired physico-chemical properties but also that compounds containing substructures that could be reactive or unstable should be removed or at least flagged. Several of the rules commonly used to filter collections will be introduced below.

Collection Cleaning and Filtering Using Physico-chemical Properties: Key Concepts

Based on the definition of RO5, presented above, several studies "re-visited" that rule of thumb and noted that the RO5 essentially dealt with molecules that were already almost at a final stage of optimization or were already marketed. In a situation where one would like to find hit compounds, it should be desirable to start with smaller molecules that could grow in size, in lipophilicity, etc. Indeed, a hit or a lead arising from HTS or *in silico - in vitro* screening generally requires a significant increase in potency often achieved through the creation of additional interactions and hence the increase in MW. As MW increases, the compounds often tend to become more greasy with an increased log P, and possibly contains more flexible bonds (*i.e.*, known to create permeability problems). Along this line of reasoning and through looking at smaller compounds, or in investigating fragments or evolution of molecules from the hit stage to the optimized lead, several authors proposed more stringent filtering criteria. For example, Congreve *et al.* [56] proposed the "Rule of Three", in which the molecular weight is < 300, the number of hydrogen bond donors is ≤ 3, the number of hydrogen bond acceptors is ≤ 3 and log P is ≤ 3. This study and several others [57, 58], also suggested that the number of flexible bonds should be ≤ 3 and that the polar surface area (PSA) [59] should be $\leq 60\text{Å}^2$. Using such easily computed parameters, it is possible to build a generic compound collection. Indeed, the resulting compounds with these types of filter will most likely possess medium potency against the desired target but can go through several optimization cycles and thus grow in size and lipophilicity while still being compatible with oral administration.

More specifically and with regard to *in silico* Absorption, Distribution, Excretion, Metabolism and Toxicity (ADMET) predictions, it is commonly accepted that absorption, distribution and excretion are dependent on similar descriptors while metabolism and toxicity are of a quite different nature and depend on numerous factors. Absorption, solubility, and permeability are all related properties and interdependent and are complex to simulate *in silico*. One property that is always examined with care is lipophilicity, since it is directly or indirectly implicated in many biological processes, from crossing the membrane to metabolism and anti-target interactions (*e.g.*, binding to undesired targets and possibly creating severe secondary effects). Lipophilicity is often expressed as a partition coefficient P (or distribution coefficient log P or log D) between octanol and water. The logarithm of the ratio P of concentrations of unionized compound between these two solutions is called log P, while the log D value is more general and can also be used for charged molecules and is thus pH-dependent. Poor aqueous solubility is also one of the major causes for low systemic exposure and, consequently, lack of *in vivo* activity. Thus, it is important to predict solubility, reflected by the log S value, but it should be remembered that this term is notoriously difficult to predict [60-62].

Several well-known commercial, open-source and free packages can be used to perform filtering a collection. In general, it is valuable to select user-customizable tools in order to tune the parameters for a given project and avoid black-boxes, where mathematical models or training sets are not available (see Table **2**) available in order to evaluate the applicability domain of the method. For example, several free tools have been developed to prepare chemical libraries such as FAFDrugs 2 [63] or ToxMatch [64] or ScreeningAssistant [65] (Table **2**). To help users, numerous online tools are also available such as PreADMET [66], FAFDrugs [67], Molinspiration [68] or OSIRIS [69]. The methods implemented are usually different and complement each other, as such it is recommended to use several packages prior to order or synthesize new compounds. These tools can also be used to "clean" virtual compound collections.

Table 2: Examples of *in silico* packages used to prepare compound collections.

Software	Description	URL	Ref.
Commercial packages			
ChemAxon Calculator Plugin Instant JChem	Tools' package for physico-chemical predictions Chemical databases manager	http://www.chemaxon.com/product/calc_pl_1 and.html http://www.chemaxon.com/product/ijc.html	[70]
OpenEye Filter	Filtering process based on ADMET predictions	www.eyesopen.com	[71]
Schrödinger QikProp	Filtering process based on ADMET predictions	https://www.schrodinger.com	[72]

Table 2: cont....

Discovery Studio 2.0	Full package including ADMET, PK/PD predictions.	www.accelrys.com	[73]
PreADMET	Full package including ADMET, PK/PD predictions.	http://preadmet.bmdrc.org/index.php	[66]
MOE	Filtering process based on ADMET predictions. 3D Generation.	http://www.chemcomp.com	[74]
Free Tools			
FAFDrugs 2	Customizable tool for preparing databases based on compounds physico-chemical calculations and chemical substrucures searching.	http://www.mti.univ-paris-diderot.fr/fr/downloads.html	[63]
Screening-Assistant	Chemical databases manager which highlights non drug-like (physico-chemical) predicted molecules.	http ://www.univ-orleans.fr/icoa/ screeningassistant/	[65]
Ligand Info	Designing tool based similarity searching and clustering process.	http://Ligand.info	[37]
ToxMatch	Flexible and free application useful to help data-mining for toxicity prediction through chemical similarity indices.	http://ecb.jrc.ec.europa.eu/qsar/qsar-tools/index.php?c=TOXMATCH	[64]
ToxTree	Mutagenicity and carcinogenicity predictions.	http://toxtree.sourceforge.net/	[75]
XLOGP3	Basic drug-likeness calculations (Lipinski, tPSA...).	http://www.sioc-ccbg.ac.cn/software/xlogp3/	[76]
Free Online Tools			
ChemAxon Calculator Plugin	Draw one molecule and get the molecular and drug-like properties	http://www.chemaxon.com/marvin/sketch/index.jsp	[70]
FAFDrugs	Online version of basic drug-likeness predictions of FAFDrugs2.	http://bioserv.rpbs.univ-paris-diderot.fr/ Help/FAFDrugs.html	[67]
Ligand Info	Online version of the free tool.	http://Ligand.info	[37]
Molinspiration	Draw one molecule and get the molecular and drug-like properties.	http://www.molinspiration.com/cgi-bin/properties	[68]
MolSoft	Draw one molecule and get the molecular and drug-like properties.	http://www.molsoft.com/mprop/	[77]
PreADMET	Online version of the commercial package.	http://preadmet.bmdrc.org/index.php	[66]
OSIRIS	Property explorer for toxicity and drug-likeness prediction.	http://www.organic-chemistry.org/prog/peo/	[69]

Common filtering protocols are often variations of the Lipinski's RO5, they can include a limit on the number of rotatable bonds, on polar surface area...etc, as further depicted in Fig. **3**. Several other related rules can also be considered and are described in these articles [50, 55-57, 78].

In practice, assuming that a user is downloading a compound collection from the internet with the aim of screening a new target and find initial hits, it is generally recommended to first "clean-up" the collection. This usually implies removing salts, molecules possessing unacceptable atoms, counterions and duplicates from the library, and checking the file format. Evaluation of possible errors such as wrong-entries, bad connection tables, incorrect chiral centers, etc... is also of importance. Then, simple filtering rules such as the RO5 can by applied. In addition, the compound collection must be cleared from compounds incorporating unacceptable functional groups, reactive species or moieties known to usually be problematic for developments through the clinical stage [79]. Molecules with some specific chemical substructures associated with poor chemical stability, reactivity or toxicity (*e.g.* epoxides, anilines, anhydrides...) are thus usually removed (some examples are depicted in Fig. **4**) or flagged. For instance, nitro derivatives have been reported to be hepatocarcinogens [80] and nitroaromatics can be reduced to form reactive, nitroanion radical, nitroso intermediate, and N-hydroxy derivative [81]. These reactive metabolites are usually not welcome in drug discovery projects and, as such, molecules containing nitro aromatic groups are in general removed from the library. In addition, it is of importance to remove frequents hitters [82] and chemicals aggregators (compounds which form aggregates in solution and induce a false inhibition of the target) [83]. Such compounds either perturb the assays (*e.g.*, colored or fluorescent molecules) or their activity is not specific of a target (defined as a promiscuous in this case).

- no more than one violation of the R05:
  200<MW<500,logP<5
  H-Bond Donors<5,H-Bond Acceptors<10
- no more then 50 rigid bonds
- nomore than 10 rotatable bonds
- no more than 4 system ring

- no more than 20 atoms in a system ring
- no more than 35 heavy atoms
- no more than 35 carbon atoms
- no more than 20 heteroatoms
- formal charge within 0 to 3
- sum of charge within - 2 to 2

remove compounds containing atoms other than those permitted as:
H,C,N,O,F,P,S,Cl,Br,I

Disease-Target Identification and Validation

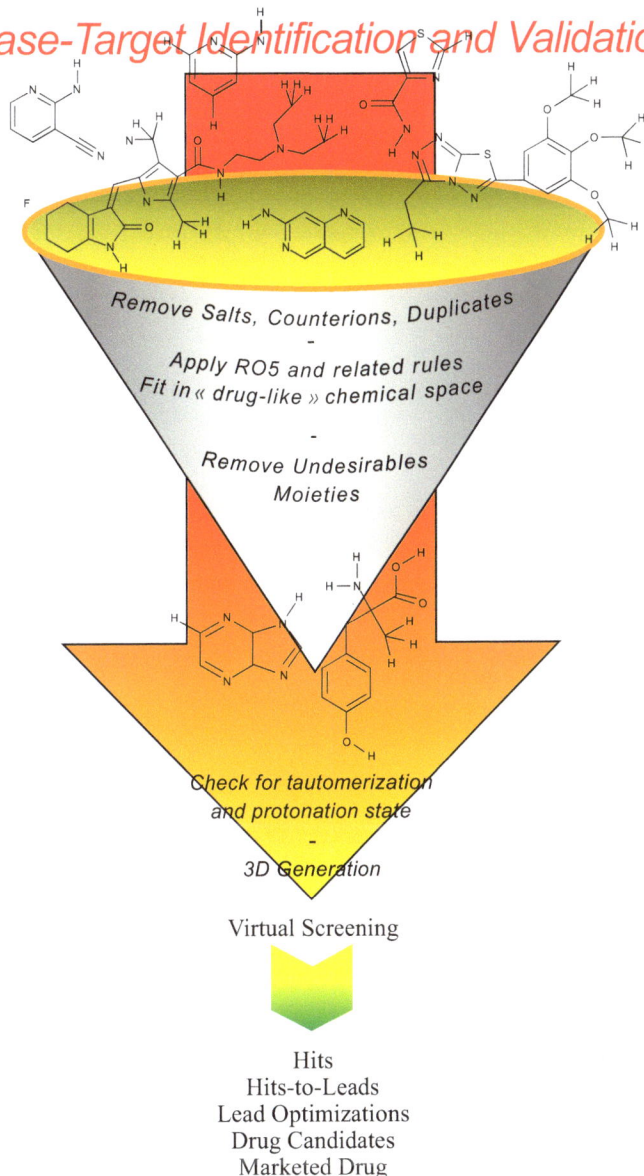

Remove Salts, Counterions, Duplicates
-
Apply RO5 and related rules
Fit in « drug-like » chemical space
-
Remove Undesirables
Moieties

Check for tautomerization
and protonation state
-
3D Generation

Virtual Screening

Hits
Hits-to-Leads
Lead Optimizations
Drug Candidates
Marketed Drug

Figure 3: Compound collection preparation: a generic strategy.

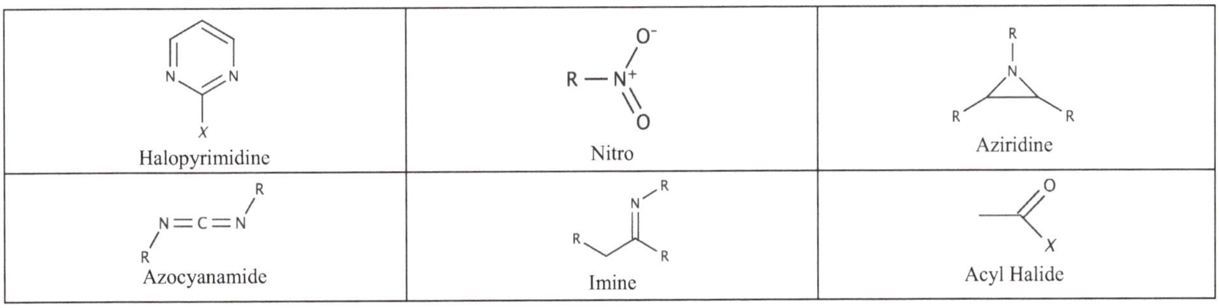

Halopyrimidine	Nitro	Aziridine
Azocyanamide	Imine	Acyl Halide

Figure 4: Examples of functional groups known to be reactive or toxic [84].

In order to illustrate the work that can be accomplished by dissecting compounds and searching for potentially reactive groups or unwanted substructure, we analyzed an in-house subset (drugs with a potential oral administration route informed on the *DrugBank* website) of *DrugBank* [15] (FDA approved small compounds) with the standalone FAFDrug2 package. Substructure matching results are reported in Fig. **5**. This information can help in selecting the appropriate parameters to generate a compound library, tolerating or not, flagging or not some substructures and as such, generating a collection suited to the project needs (*e.g.*, for cancer, less stringent filtering criteria might be acceptable).

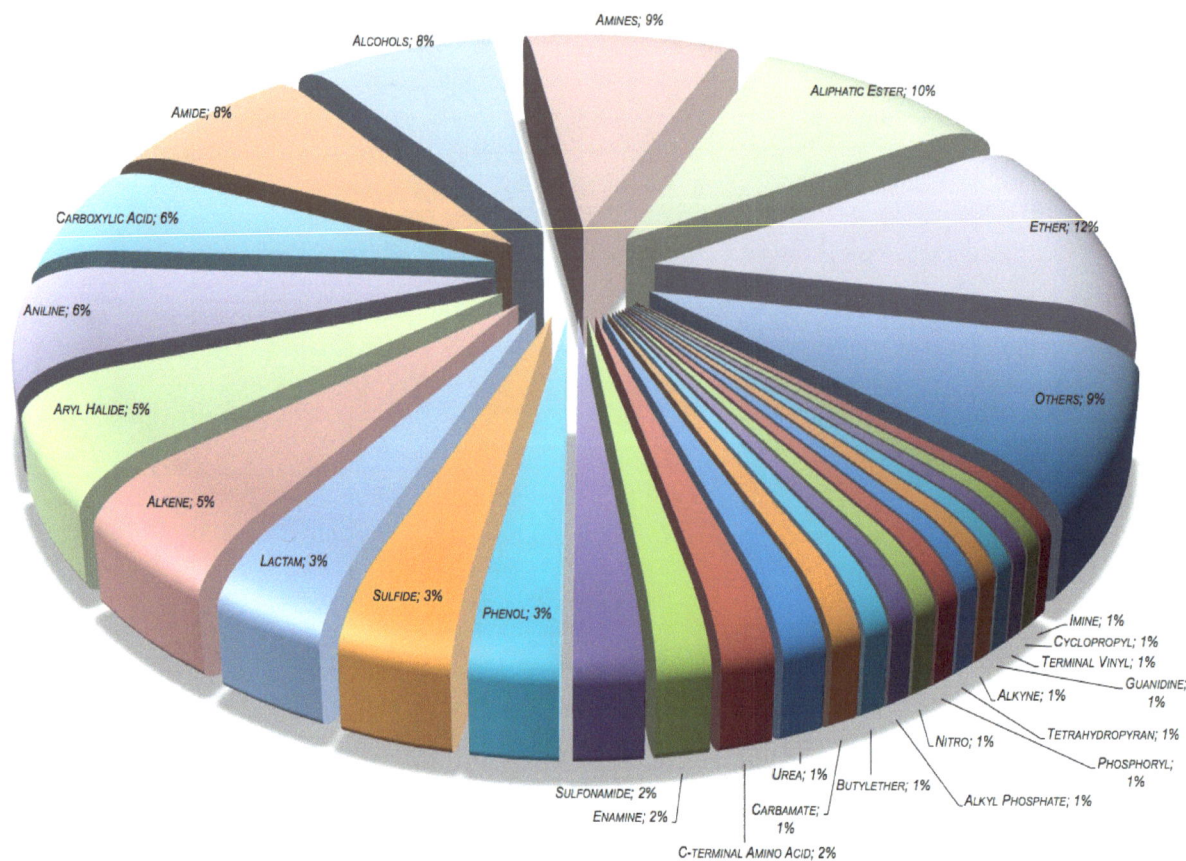

Figure 5: Pie chart of some chemical substructures detected in a set of orally bioavailable compounds selected from the DrugBank Database [15].

Generation of 3D Structures

Compound libraries need to be in 3D format for many *in silico* computations, *e.g.* 3D pharmacophore search, structure-based virtual screening, lead optimization, etc. In addition, for some computational experiments, for instance rigid-ligand docking or for 3D ligand-based screening, a multiple conformer ensemble is required (Fig. **6**). Most often, academic and commercial compound collections are in: 1D smiles [85] (.smi, simplified molecular input line entry system), 1D cansmiles [86] (.can, canonical smiles) or in 2D SDF [87]. Accurate generation of 3D structures of small chemical compounds is however not a trivial task [88] while generating conformational ensembles is also very challenging. For example, several studies benchmarking 3D generation tools [89-93] revealed that the most important parameters that play a role for a satisfactory sampling of the conformational space are: the energy window with respect to the global minimum; the RMSD value used to ensure conformational diversity; and surely the maximum number of generated conformers. Those have a major influence on the quality of the results, particularly on the RMSD value obtained between the bioactive conformation and the closest generated conformation. Over the years, various algorithms employing rule-based methods (approaches based essentially on structural data) or data-based methods [94, 95] have been developed taking into account observations about the small molecule conformational preferences observed in experimental structures [89-91, 96, 97].

Table 3: Free and commercials tools for 3D structure generation of small molecules.

Tool	Algorithm	online	free	Ref.
Balloon	Distance geometry construction and a genetic algorithm search	-	x	[98]
DG-AMMOS	Distance geometry and molecular mechanics optimization	-	x	[99]
Multiconf-DOCK	Systematic search based on DOCK anchor-first method	-	x	[100]
Frog	Fragment-based and Monte Carlo search	x		[101]
OMEGA	Fragment-based and a torsion-driving beam rule-based methods	x	-	[102]
Rubicon	Distance geometry methods to randomly sample conformations			[103]
Corina	Monocentric fragment construction	x	-	[94]
MED-3DMC	Monte Carlo search	-	-	[67]
Catalyst	Two methods of conformer sampling using a poling algorithm, FAST and BEST			[104]

Several well-established commercial packages such as Corina [94], Omega [102], Catalyst [104] or MED-3DMC [105] generate single or multiple 3D conformation of small molecules applying various approaches, including algorithms that build linker regions on the fly and combining them with pre-generated fragment libraries for the ring systems [101] and purely stochastic methods [98, 105]. In addition, several online services provide direct 2D to 3D facilities for drug-like compounds, such as OpenEye' Omega, Corina, and from academic sites such as the web-service Frog [101]. Yet, very few standalone tools generating single (or multiple) 3D conformation for a large number of compounds are freely available. Balloon [98] which uses a multi-objective genetic algorithm approach is freely available. Recently an open source program DG-AMMOS [99] (http://www.mti.univ-paris-diderot.fr/fr /downloads.html) generating a 3D conformation of a small chemical compound has been developed based on a distance geometry approach and molecular mechanics optimization [106] (http://www.cs.gsu.edu/~cscrwh/ammp/ammp.html). Several free packages are also available for multiple 3D conformer generation as Multiconf-DOCK [100] (http://dock.compbio.ucsf.edu/Contributed_Code/index.htm).

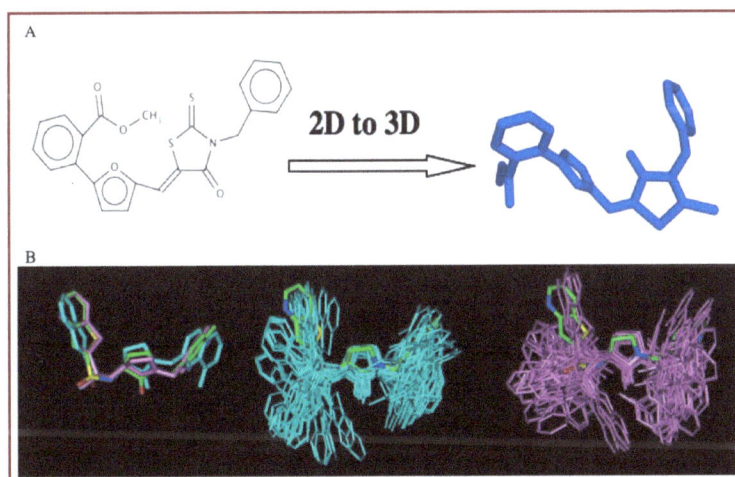

Figure 6: 3D structure generation of small chemical compounds. A. 3D conformation of a small molecule generated by the online tool Frog. B Multiple conformer ensemble of two small molecules generated by Multiconf-DOCK (magenta) and OMEGA (cyan) superimposed onto the experimental structure (all atom colors)

PROGRESS FOR PROPER COMPOUNDS' COLLECTION PREPARATION

Target-based or Focused Libraries

The success or failure of novel drug candidates can depend on the choice of the scaffolds present in the library. One of the goals when designing screening libraries is to limit *in silico* the number of compounds that will need to be tested experimentally while maximizing the hit rate and possibly the diversity of the hits [53]. For some projects, it is beneficial to focus on the targets instead of screening huge libraries. Thus, it is possible to build target-specific compound libraries starting from available compound collections using various *in silico* methods [53, 107-109]. This concept has been applied for the discovery of new inhibitors of the c-Src kinase [110]. Design and synthesis of focused genome family-targeted libraries (in particular for the GPCR proteins) can be very efficient for hit-finding, hit-to-lead or lead-optimization [111]. Libraries designed to target kinases or GPCRs are largely exploited [112] for instance, using neural network models [113]. Another possibility is the generation of focused-libraries using pharmacophoric filters based on only the atom pairs and two dimensional topological descriptors derived from active compounds [114]. In the recent years, the advances in docking, NMR and X-ray approaches applied in a large scale also suggest that focused libraries can be designed based on the receptor structure. The process for generating receptor-based libraries (using a fragment-based approach or target-specific filters) has been reviewed in [53]. Adams *et al.* identified key ligand–receptor interactions employing docking methods which were then used to guide the construction of a combinatorial library containing 700 compounds [115]. In the same line, one study [116] reported the construction of a privileged compound library for thrombin and trypsin by applying docking-scoring approach. The top-ranking compounds were synthesized, co-crystallized and their binding affinity was determined. Virtual library design for inhibitors of the human Factor Xa has been done using a chemical feature-based modeling approach to identify crucial pharmacophore patterns from 3D crystal structures of inhibitors bound to the enzyme [117]. Recently, Frecer *et al.*, derived a small highly focused combinatorial subset of analogs of triclosan, which contains virtual hits that were predicted to inhibit the Plasmodium falciparum enoyl-acyl carrier protein reductase (an important target for the design of anti-malarial agents) in the low nanomolar concentration range [118]. The above-mentioned methods can not be applied in cases where an initial set of active compounds for the selected protein target to be screened is missing. In such situations, one may create focused-libraries based exclusively on the receptor structure. Fast shape complementarity search between the receptors and ligands can be applied to generate a focused-library of smaller size. Furthermore, screening on such focused-libraries should help to achieve higher enrichment [100, 119] since all compounds screened by the flexible docking packages will show at least a good shape complementarity with the receptor, which is one of the conditions for high affinity binding [120, 121]. Oloff *et al.* have developed a novel structure-based chemoinformatics approach to search for Complimentary Ligands Based on Receptor Information (CoLiBRI) which allows a rapid pre-filtering of a large chemical database to eliminate compounds that have little chance of binding to a receptor active site [122]. Recent methods for a rapid shape complementarity search between the ligand and the receptor have been reported and can be applied as filters to create structure-based focused-libraries [100, 123].

ADMET Predictions

The efficacy of the novel drug candidates further depends on the ADMET properties. The field of *in silico* ADMET predictions is difficult to define as it encompasses many different, overlapping and inter-related areas with the aim of understanding and predicting pharmacokinetics, pharmacodynamics events, adverse effects of drugs and/or lack of efficacy. Different levels of complexity can be used but, in general, they all rely on a key assumption: the structure and physico-chemical properties of a compound can be linked to one or several ADMET events. In addition, the exact compound characteristics at some stages during the drug discovery process, become project specific. As mentioned above, when designing a collection, it can be important to use empirical guidelines, such as the RO5 and related concepts. Depending on the project, it can be important to avoid incorporating compounds that possess substructures and chemical groups that are unstable, reactive and overall inappropriate for the hit-to-lead optimization phase. This is usually accomplished by using pattern-recognition techniques. Next, different ADMET properties can be investigated. This runs from membrane permeability predictions up to toxicity. For example, it can be important at some stage to consider metabolism. Indeed, most drugs cannot be eliminated from the body without previous biotransformation to metabolites. Also, in some cases, the administrated product is not active but only the body-modified one gets efficient. Alternatively, some compounds, once metabolized, can become toxic. The key sites in the body for metabolism are the gut wall and the liver. Metabolic alteration of xenobiotics is traditionally subdivided into two main phases: phase I (oxidation, often via cytochrome P450 enzymes, reduction and hydrolysis) results in the introduction of new functional groups while in phase II (conjugation reactions), highly hydrophilic moiety such as sulfate or glucuronide is attached to make the compounds more water-soluble and to prepare for excretion through urine and bile. The most important enzymes involved in metabolism in humans are the cytochrome P450s: CYP3A4, CYP2D6, CYP2C9 and CYP2C19. Interaction of compounds with metabolizing enzymes can be in some cases predicted *in silico* [124, 125]. Further, in the pharmaceutical industry, there is a concern that close to 20% of drug attrition during preclinical and clinical development is due to toxicity issues, these ones can be directly linked to inappropriate ADME properties of a compound. Because of the relatively small number of patients enrolled in clinical trials, it is statistically difficult to detect rare adverse reactions with the current approaches (*e.g.*, animal models, experimental high-throughput early ADMET methods, small number of patients involved). Thus, all toxicities are not detected prior to and during the clinical trials causing major problem in today's drug discovery endeavors. *In silico* methods are being developed to try to facilitate the selection of more appropriate, less toxic compounds but the simulations are complex and the methods are still in development [126-129].

Numerous *in silico* methods have been developed in an attempt to assist the drug discovery process for ADMET predictions [6, 130-156]. However, because of the complexity of the biological events, ADMET properties are very difficult to predict accurately, *i.e.*, human intestinal absorption or metabolic stability, which arise from multiple physiological mechanisms are difficult to model. Nevertheless, many different approaches can be used to perform *in silico* modeling of ADMET properties ranging from empirical rules (*e.g.*, Lipinski RO5), to quantitative prediction approaches: quantitative structure-activity relationship (QSAR) and quantitative structure-property relationship (QSPR) up to classification models, and similarity searches, molecular modeling (structure-based approaches such as ligand-protein docking, pharmacophore modeling, quantum mechanics) and physiologically based pharmacokinetic (PBPK) modeling [157-160]. The level of complexity of the mathematical modeling varies greatly among the methods. In general, these approaches require a training set, molecules, for which one has experimental data about the modeled event and a validation set. The model can then involve a series of simple, statistically weighted sum of some physico-chemical descriptors (*e.g.*, a combination of log P, PSA, MW, etc.) or make use of very complex descriptors and of mathematical models that do not allow to directly translate the predictions into suggestions for chemical optimization. Overall, all models are locally valid within the structural space of the training but become questionable as global property filters without careful investigation of the applicability domain of the methods. Following the journey of a drug in the human body, the key events that need to be addressed can be solubility, absorption and permeability. Predictions can be performed by calculating the lipophilicity [60] of the molecule, and as seen above, by different empirical rules and via machine learning approaches. Next, the distribution of the drugs throughout the body determines whether the molecule will elicit a pharmacological response or not. Distribution can be predicted through predicting if a drug can reach the brain (blood–brain barrier, BBB) [161-163], permeability [60, 164-168], plasma protein binding (PPB) [147, 169, 170], volume of distribution and by estimating binding to transporters proteins [149, 171]. The BBB is formed by the brain capillary endothelium and organized in tight junctions that surround the cell margin circumferentially. This results in the brain barrier being essentially impermeable to hydrophilic compounds [172]. PPB, is a phenomenon involving many proteins (low-density lipoprotein, high-density lipoprotein, α1-acid glycoprotein, myosin, actin, etc.) but especially the human serum albumin which is present in many tissues and is responsible for unspecific drug binding. As only the free drug concentration determines the pharmaceutical activity, non-specific protein binding should be carefully considered.

The toxicity effects of a compound [129, 173], can be roughly classified in four categories according to the pathological effect induced [174]: immunological hypersensitivity, causing down-regulation of the immune system or producing autoimmunity to native proteins; altered phenotype/function relating to cell alterations; cell death (apoptosis and necrosis) and tissue response; cancer which is associated with cell death. Additionally, several pharmacophore-induced toxicities have been demonstrated such as cytochrome P450s (CYPs) induction/inhibition and drug-drug interactions [175-177] or hERG binding [178-180]. In these cases, and as compared to other types of non-target-specific toxicity, pharmacophore and/or docking-scoring models can be developed. To remove these unwanted compounds, it is in some cases possible to search for well-defined substructures, but in some other situations, the exact structures responsible of the toxicity are difficult to translate into a structure/pattern-matching algorithm. Thus, methods trained on experimental toxicity data can be used. In this context, numerous *in silico* methods have been developed and can applied for the prediction of some ADMET properties of the drug candidates [140, 181, 182].

Protonation, Tautomers and Stereoisomers

Accurate representation of the protonated and tautomeric forms of small molecules is extremely important for a proper preparation of library for lead identification but challenging. Assigning the correct protonated state of a compound is difficult because the protonation can change dynamically depending on the ligand and binding site environment. When dealing with large chemical libraries for virtual screening, where manual preparation of each ligand is not possible, an automated ligand preparation protocol is particularly useful. One may generate all relevant protonated and tautomeric states by predictions based on theoretical considerations. Similarly, it might be necessary to generate all physically relevant stereoisomers of a compound arising from stereocenter descriptions. Several tools allow to predict such states. Different protonation states, stereoisomers, and tautomers can be generated with the automated tool, called SPORES (Structure PrOtonation and Recognition System) [183], the routine LIGPREP [184] can also be employed to prepare small molecules' structures. The algorithm is able to create up to eight different tautomers for the molecules, to set up stereoisomers, and to change the ionization state of the molecule according to a given pH value. The Chemical Computing Group tool PROTONATE 3D [185] generates tautomers and rotamers, and determines an accurate protonation for polar groups in the binding site using a free energy optimization. I-interpret [186] allows to generate by default the neutral form of all functional groups but this behavior can be changed manually by the user. In addition, small molecules can be protonated at a given pH using the open source toolkit OpenBabel [187] version 2.0.2 or using the Hgene tool of the myPresto package [188]. These predictions are usually based on chemistry rules and may not be always accurate. Multiple protonated states can also be considered in chemical libraries. However, some docking and virtual screening experiments demonstrated that often the docking algorithms have problems in identifying the correct protomer for each complex [189]. It has been shown that the correct protonation leads to achieving the highest success rates in some cases. Correct protonation is thus important and possible to achieved by using more accurate theoretical methods for pH-dependent pKa computations [190, 191]. For instance the effects of protonation, tautomeric (protonation and tautomeric states enumerated in the pH range 2-14), stereochemical, and conformational states have been investigated [51] on the prioritization of known active ligands from a database containing both known actives and inactives. Recently, the ligands protonation states at pH 4.5 (where optimal BACE1 activity can be reached) and pH 6.6 have been predicted for BACE1 [192] and the effect of ligand protonation on screening efficacy has been estimated over the comparison of four docking programs (GOLD, FlexX-Pharm, eHiTS, and Surflex). It has been shown that dealing with all possible protomers does not always improve screening efficacy [192]. Further, the presence of different tautomeric forms can affect the results of virtual screening, ligand-based or structure-based, fingerprint searches, and physicochemical properties prediction. Recently, a tool named TauThor has been reported and is able to enumerate tautomers and predict tautomer stability in the aqueous medium [193]. It is based on a recursive process that generates tautomers according to the scheme HX-Y=Z <-> X=Y-ZH and a pKa based-method that utilizes pKa values predicted by MoKa [194]. To evaluate the impact of tautomerism on chemical databases, TauThor has been used to enumerate tautomers and predict tautomer stability for 683,862 structures from four databases. Results indicated that 29% of the compounds are potentially tautomeric. The TAUTOMER utility from the Molecular Networks GmBH can also be used to generate relevant tautomeric states of molecules in the database. Overall, these states are very difficult to assess accurately but they have to be considered at some stages, during compound optimization.

CONCLUSION

It is now well documented that compounds that possess drug-like or lead-like characteristics should be carefully selected to prepare a compound collection, most often prior to the screening process. Different levels of abstraction can be used to filter a compound collection, from applying empirical rules up to complex mathematical modeling of biological events. Many computer tools are available to assist the library design process. In all cases, the right balance between molecular diversity, drug-likeness, ADMET properties, focused banks has to be found. For hit finding, a generic library could be used but, rapidly, as the compounds evolve toward the optimization phase, the prediction tools and filtering parameters should be tuned according to the project, the target and the pathological conditions being addressed.

ACKNOWLEDGEMENTS

Supports from the Inserm institute and University Paris Diderot are greatly appreciated.

REFERENCES

[1]   Schmid, E.F.; Smith, D.A. Pharmaceutical R&D in the spotlight: why is there still unmet medical need? *Drug Discov. Today* **2007,** *12*(23-24), 998-1006.

[2]   Smith, D.A.; Schmid, E.F. Drug withdrawals and the lessons within. *Curr. Opin. Drug Discov. Devel.* **2006,** *9*(1), 38-46.

[3]   Dickson, M.; Gagnon, J.P. Key factors in the rising cost of new drug discovery and development. *Nat. Rev. Drug Discov.*

2004, *3*(5), 417-29.

[4]   DiMasi, J.A.; Caglarcan, E.; Wood-Armany, M. Emerging role of pharmacoeconomics in the research and development decision-making process. *Pharmacoeconomics* **2001,** *19*(7), 753-66.

[5]   Sirois, S.; Hatzakis, G.; Wei, D.; Du, Q.; Chou, K.C. Assessment of chemical libraries for their druggability. *Comput. Biol. Chem.* **2005,** *29*(1), 55-67.

[6]   Vistoli, G.; Pedretti, A.; Testa, B. Assessing drug-likeness--what are we missing? *Drug Discov. Today* **2008,** *13*(7-8), 285-94.

[7]   Verheij, H.J. Leadlikeness and structural diversity of synthetic screening libraries. *Mol. Divers* **2006,** *10*(3), 377-88.

[8]   Milne, G.M. Pharmaceutical productivity – the imperative for new paradigms. *Ann. Rep. Med. Chem.* **2002,** *38*, 383-396.

[9]   Oprea, T.I.; Zamora, I.; Ungell, A.L. Pharmacokinetically based mapping device for chemical space navigation. *J. Comb. Chem.* **2002,** *4*(4), 258-66.

[10]   Wenlock, M.C.; Austin, R.P.; Barton, P.; Davis, A.M.; Leeson, P.D. A comparison of physiochemical property profiles of development and marketed oral drugs. *J. Med. Chem.* **2003,** *46*(7), 1250-6.

[11]   Blake, J.F. Identification and evaluation of molecular properties related to preclinical optimization and clinical fate. *Med. Chem.* **2005,** *1*(6), 649-55.

[12]   Leeson, P.D.; Springthorpe, B. The influence of drug-like concepts on decision-making in medicinal chemistry. *Nat. Rev. Drug Discov.* **2007,** *6*(11), 881-90.

[13]   Villoutreix, B.O.; Renault, N.; Lagorce, D.; Sperandio, O.; Montes, M.; Miteva, M.A. Free resources to assist structure-based virtual ligand screening experiments. *Curr. Protein Pept. Sci.* **2007,** *8*(4), 381-411.

[14]   Williams, A.J. A perspective of publicly accessible/open-access chemistry databases. *Drug Discov. Today* **2008,** *13*(11-12), 495-501.

[15]   Wishart, D.S.; Knox, C.; Guo, A.C.; Cheng, D.; Shrivastava, S.; Tzur, D.; Gautam, B.; Hassanali, M. DrugBank: a knowledgebase for drugs, drug actions and drug targets. *Nucleic Acids Res* **2008,** *36*(Database issue), D901-6.

[16]   Lazo, J.S. Roadmap or roadkill: a pharmacologist's analysis of the NIH Molecular Libraries Initiative. *Mol. Interv.* **2006,** *6*(5), 240-3.

[17]   Xie, X.Q.; Chen, J.Z. Data mining a small molecule drug screening representative subset from NIH PubChem. *J. Chem. Inf. Model.* **2008,** *48*(3), 465-75.

[18]   Olah, M.; Mracec, M.; Ostopovici, L.; Rad, R.; Bora, A.; Hadaruga, N.; Olah, I.; Banda, M.; Simon, Z.; Mracec, M.; Oprea, T.I. WOMBAT: world of molecular bioactivity, in chemoinformatics in drug discovery. In *Chemoinformatics in Drug Discovery*, Wiley-VCH: New York, 2004.

[19]   Judson, R.; Richard, A.; Dix, D.; Houck, K.; Elloumi, F.; Martin, M.; Cathey, T.; Transue, T.R.; Spencer, R.; Wolf, M. ACToR--Aggregated Computational Toxicology Resource. *Toxicol. Appl. Pharmacol.* **2008,** *233*(1), 7-13.

[20]   He, Q.Y.; He, Q.Z.; Deng, X.C.; Yao, L.; Meng, E.; Liu, Z.H.; Liang, S P. ATDB: a uni-database platform for animal toxins. *Nucleic Acids Res.* **2008,** *36*(Database issue), D293-7.

[21]   Chang, A.; Scheer, M.; Grote, A.; Schomburg, I.; Schomburg, D. BRENDA, AMENDA and FRENDA the enzyme information system: new content and tools in 2009. *Nucleic Acids Res.* **2009,** *37*(Database issue), D588-92.

[22]   Ott, M.A.; Vriend, G. Correcting ligands, metabolites, and pathways. *BMC Bioinformatics* **2006,** *7*, 517.

[23]   Degtyarenko, K.; de Matos, P.; Ennis, M.; Hastings, J.; Zbinden, M.; McNaught, A.; Alcantara, R.; Darsow, M.; Guedj, M.; Ashburner, M. ChEBI: a database and ontology for chemical entities of biological interest. *Nucleic Acids Res.* **2008,** *36*(Database issue), D344-50.

[24]   Seiler, K.P.; George, G.A.; Happ, M.P.; Bodycombe, N.E.; Carrinski, H.A.; Norton, S.; Brudz, S.; Sullivan, J.P.; Muhlich, J.; Serrano, M.; Ferraiolo, P.; Tolliday, N.J.; Schreiber, S.L.; Clemons, P.A. ChemBank: a small-molecule screening and cheminformatics resource database. *Nucleic Acids Res.* **2008,** *36*(Database issue), D351-9.

[25]   Naamati, G.; Askenazi, M.; Linial, M. ClanTox: a classifier of short animal toxins. *Nucleic Acids Res.* **2009,** *37*(Web Server issue), W363-8.

[26]   Zhang, J.X.; Huang, W.J.; Zeng, J.H.; Huang, W.H.; Wang, Y.; Zhao, R.; Han, B.C.; Liu, Q.F.; Chen, Y.Z.; Ji, Z.L. DITOP: drug-induced toxicity related protein database. *Bioinformatics* **2007,** *23*(13), 1710-2.

[27]   Richard, A.M.; Williams, C.R. Distributed structure-searchable toxicity (DSSTox) public database network: a proposal. *Mutat. Res.* **2002,** *499*(1), 27-52.

[28]   Wishart, D.S.; Knox, C.; Guo, A. C.; Shrivastava, S.; Hassanali, M.; Stothard, P.; Chang, Z.; Woolsey, J. DrugBank: a comprehensive resource for *in silico* drug discovery and exploration. *Nucleic Acids Res.* **2006,** *34*(Database issue), D668-72.

[29]   Huang, N.; Shoichet, B.K.; Irwin, J.J. Benchmarking sets for molecular docking. *J. Med. Chem.* **2006,** *49*(23), 6789-801.

[30]   eMolecules, Del Mar, CA, USA http://www.emolecules.com

[31] Warr, W.A. ChEMBL. An interview with John Overington, team leader, chemogenomics at the European Bioinformatics Institute Outstation of the European Molecular Biology Laboratory (EMBL-EBI). *J. Comput. Aided Mol. Des.* **2009**, *23*(4), 195-8.

[32] Blum, L.C.; Reymond, J.L. 970 million druglike small molecules for virtual screening in the chemical universe database GDB-13. *J. Am. Chem. Soc.* **2009**, *131*(25), 8732-3.

[33] Singh, M.K.; Srivastava, S.; Raghava, G.P.; Varshney, G.C. HaptenDB: a comprehensive database of haptens, carrier proteins and anti-hapten antibodies. *Bioinformatics* **2006**, *22*(2), 253-5.

[34] Wishart, D.S.; Knox, C.; Guo, A.C.; Eisner, R.; Young, N.; Gautam, B.; Hau, D.D.; Psychogios, N.; Dong, E.; Bouatra, S.; Mandal, R.; Sinelnikov, I.; Xia, J.; Jia, L.; Cruz, J.A.; Lim, E.; Sobsey, C.A.; Shrivastava, S.; Huang, P.; Liu, P.; Fang, L.; Peng, J.; Fradette, R.; Cheng, D.; Tzur, D.; Clements, M.; Lewis, A.; De Souza, A.; Zuniga, A.; Dawe, M.; Xiong, Y.; Clive, D.; Greiner, R.; Nazyrova, A.; Shaykhutdinov, R.; Li, L.; Vogel, H.J.; Forsythe, I. HMDB: a knowledgebase for the human metabolome. *Nucleic Acids Res.* **2009**, *37*(Database issue), D603-10.

[35] Kanehisa, M. The KEGG database. *Novartis Found Symp.* **2002**, *247*, 91-101; discussion 101-3, 119-28, 244-52.

[36] Feng, Z.; Chen, L.; Maddula, H.; Akcan, O.; Oughtred, R.; Berman, H.M.; Westbrook, J. Ligand Depot: a data warehouse for ligands bound to macromolecules. *Bioinformatics* **2004**, *20*(13), 2153-5.

[37] von Grotthuss, M.; Pas, J.; Rychlewski, L. Ligand-Info, searching for similar small compounds using index profiles. *Bioinformatics* **2003**, *19*(8), 1041-2.

[38] MDPI Database http://www.mdpi.org/database.htm

[39] Boyer, S.; Arnby, C.H.; Carlsson, L.; Smith, J.; Stein, V.; Glen, R.C. Reaction site mapping of xenobiotic biotransformations. *J. Chem. Inf. Model.* **2007**, *47*(2), 583-90.

[40] Sitzmann, M.; Filippov, I.V.; Nicklaus, M.C. Internet resources integrating many small-molecule databases. *SAR QSAR Environ. Res.* **2008**, *19*(1-2), 1-9.

[41] Wheeler, D.L.; Barrett, T.; Benson, D.A.; Bryant, S.H.; Canese, K.; Chetvernin, V.; Church, D.M.; DiCuccio, M.; Edgar, R.; Federhen, S.; Geer, L.Y.; Helmberg, W.; Kapustin, Y.; Kenton, D.L.; Khovayko, O.; Lipman, D.J.; Madden, T.L.; Maglott, D.R.; Ostell, J.; Pruitt, K.D.; Schuler, G.D.; Schriml, L.M.; Sequeira, E.; Sherry, S.T.; Sirotkin, K.; Souvorov, A.; Starchenko, G.; Suzek, T.O.; Tatusov, R.; Tatusova, T.A.; Wagner, L.; Yaschenko, E. Database resources of the National Center for Biotechnology Information. *Nucleic Acids Res.* **2006**, *34*(Database issue), D173-80.

[42] Hendlich, M.; Bergner, A.; Gunther, J.; Klebe, G. Relibase: design and development of a database for comprehensive analysis of protein-ligand interactions. *J. Mol. Biol.* **2003**, *326*(2), 607-20.

[43] Dunkel, M.; Fullbeck, M.; Neumann, S.; Preissner, R. SuperNatural: a searchable database of available natural compounds. *Nucleic Acids Res.* **2006**, *34*(Database issue), D678-83.

[44] Goede, A.; Dunkel, M.; Mester, N.; Frommel, C.; Preissner, R. SuperDrug: a conformational drug database. *Bioinformatics* **2005**, *21*(9), 1751-3.

[45] Michalsky, E.; Dunkel, M.; Goede, A.; Preissner, R. SuperLigands - a database of ligand structures derived from the Protein Data Bank. *BMC Bioinformatics* **2005**, *6*, 122.

[46] Schmidt, U.; Struck, S.; Gruening, B.; Hossbach, J.; Jaeger, I.S.; Parol, R.; Lindequist, U.; Teuscher, E.; Preissner, R. SuperToxic: a comprehensive database of toxic compounds. *Nucleic Acids Res.* **2009**, *37*(Database issue), D295-9.

[47] Wexler, P. TOXNET: the National Library of Medicine's toxicology database. *Am. Fam. Physician.* **1995**, *52*(6), 1677-8.

[48] Briggs, K.A. Vitic - A data source for (quantitative) structure–activity relationship modelling. *Toxicology* **2007**, *234*(2-3), 113-114.

[49] Irwin, J.J.; Shoichet, B.K. ZINC--a free database of commercially available compounds for virtual screening. *J. Chem. Inf. Model.* **2005**, *45*(1), 177-82.

[50] Olah, M.M.; Bologa, C.G.; Oprea, T.I. Strategies for compound selection. *Curr. Drug Discov. Technol.* **2004**, *1*(3), 211-20.

[51] Knox, A.J.; Meegan, M.J.; Carta, G.; Lloyd, D.G. Considerations in compound database preparation--"hidden" impact on virtual screening results. *J. Chem. Inf. Model.* **2005**, *45*(6), 1908-19.

[52] Goodnow, R.A., Jr.; Guba, W.; Haap, W. Library design practices for success in lead generation with small molecule libraries. *Comb. Chem. High Throughput Screen.* **2003**, *6*(7), 649-60.

[53] Orry, A.J.; Abagyan, R.A.; Cavasotto, C.N. Structure-based development of target-specific compound libraries. *Drug Discov. Today* **2006**, *11*(5-6), 261-6.

[54] Frimurer, T.M.; Bywater, R.; Naerum, L.; Lauritsen, L.N.; Brunak, S. Improving the odds in discriminating "drug-like" from "non drug-like" compounds. *J. Chem.Inf. Comput. Sci.* **2000**, *40*(6), 1315-24.

[55] Lipinski, C.A.; Lombardo, F.; Dominy, B.W.; Feeney, P.J. Experimental and computational approaches to estimate solubility and permeability in drug discovery and development settings. *Adv. Drug Deliv. Rev.* **2001**, *46*(1-3), 3-26.

[56] Congreve, M.; Carr, R.; Murray, C.; Jhoti, H. A 'rule of three' for fragment-based lead discovery? *Drug Discov. Today*

2003, *8*(19), 876-7.

[57] Veber, D.F.; Johnson, S.R.; Cheng, H.Y.; Smith, B.R.; Ward, K.W.; Kopple, K.D. Molecular properties that influence the oral bioavailability of drug candidates. *J. Med. Chem.* **2002**, *45*(12), 2615-23.

[58] Clark, D.E.; Pickett, S.D. Computational methods for the prediction of 'drug-likeness'. *Drug Discov. Today* **2000**, *5*(2), 49-58.

[59] Ertl, P.; Rohde, B.; Selzer, P. Fast calculation of molecular polar surface area as a sum of fragment-based contributions and its application to the prediction of drug transport properties. *J. Med. Chem.* **2000**, *43*(20), 3714-7.

[60] Di, L.; Kerns, E.H. Profiling drug-like properties in discovery research. *Curr. Opin. Chem. Biol.* **2003**, *7*(3), 402-8.

[61] Jorgensen, W.L.; Duffy, E.M. Prediction of drug solubility from structure. *Adv. Drug Deliv. Rev.* **2002**, *54*(3), 355-66.

[62] Kerns, E.H.; Di, L. Pharmaceutical profiling in drug discovery. *Drug Discov. Today* **2003**, *8*(7), 316-23.

[63] Lagorce, D.; Sperandio, O.; Galons, H.; Miteva, M.A.; Villoutreix, B.O. FAF-Drugs2: free ADME/tox filtering tool to assist drug discovery and chemical biology projects. *BMC Bioinformatics* **2008**, *9*, 396.

[64] Patlewicz, G.; Jeliazkova, N.; Gallegos Saliner, A.; Worth, A.P. Toxmatch-a new software tool to aid in the development and evaluation of chemically similar groups. *SAR QSAR Environ. Res.* **2008**, *19*(3-4), 397-412.

[65] Monge, A.; Arrault, A.; Marot, C.; Morin-Allory, L. Managing, profiling and analyzing a library of 2.6 million compounds gathered from 32 chemical providers. *Mol. Divers* **2006**, *10*(3), 389-403.

[66] PreADMET. http://preadmet.bmdrc.org/index.php

[67] Miteva, M.A.; Violas, S.; Montes, M.; Gomez, D.; Tuffery, P.; Villoutreix, B.O. FAF-Drugs: free ADME/tox filtering of compound collections. *Nucleic Acids Res* **2006**, *34*(Web Server issue), W738-44.

[68] Molinspiration Calculation of Molecular Properties and Drug-likeness. http://www.molinspiration.com/cgi-bin/properties.

[69] OSIRIS Property Explorer. http://www.organic-chemistry.org/prog/peo/.

[70] ChemAxon. http://www.chemaxon.com

[71] FILTER, OpenEye Scientific Software. http://www.eyesopen.com/products/applications/filter.html

[72] Schrödinger, QikProp. http://www.schrodinger.com/ProductDescription.php?mID=6&sID=10

[73] Accelrys, Discovery Studio 2.0 http://accelrys.com/products/discovery-studio/

[74] MOE, Chemical Computing Group http://www.chemcomp.com

[75] ToxTree Ideaconsult Ltd. http://toxtree.sourceforge.net/

[76] Cheng, T.; Zhao, Y.; Li, X.; Lin, F.; Xu, Y.; Zhang, X.; Li, Y.; Wang, R.; Lai, L. Computation of octanol-water partition coefficients by guiding an additive model with knowledge. *J. Chem. Inf. Model.* **2007**, *47*(6), 2140-8.

[77] MolSoft Calculation of Molecular Properties and Drug-likeness. http://www.molsoft.com/mprop/.

[78] Oprea, T.I. Property distribution of drug-related chemical databases. *J. Comput. Aided Mol. Des.* **2000**, *14*(3), 251-64.

[79] Rishton, G.M. Reactive compounds and *in vitro* fake positives in HTS. *Drug Discov. Today* **1997**, *2*(9), 382-384.

[80] Gultekin, F.; Hicyilmaz, H. Renal deterioration caused by carcinogens as a consequence of free radical mediated tissue damage: a review of the protective action of melatonin. *Arch. Toxicol.* **2007**, *81*(10), 675-81.

[81] Metosh-Dickey, C.A.; Mason, R.P.; Winston, G.W. Nitroarene reduction and generation of free radicals by cell-free extracts of wild-type, and nitroreductase-deficient and -enriched Salmonella typhimurium strains used in the umu gene induction assay. *Toxicol. Appl. Pharmacol.* **1999**, *154*(2), 126-34.

[82] Roche, O.; Schneider, P.; Zuegge, J.; Guba, W.; Kansy, M.; Alanine, A.; Bleicher, K.; Danel, F.; Gutknecht, E.M.; Rogers-Evans, M.; Neidhart, W.; Stalder, H.; Dillon, M.; Sjogren, E.; Fotouhi, N.; Gillespie, P.; Goodnow, R.; Harris, W.; Jones, P.; Taniguchi, M.; Tsujii, S.; von der Saal, W.; Zimmermann, G.; Schneider, G. Development of a virtual screening method for identification of "frequent hitters" in compound libraries. *J. Med. Chem.* **2002**, *45*(1), 137-42.

[83] McGovern, S.L.; Caselli, E.; Grigorieff, N.; Shoichet, B.K. A common mechanism underlying promiscuous inhibitors from virtual and high-throughput screening. *J. Med. Chem.* **2002**, *45*(8), 1712-22.

[84] Rishton, G.M. Nonleadlikeness and leadlikeness in biochemical screening. *Drug Discov. Today* **2003**, *8*(2), 86-96.

[85] Weininger, D. SMILES, a chemical language and information system. 1. Introduction to methodology and encoding rules. *J. Chem. Inf. Comput. Sci.* **1988**, *28*(1), 31-36.

[86] Weininger, D. SMILES. 2. Algorithm for generation of unique SMILES notation. *J. Chem. Inf. Comput. Sci.* **1989**, *29*(2), 97-101.

[87] Dalby, A. Description of several chemical structure file formats used by computer programs developed at Molecular Design Limited. *J. Chem. Inf. Comput. Sci.* **1992**, *32*, 244-255.

[88] Sadowski, J.; Gasteiger, J. From atoms and bonds to three-dimensional atomic coordinates: automatic model builders. *Chem. Rev.* **1993**, *93*, 2567-2581.

[89] Bostrom, J. Reproducing the conformations of protein-bound ligands: a critical evaluation of several popular conformational searching tools. *J. Comput. Aided Mol. Des.* **2001**, *15*(12), 1137-52.

[90] Bostrom, J.; Greenwood, J.R.; Gottfries, J. Assessing the performance of OMEGA with respect to retrieving bioactive conformations. *J. Mol. Graph. Model.* **2003**, *21*(5), 449-62.

[91] Kirchmair, J.; Wolber, G.; Laggner, C.; Langer, T. Comparative performance assessment of the conformational model generators omega and catalyst: a large-scale survey on the retrieval of protein-bound ligand conformations. *J. Chem. Inf. Mode.l* **2006**, *46*(4), 1848-61.

[92] Perola, E.; Charifson, P.S. Conformational analysis of drug-like molecules bound to proteins: an extensive study of ligand reorganization upon binding. *J. Med. Chem.* **2004**, *47*(10), 2499-510.

[93] Renner, S.; Schwab, C.H.; Gasteiger, J.; Schneider, G. Impact of conformational flexibility on three-dimensional similarity searching using correlation vectors. *J Chem Inf Model* **2006**, *46*(6), 2324-32.

[94] Sadowski, J.; Gasteiger, J.; Klebe, G. Comparison of automatic three-dimensional model builders using 639 X-ray structures. *J. Chem. Inf. Comput. Sci.* **1994**, *34*(4), 1000-1008.

[95] Steinbeck, C.; Hoppe, C.; Kuhn, S.; Floris, M.; Guha, R.; Willighagen, E.L. Recent developments of the chemistry development kit (CDK) - an open-source java library for chemo- and bioinformatics. *Curr. Pharm. Des.* **2006**, *12*(17), 2111-20.

[96] Brameld, K.A.; Kuhn, B.; Reuter, D.C.; Stahl, M. Small molecule conformational preferences derived from crystal structure data. A medicinal chemistry focused analysis. *J. Chem. Inf. Model.* **2008**, *48*(1), 1-24.

[97] Kirchmair, J.; Ristic, S.; Eder, K.; Markt, P.; Wolber, G.; Laggner, C.; Langer, T. Fast and efficient *in silico* 3D screening: toward maximum computational efficiency of pharmacophore-based and shape-based approaches. *J. Chem. Inf. Model.* **2007**, *47*(6), 2182-96.

[98] Vainio, M.J.; Johnson, M.S. Generating conformer ensembles using a multiobjective genetic algorithm. *J. Chem. Inf. Model.* **2007**, *47*(6), 2462-74.

[99] Lagorce, D.; Pencheva, T.; Villoutreix, B.O.; Miteva, M.A. DG-AMMOS: A New tool to generate 3D conformation of small molecules using Distance Geometry and Automated Molecular Mechanics Optimization for *in silico* Screening. *BMC Chemical Biology* **2009**, *9*, 6.

[100] Sauton, N.; Lagorce, D.; Villoutreix, B.O.; Miteva, M.A. MS-DOCK: accurate multiple conformation generator and rigid docking protocol for multi-step virtual ligand screening. *BMC Bioinformatics* **2008**, *9*, 184.

[101] Leite, T. B.; Gomes, D.; Miteva, M.A.; Chomilier, J.; Villoutreix, B.O.; Tuffery, P. Frog: a FRee Online druG 3D conformation generator. *Nucleic Acids Res.* **2007**, *35*(Web Server issue), W568-72.

[102] OMEGA OpenEye Scientific Software. http://www.eyesopen.com/products/ applications/omega.html

[103] RUBICON Daylight Chemical Information Systems Inc. http://www.daylight.com

[104] Kirchmair, J.; Laggner, C.; Wolber, G.; Langer, T. Comparative analysis of protein-bound ligand conformations with respect to catalyst's conformational space subsampling algorithms. *J. Chem. Inf. Model.* **2005**, *45*(2), 422-30.

[105] Sperandio, O.; Souaille, M.; Delfaud, F.; Miteva, M.A.; Villoutreix, B.O. MED-3DMC: a new tool to generate 3D conformation ensembles of small molecules with a Monte Carlo sampling of the conformational space. *Eur. J. Med. Chem.* **2009**, *44*(4), 1405-9.

[106] Pencheva, T.; Lagorce, D.; Pajeva, I.; Villoutreix, B.O.; Miteva, M.A. AMMOS: Automated Molecular Mechanics Optimization tool for *in silico* Screening. *BMC Bioinformatics* **2008**, *9*, 438.

[107] Lamb, M.L.; Burdick, K.W.; Toba, S.; Young, M.M.; Skillman, A.G.; Zou, X.; Arnold, J.R.; Kuntz, I.D. Design, docking, and evaluation of multiple libraries against multiple targets. *Proteins* **2001**, *42*(3), 296-318.

[108] Eksterowicz, J.E.; Evensen, E.; Lemmen, C.; Brady, G.P.; Lanctot, J.K.; Bradley, E.K.; Saiah, E.; Robinson, L.A.; Grootenhuis, P.D.; Blaney, J.M. Coupling structure-based design with combinatorial chemistry: application of active site derived pharmacophores with informative library design. *J. Mol. Graph. Model.* **2002**, *20*(6), 469-77.

[109] Zhang, Y.J.; Wang, Z.; Sprous, D.; Nabioullin, R. *In silico* design and synthesis of piperazine-1-pyrrolidine-2,5-dione scaffold-based novel malic enzyme inhibitors. *Bioorg. Med. Chem. Lett.* **2006**, *16*(3), 525-8.

[110] Maly, D.J.; Choong, I.C.; Ellman, J.A. Combinatorial target-guided ligand assembly: identification of potent subtype-selective c-Src inhibitors. *Proc. Natl. Acad. Sci. U S A* **2000**, *97*(6), 2419-24.

[111] Crossley, R. The design of screening libraries targeted at G-protein coupled receptors. *Curr. Top. Med. Chem.* **2004**, *4*(6), 581-8.

[112] Klabunde, T.; Hessler, G. Drug design strategies for targeting G-protein-coupled receptors. *Chembiochem* **2002**, *3*(10), 928-44.

[113] Manallack, D.T.; Pitt, W.R.; Gancia, E.; Montana, J.G.; Livingstone, D.J.; Ford, M.G.; Whitley, D.C. Selecting screening candidates for kinase and G protein-coupled receptor targets using neural networks. *J. Chem. Inf. Comput. Sci.* **2002**, *42*(5), 1256-62.

[114] Kellenberger, E.; Springael, J.Y.; Parmentier, M.; Hachet-Haas, M.; Galzi, J.L.; Rognan, D. Identification of nonpeptide

CCR5 receptor agonists by structure-based virtual screening. *J. Med. Chem.* **2007**, *50*(6), 1294-303.

[115] Adams, C.; Aldous, D.J.; Amendola, S.; Bamborough, P.; Bright, C.; Crowe, S.; Eastwood, P.; Fenton, G.; Foster, M.; Harrison, T.K.; King, S.; Lai, J.; Lawrence, C.; Letallec, J.P.; McCarthy, C.; Moorcroft, N.; Page, K.; Rao, S.; Redford, J.; Sadiq, S.; Smith, K.; Souness, J.E.; Thurairatnam, S.; Vine, M.; Wyman, B. Mapping the kinase domain of Janus Kinase 3. *Bioorg. Med. Chem. Lett.* **2003**, *13*(18), 3105-10.

[116] DeSimone, R.W.; Currie, K.S.; Mitchell, S.A.; Darrow, J.W.; Pippin, D.A. Privileged structures: applications in drug discovery. *Comb. Chem. High. Throughput Screen.* **2004**, *7*(5), 473-94.

[117] Krovat, E.M.; Fruhwirth, K.H.; Langer, T. Pharmacophore identification, *in silico* screening, and virtual library design for inhibitors of the human factor Xa. *J. Chem. Inf. Model.* **2005**, *45*(1), 146-59.

[118] Frecer, V.; Megnassan, E.; Miertus, S. Design and *in silico* screening of combinatorial library of antimalarial analogs of triclosan inhibiting Plasmodium falciparum enoyl-acyl carrier protein reductase. *Eur. J. Med. Chem.* **2009**, *44*, 3009-19.

[119] Miteva, M.A.; Lee, W.H.; Montes, M.O.; Villoutreix, B.O. Fast structure-based virtual ligand screening combining FRED, DOCK, and Surflex. *J. Med. Chem.* **2005**, *48*(19), 6012-22.

[120] Kahraman, A.; Morris, R.J.; Laskowski, R.A.; Thornton, J.M. Shape variation in protein binding pockets and their ligands. *J. Mol. Biol.* **2007**, *368*(1), 283-301.

[121] Meek, P.J.; Liu, Z.; Tian, L.; Wang, C.Y.; Welsh, W.J.; Zauhar, R.J. Shape Signatures: speeding up computer aided drug discovery. *Drug Discov. Today* **2006**, *11*(19-20), 895-904.

[122] Oloff, S.; Zhang, S.; Sukumar, N.; Breneman, C.; Tropsha, A. Chemometric analysis of ligand receptor complementarity: identifying Complementary Ligands Based on Receptor Information (CoLiBRI). *J. Chem. Inf. Model.* **2006**, *46*(2), 844-51.

[123] Yamagishi, M.E.; Martins, N.F.; Neshich, G.; Cai, W.; Shao, X.; Beautrait, A.; Maigret, B. A fast surface-matching procedure for protein-ligand docking. *J. Mol. Model.* **2006**, *12*(6), 965-72.

[124] Hennemann, M.; Friedl, A.; Lobell, M.; Keldenich, J.; Hillisch, A.; Clark, T.; Goller, A.H. CypScore: Quantitative prediction of reactivity toward cytochromes P450 based on semiempirical molecular orbital theory. *ChemMedChem* **2009**, *4*(4), 657-69.

[125] Dickins, M.; van de Waterbeemd, H. Simulation models for drug disposition and drug interactions. *Drug Discov. Today* **2004**, *2*(1), 38-45.

[126] Valerio, L.G., Jr. *In silico* toxicology for the pharmaceutical sciences. *Toxicol. Appl. Pharmacol.* **2009**, *241*, 356-70.

[127] Nigsch, F.; Macaluso, N.J.; Mitchell, J. B.; Zmuidinavicius, D. Computational toxicology: an overview of the sources of data and of modelling methods. *Expert Opin. Drug Metab. Toxicol.* **2009**, *5*(1), 1-14.

[128] Nassar, A.E.; Kamel, A.M.; Clarimont, C. Improving the decision-making process in structural modification of drug candidates: reducing toxicity. *Drug Discov. Today* **2004**, *9*(24), 1055-64.

[129] Merlot, C. *In silico* methods for early toxicity assessment. *Curr. Opin. Drug Discov. Devel.* **2008**, *11*(1), 80-5.

[130] Walters, W.P.; Murcko, M.A. Prediction of 'drug-likeness'. *Adv. Drug Deliv. Rev.* **2002**, *54*(3), 255-71.

[131] van de Waterbeemd, H. Physicochemical concepts in drug design. *Exs* **2003**, (93), 243-57.

[132] Van de Waterbeemd, H. From *in vivo* to *in vitro*/*in silico* ADME: progress and challenges. *Expert Opin. Drug Metab. Toxicol.***2005**, *1*(1), 1-4.

[133] van de Waterbeemd, H.; Jones, B.C. Predicting oral absorption and bioavailability. *Prog. Med. Chem.* **2003**, *41*, 1-59.

[134] van De Waterbeemd, H.; Smith, D.A.; Beaumont, K.; Walker, D.K. Property-based design: optimization of drug absorption and pharmacokinetics. *J. Med. Chem.* **2001**, *44*(9), 1313-33.

[135] Beresford, A.P.; Segall, M.; Tarbit, M.H. *In silico* prediction of ADME properties: are we making progress? *Curr. Opin. Drug Discov. Devel.* **2004**, *7* (1), 36-42.

[136] Hutter, M.C. *In silico* prediction of drug properties. *Curr. Med. Chem.* **2009**, *16*(2), 189-202.

[137] Davis, A.M.; Riley, R.J. Predictive ADMET studies, the challenges and the opportunities. *Curr. Opin. Chem. Biol.* **2004**, *8*(4), 378-86.

[138] Keseru, G.M.; Makara, G.M. The influence of lead discovery strategies on the properties of drug candidates. *Nat. Rev. Drug Discov.* **2009**, *8*(3), 203-12.

[139] Benigni, R.; Bossa, C. Structure alerts for carcinogenicity, and the Salmonella assay system: a novel insight through the chemical relational databases technology. *Mutat. Res.* **2008**, *659*(3), 248-61.

[140] Muster, W.; Breidenbach, A.; Fischer, H.; Kirchner, S.; Muller, L.; Pahler, A. Computational toxicology in drug development. *Drug Discov. Today* **2008**, *13*(7-8), 303-10.

[141] Axerio-Cilies, P.; Castaneda, I.P.; Mirza, A.; Reynisson, J. Investigation of the incidence of "undesirable" molecular moieties for high-throughput screening compound libraries in marketed drug compounds. *Eur. J. Med. Chem.* **2009**, *44*(3), 1128-34.

[142] DeWitte, R.S.; Robins, R.H. A hierarchical screening methodology for physicochemical/ADME/Tox profiling. *Expert*

Opin. Drug Metab. Toxicol. **2006**, *2*(5), 805-17.

[143]   Muegge, I. Selection criteria for drug-like compounds. *Med. Res. Rev.* **2003**, *23*(3), 302-21.

[144]   Li, A.P. Screening for human ADME/Tox drug properties in drug discovery. *Drug Discov. Today* **2001**, *6*(7), 357-366.

[145]   Li, A.P.; Segall, M. Early ADME/Tox studies and *in silico* screening. *Drug Discov. Today* **2002**, *7*(1), 25-7.

[146]   Rishton, G.M. Failure and success in modern drug discovery: guiding principles in the establishment of high probability of success drug discovery organizations. *Med. Chem.* **2005**, *1*(5), 519-27.

[147]   Colmenarejo, G. *In silico* prediction of drug-binding strengths to human serum albumin. *Med. Res. Rev.* **2003**, *23*(3), 275-301.

[148]   Kramer, J.A.; Sagartz, J.E.; Morris, D.L. The application of discovery toxicology and pathology towards the design of safer pharmaceutical lead candidates. *Nat. Rev. Drug Discov.* **2007**, *6*(8), 636-49.

[149]   Szakacs, G.; Varadi, A.; Ozvegy-Laczka, C.; Sarkadi, B. The role of ABC transporters in drug absorption, distribution, metabolism, excretion and toxicity (ADME-Tox). *Drug Discov. Today* **2008**, *13*(9-10), 379-93.

[150]   Oprea, T.I.; Allu, T.K.; Fara, D.C.; Rad, R.F.; Ostopovici, L.; Bologa, C.G. Lead-like, drug-like or "Pub-like": how different are they? *J. Comput. Aided Mol. Des.* **2007**, *21*(1-3), 113-9.

[151]   Oprea, T.I.; Davis, A.M.; Teague, S.J.; Leeson, P.D. Is there a difference between leads and drugs? A historical perspective. *J. Chem. Inf. Comput. Sci.* **2001**, *41*(5), 1308-15.

[152]   Dearden, J.C. *In silico* prediction of ADMET properties: how far have we come? *Expert Opin. Drug Metab. Toxicol.* **2007**, *3*(5), 635-9.

[153]   Hou, T.; Wang, J. Structure-ADME relationship: still a long way to go? *Expert Opin. Drug Metab. Toxicol.* **2008**, *4*(6), 759-70.

[154]   Xu, J.; Stevenson, J. Drug-like index: a new approach to measure drug-like compounds and their diversity. *J. Chem. Inf. Comput. Sci.* **2000**, *40*(5), 1177-87.

[155]   Fostel, J. Predictive ADME-Tox London, UK, 27 -- 28 April 2005. *Expert Opin. Drug Metab. Toxicol.* **2005**, *1*(3), 565-70.

[156]   Summerfield, S.; Jeffrey, P. Discovery DMPK: changing paradigms in the eighties, nineties and noughties. *Expert Opin. Drug Discov.* **2009**, *4*, 207-218.

[157]   Yamashita, F.; Hashida, M. *In silico* approaches for predicting ADME properties of drugs. *Drug Metab. Pharmacokinet.* **2004**, *19*(5), 327-38.

[158]   Butina, D.; Segall, M.D.; Frankcombe, K. Predicting ADME properties *in silico*: methods and models. *Drug Discov. Today* **2002**, *7*(11 Suppl), S83-8.

[159]   Gedeck, P.; Lewis, R.A. Exploiting QSAR models in lead optimization. *Curr. Opin. Drug Discov. Devel.* **2008**, *11*(4), 569-75.

[160]   Gedeck, P.; Rohde, B.; Bartels, C. QSAR--how good is it in practice? Comparison of descriptor sets on an unbiased cross section of corporate data sets. *J. Chem. Inf. Model.* **2006**, *46*(5), 1924-36.

[161]   Clark, D.E. Prediction of intestinal absorption and blood-brain barrier penetration by computational methods. *Comb. Chem. High Throughput Screen.* **2001**, *4*(6), 477-96.

[162]   Mensch, J.; Oyarzabal, J.; Mackie, C.; Augustijns, P. In vivo, *in vitro* and *in silico* methods for small molecule transfer across the BBB. *J. Pharm. Sci.* **2009**, *98*, 4429-68.

[163]   Ekins, S.; Tropsha, A. A turning point for blood-brain barrier modeling. *Pharm. Res.* **2009**, *26*(5), 1283-4.

[164]   Egan, W.J.; Lauri, G. Prediction of intestinal permeability. *Adv. Drug. Deliv. Rev.* **2002**, *54*(3), 273-89.

[165]   Gleeson, M.P. Generation of a set of simple, interpretable ADMET rules of thumb. *J. Med. Chem.* **2008**, *51*(4), 817-34.

[166]   Di, L.; Kerns, E.H. Solution stability--plasma, gastrointestinal, bioassay. *Curr. Drug Metab.* **2008**, *9*(9), 860-8.

[167]   Ekins, S.; Andreyev, S.; Ryabov, A.; Kirillov, E.; Rakhmatulin, E.A.; Bugrim, A.; Nikolskaya, T. Computational prediction of human drug metabolism. *Expert Opin. Drug Metab. Toxicol.* **2005**, *1*(2), 303-24.

[168]   Ekins, S.; Nikolsky, Y.; Nikolskaya, T. Techniques: application of systems biology to absorption, distribution, metabolism, excretion and toxicity. *Trends Pharmacol. Sci.* **2005**, *26*(4), 202-9.

[169]   Colmenarejo, G.; Alvarez-Pedraglio, A.; Lavandera, J.L. Cheminformatic models to predict binding affinities to human serum albumin. *J. Med. Chem.* **2001**, *44*(25), 4370-8.

[170]   Kratochwil, N.A.; Huber, W.; Muller, F.; Kansy, M.; Gerber, P.R. Predicting plasma protein binding of drugs: a new approach. *Biochem. Pharmacol.* **2002**, *64*(9), 1355-74.

[171]   Fenner, K.S.; Troutman, M.D.; Kempshall, S.; Cook, J.A.; Ware, J.A.; Smith, D.A.; Lee, C.A. Drug-drug interactions mediated through P-glycoprotein: clinical relevance and *in vitro*-in vivo correlation using digoxin as a probe drug. *Clin. Pharmacol. Ther.* **2009**, *85*(2), 173-81.

[172]   Eyal, S.; Hsiao, P.; Unadkat, J.D. Drug interactions at the blood-brain barrier: fact or fantasy? *Pharmacol. Ther.* **2009**, *123*(1), 80-104.

[173] Erve, J.C. Chemical toxicology: reactive intermediates and their role in pharmacology and toxicology. *Expert Opin. Drug Metab. Toxicol.* **2006,** *2,* 923-46.

[174] Liebler, D.C.; Guengerich, F.P. Elucidating mechanisms of drug-induced toxicity. *Nat. Rev. Drug Discov.* **2005,** *4*(5), 410-20.

[175] Tompkins, L.M.; Wallace, A.D. Mechanisms of cytochrome P450 induction. *J. Biochem. Mol.Toxicol.* **2007,** *21*(4), 176-81.

[176] Lin, J.H. CYP induction-mediated drug interactions: *in vitro* assessment and clinical implications. *Pharm. Res.* **2006,** *23*(6), 1089-116.

[177] Pelkonen, O.; Turpeinen, M.; Hakkola, J.; Honkakoski, P.; Hukkanen, J.; Raunio, H. Inhibition and induction of human cytochrome P450 enzymes: current status. *Arch. Toxicol.* **2008,** *82*(10), 667-715.

[178] Gepp, M.M.; Hutter, M.C. Determination of hERG channel blockers using a decision tree. *Bioorg. Med. Chem.* **2006,** *14*(15), 5325-32.

[179] Mitcheson, J.S. hERG potassium channels and the structural basis of drug-induced arrhythmias. *Chem. Res. Toxicol.* **2008,** *21*(5), 1005-10.

[180] Raschi, E.; Ceccarini, L.; De Ponti, F.; Recanatini, M. hERG-related drug toxicity and models for predicting hERG liability and QT prolongation. *Expert Opin. Drug Metab. Toxicol.* **2009,** *5*(9), 1005-21.

[181] Greene, N. Computer systems for the prediction of toxicity: an update. *Adv. Drug Deliv. Rev.* **2002,** *54*(3), 417-31.

[182] van de Waterbeemd, H.; Gifford, E. ADMET *in silico* modelling: towards prediction paradise? *Nat. Rev. Drug Discov.* **2003,** *2*(3), 192-204.

[183] ten Brink, T.; Exner, T.E. Influence of protonation, tautomeric, and stereoisomeric states on protein-ligand docking results. *J. Chem. Inf. Model.* **2009,** *49*(6), 1535-46.

[184] LIGPREP *Maestro, Version 7.5*, Schrodinger, LLC: New York, 2006.

[185] Labute, P. Protonate3D: assignment of ionization states and hydrogen coordinates to macromolecular structures. *Proteins* **2009,** *75*(1), 187-205.

[186] Zhao, Y.; Cheng, T.; Wang, R. Automatic perception of organic molecules based on essential structural information. *J. Chem. Inf. Model.* **2007,** *47*(4), 1379-85.

[187] OpenBabel 2.1.1 http://openbabel.org/wiki/Main_Page

[188] myPresto. http://medals.jp/myPresto/index.html.

[189] Ten Brink, T.; Exner, T.E. Influence of Protonation, Tautomeric, and Stereoisomeric States on Protein-Ligand Docking Results. *J. Chem. Inf. Model.* **2009,** *49*, 1535-46.

[190] Mitra, R.; Shyam, R.; Mitra, I.; Miteva, M.A.; Alexov, E. Calculating the Protonation States of Proteins and Small Molecules: Implications to Ligand-Receptor Interactions. *Current Computer-Aided Drug Design* **2008,** *4*, 169-179.

[191] Lee, A.C.; Crippen, G.M. Predicting pKa. *J. Chem. Inf. Model.* **2009,** *49*(9), 2013-33.

[192] Polgar, T.; Magyar, C.; Simon, I.; Keseru, G.M. Impact of ligand protonation on virtual screening against beta-secretase (BACE1). *J. Chem. Inf. Model.* **2007,** *47*(6), 2366-73.

[193] Milletti, F.; Storchi, L.; Sforna, G.; Cross, S.; Cruciani, G. Tautomer enumeration and stability prediction for virtual screening on large chemical databases. *J. Chem. Inf. Model.* **2009,** *49*(1), 68-75.

[194] Milletti, F.; Storchi, L.; Sforna, G.; Cruciani, G. New and original pKa prediction method using grid molecular interaction fields. *J. Chem. Inf. Model.* **2007,** *47*(6), 2172-81.

CHAPTER 2

Structure-Based Virtual Screening

Olivier Sperandio, Bruno O. Villoutreix and Maria A. Miteva*

MTI, Inserm UMR-S 973, University Paris Diderot, 35 Rue Helene Brion 75013 Paris, France; Fax: +331 57 27 83 72, maria.miteva@univ-paris-diderot.fr

Abstract: The number of promising macromolecular targets involved in drug discovery programs has considerably increased because of the recent advances in human genomics and proteomics. In this respect, modern techniques like virtual screening in combination with high-throughput screening are well established approaches to assist identification of novel lead compounds. In particular, structure-based virtual screening is widely applied to search for new hit molecules among a large number of chemical compounds against therapeutically relevant protein targets with known three-dimensional structures. Here, we introduce the main techniques and programs applied for structure-based virtual screening with a focus on the molecular docking–scoring methodology and we discuss key issues that still need to be improved. In addition, we give recent examples of successful applications of hierarchical structure-based virtual screening methods combined with ligand-based ones.

INTRODUCTION

The number of promising macromolecular targets involved in drug discovery programs has considerably increased thanks to the recent advances in human genomics and proteomics. Drug discovery processes involve disease characterization, target validation, identification of hit molecules (screening), optimization of the lead compounds, pre-clinical and clinical trials. Such campaigns are very long and expensive and clearly optimizations in terms of speed and cost are required. Because it is very difficult to speed-up clinical trials, the only way to increase the drug discovery effectiveness is to optimize the initial hit/lead discovery stages. Today technologies like combinatorial chemistry and high-throughput screening (HTS) authorize biological assays of a large number of small molecules against therapeutically relevant targets (the number of drug/lead-like compounds is estimated to be at least 10^{24} [1]. However the escalating costs, from $500 000 to $1000 000 for HTS experiments of about 1 million compounds [2], highlight the need of developing novel approaches while still allowing to explore new areas of the chemical space. In this respect, virtual screening (VS), or *in silico* screening, is established as an attractive approach to handle large sets of compounds and to improve the "hit-rate" of drug discovery programs (Fig. **1**) [3, 4]. In fact, VS has become a method of choice for hit identification complementing HTS and NMR-based screens, such that experiments are only required for a small list of compounds pre-selected via computer methods [5, 6]. Numerous examples of hit molecules inhibiting enzymes or blocking/enhancing macromolecular interactions (either the catalytic or outside catalytic sites) clearly identified by virtual screening have been reported [7-13]. Moreover a recent report demonstrated that virtual screening contributed to the discovery of several compounds that have either reached the market or entered clinical trials, like for instance Aggrastat (Market), PRX-00023, PRX-03140 (Phase IIb) or PRX-08066, SC12267 (Phase IIa), PRX-07034 (Phase Ib), etc [9].

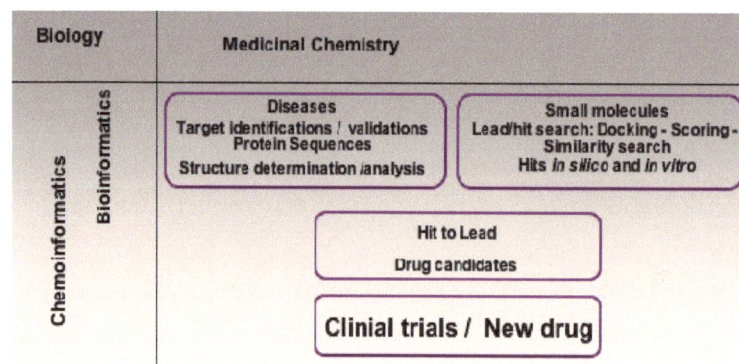

Figure 1: Combination of different approaches to assist the drug discovery.

The various VS methods directly used for hit identification can be divided into two large groups (Fig. **2**): ligand-based and structure-based [14]. The ligand-based methods widely used by medical chemists are very useful when the specific mechanism of ligand recognition by the target is not known, or when there is no sufficient structural information available [15, 16]. For ligand-based methods, the strategy relies on the principle that similar compounds have similar properties and that screening molecular databases for similar compounds to known potent ligands acting on a given target can provide new chemical entities having even greater potencies [17-21]. This can be done by similarity and substructure search, clustering, 2D-QSAR, 3D-QSAR (CoMFA, CoMSIA, etc...), pharmacophore matching or three-dimensional (3D) shape matching (*e.g.*, "lead-hopping" strategies). QSARs employ chemical descriptors that characterize a large variety of pharmacochemical properties of the small molecules, and are based on the assumption that a mathematical expression can be found to correlate these descriptors and the affinity of a compound for a given target. Fingerprint techniques are based on a signature of given fragments and as such allow a fast molecular similarity evaluation with respect to a reference molecule. For a more complete molecule presentation, pharmacophore searches are based on the definition of physicochemical properties such as hydrogen bond donors/acceptors, aromatic rings, hydrophobic groups. These techniques are very fast and thus appropriate for exploring huge chemical compounds collections.

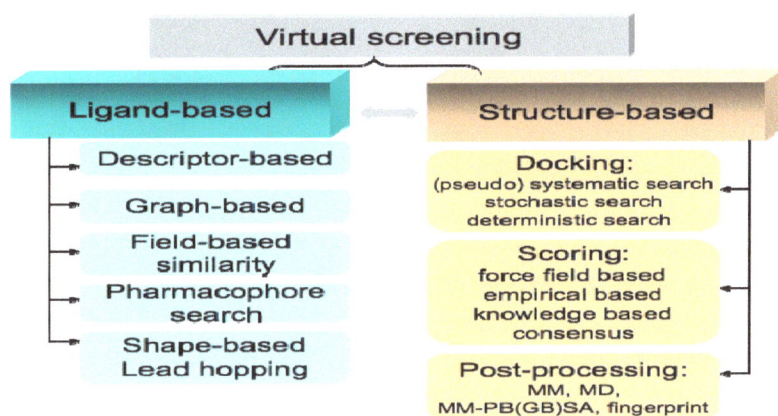

Figure 2: Main structure-based and ligand-based VS approaches for hit identification.

Virtual screening based on the 3D structure of macromolecular targets (structure-based SBVS) is widely applied to identify chemical entities that have a high likelihood of binding to a therapeutic target to elicit desired biological responses [22-25]. For SBVS methods (Fig. **3**), it is assumed that the 3D structure of the target is known either by X-ray crystallography or NMR experiments, or predicted by homology modeling [26-29] and that a compound collection in 3D is also available. The binding pocket to perform the search has to be known or has to be predicted. The principle here is to dock all the molecules present in a database into the binding pocket of the selected target and evaluate the fit between the molecules [3]. The quality of the fit is then used to rank the small molecules. Assuming that the receptor 3D structure is available, a primary challenge in lead discovery via SBVS is to predict both ligand orientation in the binding site and binding affinity; the former is often referred to as 'molecular docking' while the latter is referred to as 'scoring". Docking protocols can be described as a combination of two components; a search strategy and a simple scoring/fitting function to assess the poses prior to the real scoring step. The search algorithm should generate an optimum number of conformations for the ligands and of poses that should include the experimentally determined binding mode.

A number of reviews on computer aided drug design/VS, structure- and ligand-based methods have been published [30-37]. In other reviews we presented [24, 38] existing docking algorithms. In this chapter we will focus on the molecular docking-scoring methodology and on packages commonly used for SBVS. The complexity of molecular docking implies several approximations, from rigid body docking, to (pseudo)-flexible ligand docking (where the receptor is held rigid and the ligand is flexible) to flexible docking (where both receptor and ligand flexibility are considered). Algorithms dealing with flexibility can be divided in three types, namely systematic, stochastic and deterministic searches (*e.g.*, energy minimization and molecular dynamics MD) [39-41].

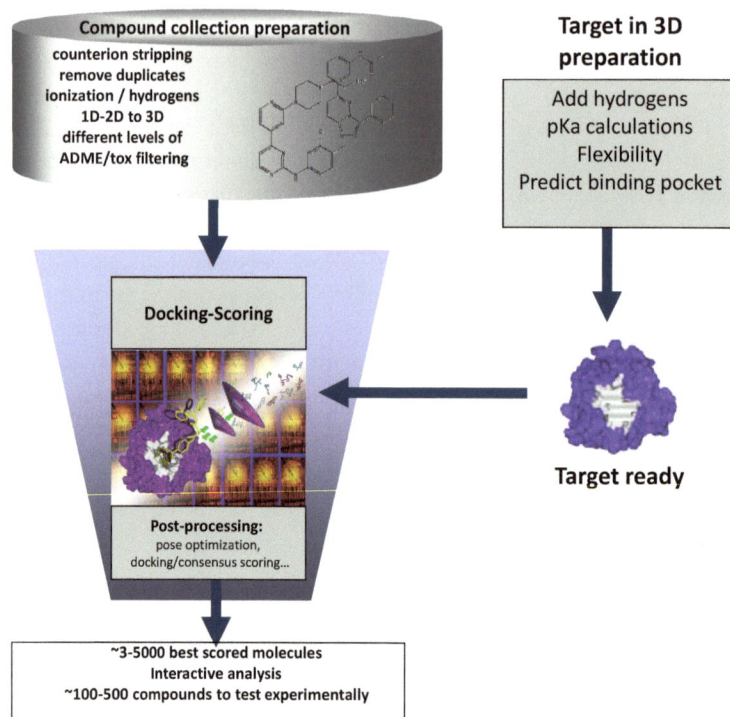

Figure 3: Structure-based VS protocol.

BINDING FREE ENERGY AND SCORING IN SBVS

The prediction of protein-ligand binding affinities, which is a basis for protein-ligand docking–scoring methodology, is extensively discussed in a number of reviews and articles [41, 42]. It is well known that molecular complexes (protein-protein or protein-ligand) are stabilized by interactions such as: van der Waals, hydrophobic, polar, ionic, and hydrogen bonds. Upon receptor-ligand binding significant solvation and entropic changes occur due to rearrangement of water molecules surrounding the unbound ligand and receptor. The stability of a complex (or the binding free energy ΔGbind) can be measured by determining the equilibrium binding constant, Keq (ΔGbind = - RT ln Keq). The binding free energy ΔGbind involving both enthalpic (ΔH) and entropic (ΔS) contributions can be presented as a difference between the free energy of the complex and the free energies of the receptor and the ligand in its unbound state. Rigorous and accurate computation of binding free energy can be done through time-consuming methods such as the free energy perturbation (FEP) and thermodynamic integration (TI). In such approaches the free energy difference between bound and unbound ligand and receptor is determined by slow intermediate changes from one state to the other. By now, these are the most precise theoretical methods but they are very time-consuming and as such, not appropriate for screening large compound collections. Along the same line, the linear interaction energy (LIE) approximation [43] has been shown to be useful for prediction of binding free energies for a small number of compounds, but in general still too computationally expensive for exploring a large number of small molecules.

More approximate models have been developed to evaluate relative binding affinities, the so-called scoring functions. They can be applied in the scoring stages in VS projects. Accurate prediction of relative binding affinities firstly depends on finding the correct binding poses. However, the required good binding modes are not sufficient for correct ligand scoring and ranking. The scoring of the ligands is a crucial step [30, 41] for the success of VS projects. Scoring functions are used (1) to evaluate different bound poses for a single ligand generated by the docking algorithm in order to select the energetically preferred pose; (2) to rank different docked ligands in order to discriminate the active compounds. Scoring functions developed during the last few years have been reviewed in [33, 37, 44]. The scoring functions commonly used are classified into three categories: force field-based, knowledge-based and empirical (Fig. **4**) [34, 44]. Some scoring functions include mixed force field and empirical terms. The force field scoring functions [45, 46] estimate the binding free energy as a sum of independent molecular mechanics force fields potentials, such as Coulomb, van der Waals, hydrogen bonding. Solvation [47, 48] and entropy [49] contributions can also be considered. It has recently been shown that free energy scoring methods by means of Molecular Mechanics-Poisson-Boltzmann

Surface Area/ Generalized Born Surface Area (MM-PBSA/GBSA) can be successful in improving the binding affinity prediction and ranking the actives [50-53]. The MM-PBSA methodology is based on calculation of the binding free energy ΔGbind of a ligand by combining molecular mechanics energy, solvation free energies with Poisson-Boltzmann (or Generalized Born) method, and entropy estimation from normal mode calculations [54-56].

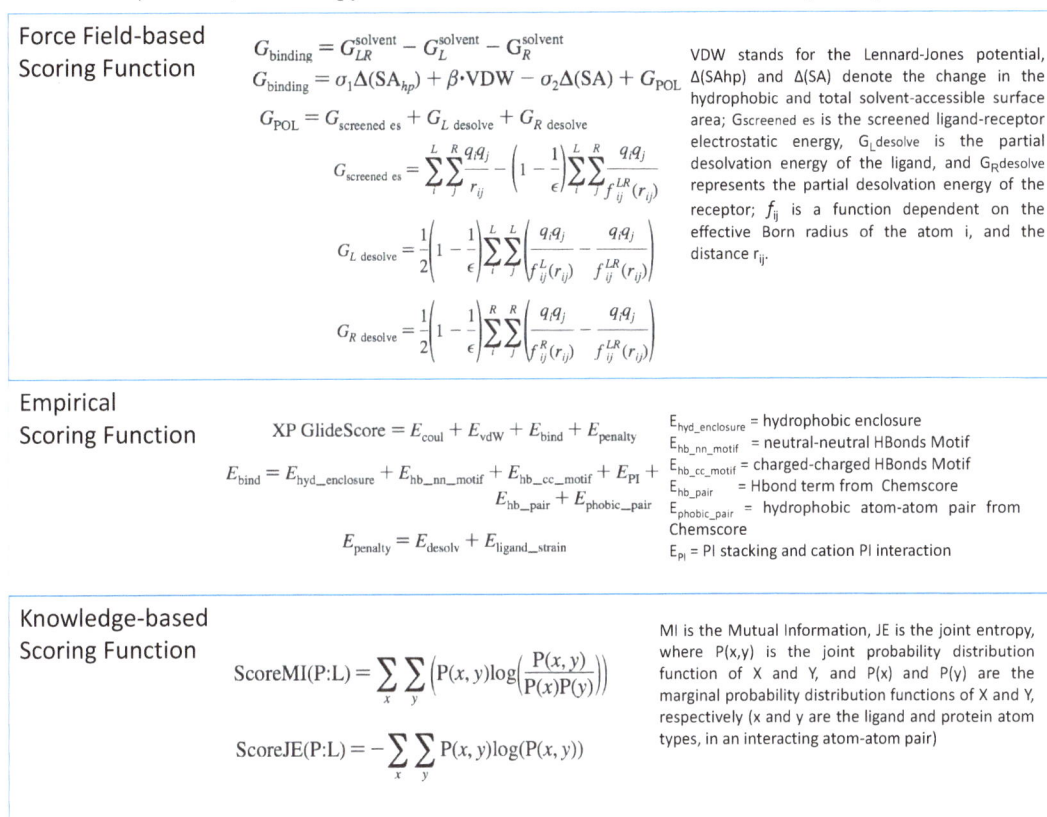

Force Field-based Scoring Function	$G_{binding} = G_{LR}^{solvent} - G_{L}^{solvent} - G_{R}^{solvent}$ $G_{binding} = \sigma_1 \Delta(SA_{hp}) + \beta \cdot VDW - \sigma_2 \Delta(SA) + G_{POL}$ $G_{POL} = G_{screened\ es} + G_{L\ desolve} + G_{R\ desolve}$ $G_{screened\ es} = \sum_i^L \sum_j^R \frac{q_i q_j}{r_{ij}} - \left(1 - \frac{1}{\epsilon}\right) \sum_i^L \sum_j^R \frac{q_i q_j}{f_{ij}^{LR}(r_{ij})}$ $G_{L\ desolve} = \frac{1}{2}\left(1 - \frac{1}{\epsilon}\right) \sum_i^L \sum_j^L \left(\frac{q_i q_j}{f_{ij}^L(r_{ij})} - \frac{q_i q_j}{f_{ij}^{LR}(r_{ij})}\right)$ $G_{R\ desolve} = \frac{1}{2}\left(1 - \frac{1}{\epsilon}\right) \sum_i^R \sum_j^R \left(\frac{q_i q_j}{f_{ij}^R(r_{ij})} - \frac{q_i q_j}{f_{ij}^{LR}(r_{ij})}\right)$	VDW stands for the Lennard-Jones potential, $\Delta(SAhp)$ and $\Delta(SA)$ denote the change in the hydrophobic and total solvent-accessible surface area; Gscreened es is the screened ligand-receptor electrostatic energy, G$_L$desolve is the partial desolvation energy of the ligand, and G$_R$desolve represents the partial desolvation energy of the receptor; f_{ij} is a function dependent on the effective Born radius of the atom i, and the distance r$_{ij}$.
Empirical Scoring Function	$XP\ GlideScore = E_{coul} + E_{vdW} + E_{bind} + E_{penalty}$ $E_{bind} = E_{hyd\_enclosure} + E_{hb\_nn\_motif} + E_{hb\_cc\_motif} + E_{PI} + E_{hb\_pair} + E_{phobic\_pair}$ $E_{penalty} = E_{desolv} + E_{ligand\_strain}$	$E_{hyd\_enclosure}$ = hydrophobic enclosure $E_{hb\_nn\_motif}$ = neutral-neutral HBonds Motif $E_{hb\_cc\_motif}$ = charged-charged HBonds Motif E_{hb_pair} = Hbond term from Chemscore E_{phobic_pair} = hydrophobic atom-atom pair from Chemscore E_{PI} = PI stacking and cation PI interaction
Knowledge-based Scoring Function	$ScoreMI(P:L) = \sum_x \sum_y \left(P(x,y)\log\left(\frac{P(x,y)}{P(x)P(y)}\right)\right)$ $ScoreJE(P:L) = -\sum_x \sum_y P(x,y)\log(P(x,y))$	MI is the Mutual Information, JE is the joint entropy, where $P(x,y)$ is the joint probability distribution function of X and Y, and $P(x)$ and $P(y)$ are the marginal probability distribution functions of X and Y, respectively (x and y are the ligand and protein atom types, in an interacting atom-atom pair)

Figure 4: Examples of a force field scoring function involving a GBSA term [48], an empirical scoring function (the Glide 4.0 XP scoring function) [57] and a knowledge based scoring function [58].

The empirical approximation [59, 60] represents the binding free energy as a weighted sum of interaction terms like hydrogen bonding and hydrophobic contacts by fitting the scoring function to experimental binding affinity data for a training set of protein-ligand complexes. These functions have proven to be successful for many protein-ligand complexes [61], although some uncertainty can be present when dealing with protein-ligand interactions not represented in the training sets. Recently new original empirical scoring functions have been developed (involving for instance improved models for hydrophobic interactions [57, 62], quantum chemical energy binding terms [63] or NMR chemical shift perturbation prediction [64]). The knowledge-based scoring functions [58, 65, 66] are exclusively derived from statistical analyses of atom-pairs frequencies in protein-ligand complexes with known 3D structures. The probability distributions of interatomic distances are converted into distance-dependent interaction energies of protein-ligand atom pairs. The several models developed differ in the definition of the set of ligand and protein atom pairs and the training set of protein-ligand complexes functions [58, 65, 66]. Other less explored knowledge-based methods propose to combine the advantage of data mining on natural structures and the atom type description provided by force fields [67]. In order to increase the efficiency in hits discovering instead on relying only on the above-mentioned scoring functions one may use in addition molecular fingerprints to prioritize specific interactions or to re-evaluate the compounds in a post-processing stage [68, 69].

The scoring functions have been shown to be the Achilles heel of structure-based methods when used indistinctively on any kind of protein targets. Nonetheless, such methodology can be very accurate to predict the binding affinity of ligands on some protein targets or dramatically deficient on others. Such that some have tried to rationalize their use either by combining them to ligand-based methods (mostly pharmacophore and/or 3D-QSAR see below) or by proposing target-specific scoring functions [70] through the definition of elaborated process that can manage to increase significantly the enrichment results on specific target proteins [71].

DOCKING ALGORITHMS

Rigid Body Docking with Multiple Conformers

Several docking programs perform rigid docking of pre-generated conformers by matching interaction points from the receptor site with ligand atoms. In the rigid body docking methods, an orientational search of the ligand in the protein-binding pocket is carried out while the receptor and the ligand remain rigid. One of the first molecular docking program for protein-small molecule interactions involving rigid body docking was DOCK developed by Kuntz and co-workers [72]. The program DOCK generates a negative image of the receptor - spheres that fill the binding pocket representing potential interaction sites. The DOCK algorithm attempts to superimpose the ligand atoms onto the centers of the spheres. Another rigid body docking program FRED (http://www.eyesopen.com) [73] applies a Gaussian shape fitting function to optimize the contact surface between the ligand and the protein which allows extremely fast rigid docking procedure. FRED filters the pose ensemble by rejecting the ones that clash with the protein using a negative image of the active site. The refined poses can then be scored using various scoring functions [74]. A full coordinate optimization of the pose via the MMFF force-field can also be performed. The refined poses can then be scored using one or more scoring functions, including MASC (Multiple Active Site Correction) corrected scoring functions [74]. The programs FLOG [75] and CLIX [76] apply grids with pre-calculated potential interaction energies with putative ligand atoms. The physical properties of the ligand atoms are divided into several atom types (neutral hydrogen-bond donors, neutral hydrogen-bond acceptors, polar, hydrophobic…). An alternative approach of continuum calculation of the binding energy is used in EUDOC [77] to score the docking poses via the three different force fields (TRIPOS, AMBER, CHARMM), while cation-pi interaction between the ligand and the target can also be taken into account.

Despite obvious limitations, rigid body docking methods are interesting because they are much faster than the flexible docking algorithms. Software such as FRED can dock up to 10 compounds per second on a standard mono-processor Linux workstation [73]. The speed and the relative accuracy of multiple conformation rigid body docking methods make them attractive [78] especially as an initial filter to remove from the docking library compounds that could not fit into the binding pocket or that have low surface complementarity with the receptor [54, 79]. One way to improve rigid body docking performance is to dock pre-generated multiple conformers of a ligand. In a modified version of DOCK [80], multiple conformations of ligands in the same frame of reference are docked as an ensemble, into a receptor binding site allowing extremely fast docking. Moreover, each ensemble of pre-generated ligand conformations can be processed into a hierarchical data structure such that atom connectivity is implicitly represented across all conformations of the ensemble [81]. Recently MS-DOCK [82] has been elaborated employing rigid body docking with DOCK 6.0 [83] on a multiple conformer state for each small molecule. MS-DOCK prioritizes the molecules that have satisfactory shape complementarity with the receptor-binding pocket using a simple contact scoring function (see for details [82]). It has been demonstrated that MS-DOCK is able to decrease the size of the initial compounds collection by 2- or 3-fold depending on the binding pocket shape and volume [82]. Multiple ligand conformers required for the rigid-body docking can be generated with well established commercial packages, such as Corina (Molecular Networks GmbH. http://www.mol-net.de), OMEGA (Openeye Scientific Software. http://www.eyesopen.com), Catalyst [84] or MED-3DMC [85] as well as with free online tools like FROG [86] (http://bioserv.rpbs.jussieu.fr/Frog.html). Yet, very few tools generating single or multiple 3D conformation for a large number of compounds are freely available. For example, the freely available Balloon program [87] (http://www.abo.fi/_mivainio/balloon/) offers the possibility to generate conformer ensembles for small molecules using a multi-objective genetic algorithm approach. We have recently developed the freely available programs DG-AMMOS [88] (http://www.mti.univ-paris-diderot.fr/en/downloads.html) and Multiconf-DOCK [82] (http://dock.compbio.ucsf.edu/Contributed_Code/index.htm) for generation of single and multiple conformers (based on systematic search).

Systematic and Incremental Construction search

Systematic (or pseudo-systematic) search algorithms try to explore all degrees of freedom. To reduce the number of evaluations to be performed, termination criteria is defined to prevent the algorithm from facing combinatorial explosion. Fragmentation/reconstruction algorithms (incremental construction methods) generally divide a ligand into small rigid cores and flexible linking parts (see Fig. **5**). The rigid core fragments are docked first into the binding site and the flexible parts are added incrementally to reconstruct the complete ligand.

The general approach of the program DOCK (http://dock.compbio.ucsf.edu/) for treatment of ligand flexibility [14, 89] (anchor-and-grow method) is divided into three main steps. First, the determination of a set of overlapping spheres in contact with the surface of the receptor site. Second, the center of these spheres is matched with the ligand atoms via the use of a graph-matching algorithm. Third, a scoring function is used to evaluate the pertinence of the docking poses. The evaluation of the ligand orientation uses a grid-based procedure in which steric and electrostatic interactions between the putative ligand and the receptor are pre-computed at each grid point. In order to score generated binding modes, as well to rank a number of ligands, several scoring functions can be used in DOCK3.5 and DOCK4.0: contact score, chemical score, energy score (Lennard-Jones van der Waals potential and Coulombic electrostatics with a distance-dependent dielectric constant). In the new versions of DOCK additional scoring functions have been developed taking into account the contribution of the solvation energy for molecular recognition: the Generalized Born/Surface Area (GB/SA) score (implemented in DOCK5) [48] and Poisson-Boltzmann/Surface Area (PB/SA) scoring function [90] (in DOCK6).

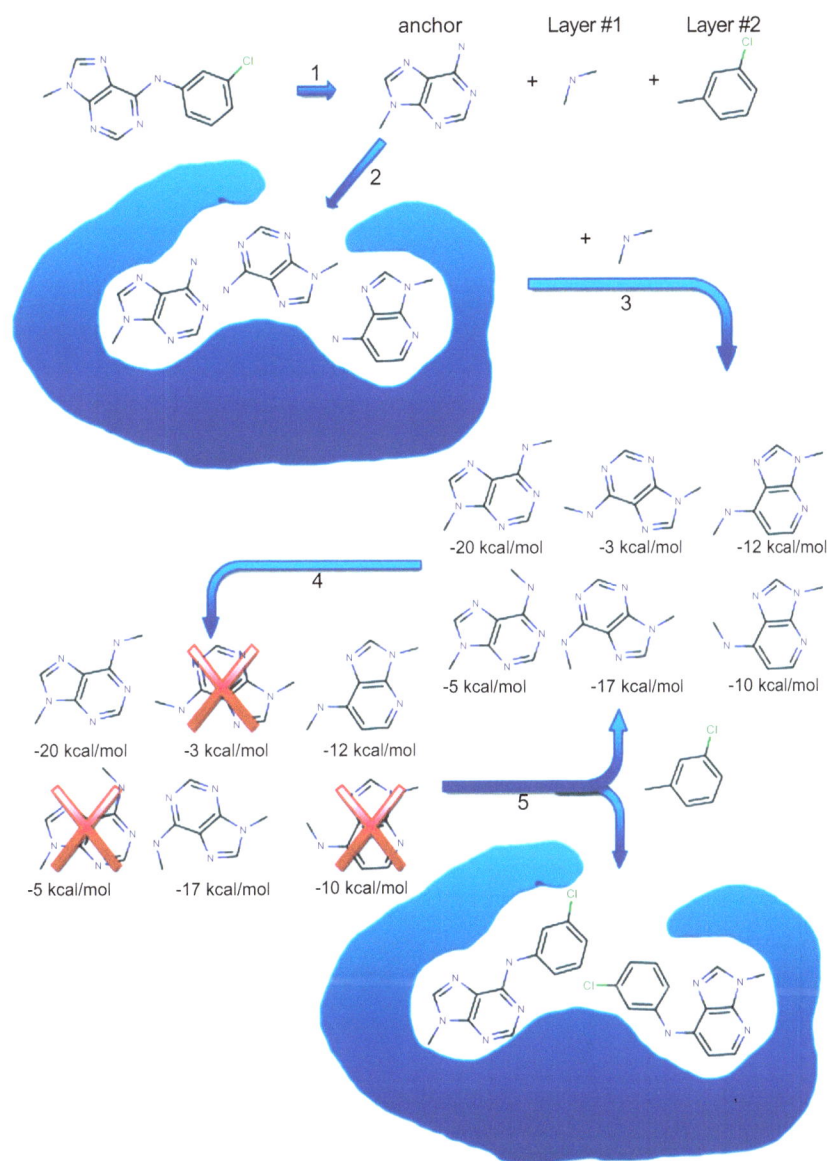

Figure 5: Incremental construction docking approach of DOCK [83]. The algorithm performs the following steps: (1) DOCK uses the ligand rotatable bonds to identify an anchor segment and overlapping rigid layer segments. (2) Rigid docking is used to generate multiple poses of the anchor within the receptor. (3) The first layer atoms are added to each anchor pose, and multiple conformations of the layer 1 atoms are generated. An energy score within the context of the receptor is computed for each conformation. (4) The partially grown conformations are ranked by their score and are spatially clustered. The least energetically favorable and spatially diverse conformations are discarded. (5) The next rigid layer is added to each remaining conformation, generating a new set of conformations. (6) Once all layers have been added, the set of completely grown conformations and orientations is returned.

Several programs based on the DOCK methodology have been developed [91, 92]. The program PhDOCK [92] is based on the pharmacophore representations of small molecules that are stored in a database. This pharmacophoric-point representation is compared to pre-defined DOCK site points in the binding region in order to orient the complex (ligand + protein). The pharmacophore representation is first used to overlay molecules based on their widest 3D pharmacophore. The basic objects of this representation are simply hydrogen-bond donors, hydrogen-bond acceptors, and ring centroids. For each orientation that provides a good match with the receptor points, the ensemble of conformers is docked into the binding site, and all members of the ensembles are scored.

The program Surflex (www.biopharmics.com) [93, 94] is based on a previously developed program named Hammerhead [95]. It uses the same concept of pocket finder and binding site-probing definition (protomols, see Fig. 6) but it is characterized by an innovative incremental construction of the ligand and recently refined scoring function. The program first creates an idealized binding site "a protomol" that serves as a template to which putative ligands or ligand fragments are aligned on the basis of molecular similarity. Each putative ligand is fragmented, resulting in 1-10 molecular fragments, each of which may have some rotatable bonds. Each fragment is then conformationally searched and each conformation of each fragment is aligned to the protomol to yield poses that maximize molecular similarity to the protomol. The scoring function terms involve hydrophobic and polar complementarity, entropic and solvation effects.

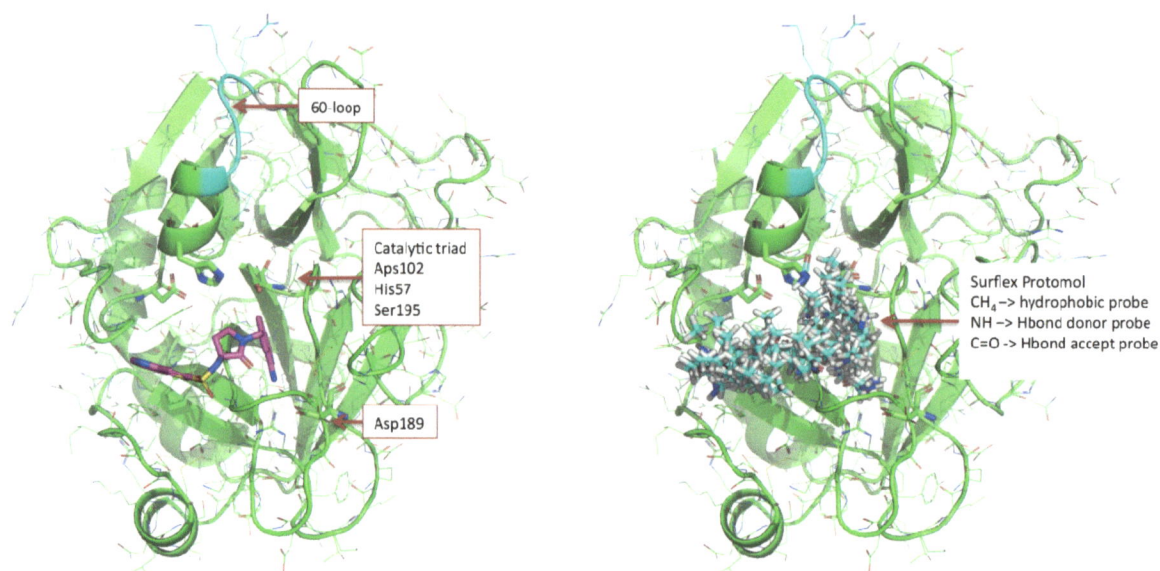

Figure 6: Surflex protomol on coaguation Factor Xa. Left panel: the structure of FXa is shown as cartoon and atomic details (green) with the 60-loop (cyan), the catalytic triad (D102, H57, S195) and Asp189 at the bottom of the P1 pocket both in sticks, along with a cocrystallized ligand (magenta). Right panel: the same orientation but with the Surflex protomol calculated from the localization of the cocrystallized ligand. It shows the hydrophobic interaction probes (CH4), the H-Bonds donor interaction probes (N-H fragments), and the H-Bonds acceptor interaction probes (C=0 fragments) that are used to incrementally dock the molecules within the binding site.

The software FlexX (http://www.biosolveit.de/) (http://www.tripos.com/) [96] docks flexible ligands into rigid receptors using an incremental approach and some concepts present in the LUDI program [97]. The approach can be divided into three areas: conformational flexibility, protein-ligand interactions and scoring. The conformational flexibility of the ligand is modeled by a discrete set of preferred torsion angles at acyclic single bonds [98] and multiple conformations for ring systems. With regard to the interaction scheme, FlexX relies on the detection of geometrically restrictive interactions such as hydrogen-bonds, specific hydrophobic interactions such as phenyl-methyl doublets, or spherical surfaces that are derived from favored interaction distances. The docking algorithm involves: base fragment selection, base fragment placement, and complex construction, where the ligand is built incrementally from the base fragment. The ranking of the ligands is performed with the modified empirical Bohm scoring function. This scoring function includes several (weighted) terms: a fixed ground term, a term taking into account the loss of entropy during ligand binding due to the hindrance of rotatable bonds in the ligand, hydrogen bonds, ionic interactions, aromatic interactions and lipophilic interactions.

eHITS [99] (http://www.simbiosys.ca/ehits/) also performs fast flexible ligand docking via systematic search. The eHiTS system generates all major docking modes that are compatible with the steric and chemistry constraints of the target cavity for each candidate structure. The output consists of multiple sets of 3D coordinates per structure with rough fitness scores that are highly configurable. eHiTS uses eHiTS_Score, a statistically derived empirical scoring function, with many novel features, including consideration of the temperature factors of crystal structures. In addition to a generic, default weight set, eHiTS_Score uses automated receptor family clustering which is used to create sets of family-specific scoring function weights which better represents those receptors.

Stochastic Search

Stochastic (or random) search algorithms involve random changes to modify the position of the ligand (translation and rotation) as well as torsion angles in order to generate different conformations. The main stochastic search methods are Monte Carlo (MC), Genetic Algorithms (GA) and Tabu search. MC can generate an ensemble of conformations that are statistically consistent at a given temperature. Random perturbations of the atomic positions are applied in order to explore the conformational space of the molecular system. An energy function evaluates whether the energy of the newly generated conformation is either lower than the one from the previous step or, if higher, is within an energy range defined by the so-called Boltzmann factor (Metropolis criteria).

The package LigandFit [100] (www.accelrys.com) has utilities to predict/define the binding site based upon the protein shape (flood-filling algorithm). A Monte Carlo method is employed for the conformational search of the ligand. During this search, bond lengths/bond angles are untouched while torsion angles are randomized. Multiple structural changes may thus occur at the same time during this step. Once a new conformation for the ligand is generated, the fitting of the compound in the binding pocket is carried out (shape similarity search procedure), eventually followed by rigid body minimization. If the shape is similar then the ligand is docked into the binding site and its binding energy is evaluated via an energy function called DockScore involving a soft 9-6 van der Waals term and an electrostatic term with a distance-dependent dielectric constant and, eventually, the internal energy of the ligand. Several scoring functions are available including, for instance, Ludi [97], LigScore [101] or PLP [102].

The method MCDOCK [103] is based on a three-stage strategy using a Monte Carlo algorithm consisting of increasingly refining the level at which the ligand is placed within the receptor site. This first stage (geometry-based docking) consists of placing the ligand inside the receptor site without major clashes by using a binary grid. A MC routine is in charge of positioning the rigid ligand into the binding site. The second stage uses another MC protocol (energy-based docking). After a global sampling that allows the eviction of bad contacts between the ligand and the receptor, a simulated annealing protocol is applied using Metropolis criteria. Finally, a MC protocol is used to prevent the system to be trapped in a local minimum. Other MC methods for SBVS are Dockvision [104], QXP [105], QUASI [106]. A new MC docking tool GlamDock [107] has been reported. The main features of the method are the energy function, which is a continuously differentiable empirical potential, and the definition of the search space, which combines internal coordinates for the conformation of the ligand, with a mapping based description of the rigid body translation and rotation.

GA are based and modeled on concepts borrowed from Darwinian evolution. Different ligand conformations and positions are generated, forming a population of solutions. This initial state of population is evolved as the lowest energy positions. This population is submitted to crossing over and random or biased mutations in order to form the next population. The algorithms maintain a selective pressure towards an optimal solution, with a randomized information exchange permitting exploration of the search space. After successive steps of evolution, the best ligand positions and conformations are kept resulting into the lowest energy ligand pose. These algorithms are used in many docking programs such as GOLD [108], AutoDock [46], DARWIN [109], FFLD [110]. For example, in GOLD, the mechanism for ligand placement is based on fitting points. The program adds fitting points to hydrogen bonding groups on protein and ligand, and maps acceptor points in the ligand on donor points in the protein and vice versa. The genetic algorithm optimizes flexible ligand dihedrals, ligand ring geometries, dihedrals of protein OH and NH groups, and the mappings of the fitting points.

The widely used open source program AutoDock3 [111] (http://www.scripps.edu/mb/olson/doc/autodock) employs a Lamarkian genetic algorithm (LGA) that incorporates a local minimization for a given fraction of the population. The LGA mixes a global search for ligand conformation and orientation, handled by a genetic algorithm switching between "genotypic space" and "phenotypic space, with an adaptive local search to perform energy minimization. The scoring function of AutoDock3.0 uses a pairwise sum of energetic terms with parameters for van der Waals, hydrogen bonding and distance-dependent dielectric electrostatics, as well as conformational torsional restriction entropy and empirical solvation terms. An easy-to-use applications to run AutoDock have been recently reported and should indeed help the use of this program for screening large chemical compounds' libraries [112-114]. The new version of AutoDock v.4 also includes receptor side chain flexibility. On the basis of AutoDock, the new program PSO@AUTODOCK has been recently developed [115] as a tool for fast flexible molecular docking based on a Particle Swarm Optimization (PSO) algorithm.

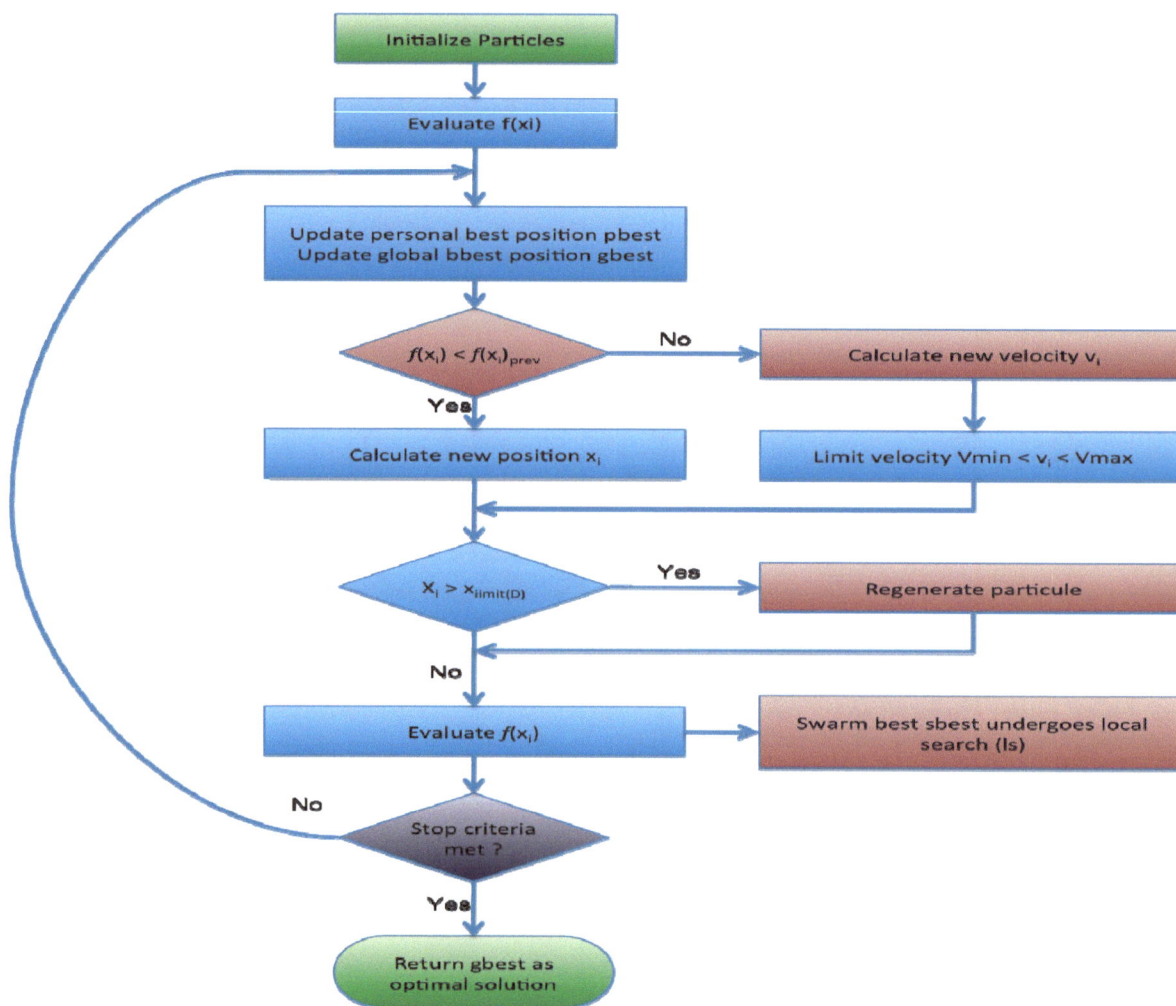

Figure 7: PSO@AUTODOCK: Pseudo-code of the particle swarm optimization algorithm.

An innovative molecular docking algorithm and three specialized scoring functions (for binding energy predictions, energy-ranking of docked ligand poses; and rank-ordering of active and inactive compounds) have been recently introduced in the Lead Finder docking software [116]. Lead Finder's algorithm for ligand docking combines the classical genetic algorithm with various local optimization procedures and resourceful exploitation of the knowledge generated during docking process. Lead Finder's scoring functions are based on a molecular mechanics function, which accounts for different types of energy contributions scaled with empiric coefficients.

Tabu search combines a minimization procedure with restrictions on the search path, such that the solution is forced into previously unexplored regions of the search space. It proceeds stepwise from an initial solution, while maintaining a list of previous solutions. The list of previous solutions provides both a ranking of solutions and a partial record of explored regions of the search space. The Tabu algorithm generates a set of N new solutions from the previous solution, and one of the N solutions is kept. A solution is added to the list if it is the best solution so far, or the solution explores a new region of the search space. A Tabu search algorithm is for example implemented in the PRO_LEADS package [117]. PSI-DOCK [118] is a recent SBVS package that makes use of a unique two-step searching/docking strategy. In the first step, it quickly explores the possible binding poses of the ligands using a Tabu enhanced genetic algorithm with a rapid shape complementarity scoring function.

A new docking program which is based on iterative stochastic elimination (ISE) algorithm [119] has been shown to provide high quality binding modes. ISE is a generic discrete combinatorial optimization method that iteratively eliminates values which contribute consistently to highest energy conformations. The docking algorithm PLANTS [120] is based on another stochastic optimization algorithm called ant colony optimization (ACO). ACO is inspired by the behavior of real ants finding a shortest path between their nest and a food source. In PLANTS, the flexibility of the protein is partially considered by optimizing the positions of some groups that could be involved in hydrogen bonding. Several scoring functions have been implemented proposing rescoring capability.

Deterministic Search

In the deterministic search, the initial state determines the change that can be made to generate the next state, which generally has to be energetically preferred as compared with the initial state. One problem can be a case where the simulation can be trapped in local minima. Deterministic methods for SBVS are energy minimization and molecular dynamics (MD) methods accounting for the flexibility of the receptor [121, 122, 40, 123]. Molecular dynamics as compared to MC simulations, cannot cross easily high-energy barriers within reasonable simulation time and at room temperature [41, 30]. Several simulation methods that could be useful for SBVS have been developed to overcome more rapidly the energy barriers, for example using simulated annealing molecular dynamics (SDOCKER [124]). Some authors propose to carry out time-consuming MD simulations at the final steps of the SBVS process on a smaller list of pre-selected compound in order to sample different conformations of the protein-ligand complex and to better predict the binding affinities [125, 54]. Approaches that combine several methods are emerging, like for instance, docking with FlexX and applying molecular dynamics and quantum mechanics and molecular mechanics methods [126] or molecular dynamics with Quantum-Refined Force-Field [127].

Flexible Receptor

One of the main challenges for the SBVS methodology today is to take into account conformational changes of the receptor upon ligand binding [128]. Indeed, receptor flexibility is neglected during VS experiments [30]. However, in many receptor-ligand interactions, significant conformational changes can occur upon binding [128-130], for instance the induced-fit of protein kinases upon inhibitor binding [129], HIV-1 protease [131], or at protein-protein interfaces [132, 133]. On the other hand, an incorrect inclusion of the receptor flexibility can lead to worse discrimination of the active compounds [134-136] due to an increase in the number of false positives. Various strategies have been proposed to take into consideration this critical issue reviewed in a number of papers as well as in Chapter 6 of this eBook. Here will only provide one of the methods reported to be efficient. Cavasotto and Abagyan [129] proposed the ICM-flexible (www.molsoft.com) receptor-docking algorithm (IFREDA) to account for protein flexibility during virtual screening. The ICM program is based on Monte Carlo simulation that relies on global optimization of the energy function of the flexible ligand in the receptor field via a grid receptor approach (receptor side chains/main chain flexibility). A Monte Carlo minimization procedure in the internal coordinate space is employed to search for the global minimum of the energy function. Each step of the algorithm consists of a random change of two types, torsional or positional, followed by a local minimization. Torsional changes of amino acid side chains at the interface can also be applied using a biased probability methodology. The SBVS scoring function used in ICM consists of internal force field energy of the ligand and the ligand/receptor interaction energy with eventually a term to account for the size of the binding site/ligand. The ligand/receptor interaction energy includes several weighted terms: van der Waals, a hydrophobicity term based on the solvent accessible surface buried upon binding, an electrostatic solvation term calculated using a boundary-element solution of the Poisson equation, hydrogen bond interactions and an entropic term proportional to the number of flexible torsions in the ligand.

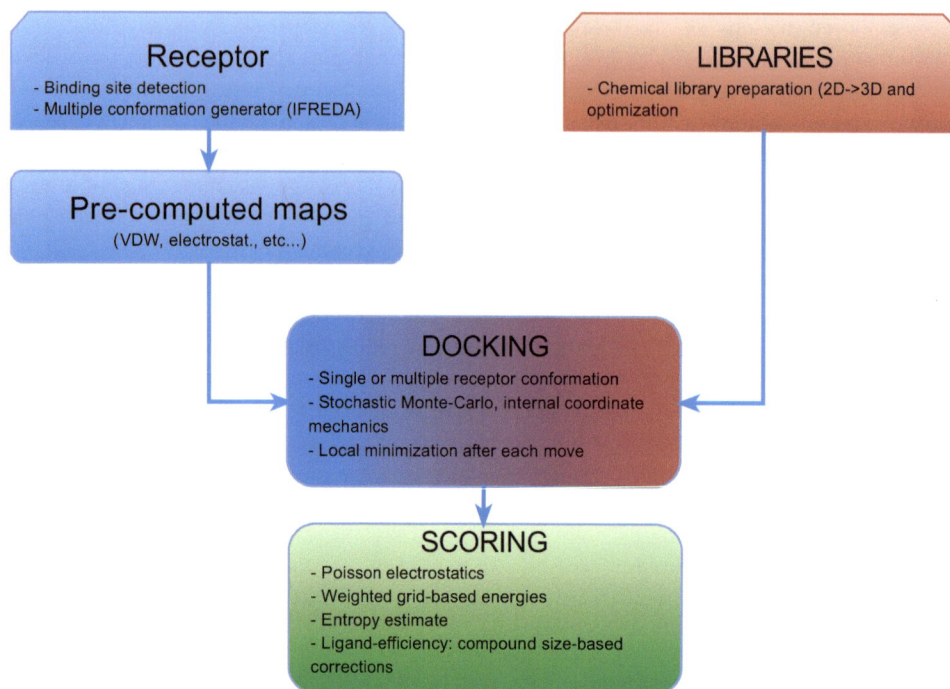

Figure 8: ICM IFREDA flowchart.

NEW STRATEGIES IN SBVS FOR LEAD DISCOVERY

Various concepts have been proposed to increase the efficacy of SBVS campaigns. One of the most exploited one is to combine different methods in order to increase the chances to find the most promising hit molecules.

Hierarchical Protocols

Recently attractive approaches including multi-step funnel protocols of different levels of docking-scoring filtering have been applied for lead discovery [54, 57, 79, 137]. It was demonstrated that such procedures can significantly improve the speed and the quality of SBVS process [138-140]). Several of these protocols start with pharmacophoric constrains or a geometrical matching of the protein target binding site and the ligands (see Fig. **9A**). The following more time consuming filtering steps usually involve flexible ligand docking and eventually partial receptor flexibility and different levels of precision in the estimation of binding energies with final free energy calculations involving estimation of van der Waals, Coulombic interactions, and changes in solvation and entropy due to the ligand binding [55]. Further, multi-step procedures can end with molecular mechanics, molecular dynamics or Monte Carlo. For example, recent hierarchical database screenings using a pharmacophore model, rigid body docking, solvation docking, and MM-PBSA have been shown to be useful in order to predict more precisely the binding energies [54, 56]. Another fast hierarchical docking approach, HierVLS method [137] starts with a coarse grain conformational search over a large number of configurations filtered with a fast but crude energy function, followed by a succession of finer grain levels, using more accurate but more expensive descriptions of the ligand-protein-solvent interactions. The method MPSim-Dock [141] combines elements of DOCK and MD methods available in the software MPSim. The widely known program Glide (http://www.schrodinger.com/) [142] (Fig. **9B**) also uses hierarchical filters to explore plausible docking poses for a given ligand within the receptor site. The shape and properties of the receptor are represented on a grid by several different sets of fields and calculations become progressively more accurate as the docking proceed. A set of initial ligand conformations is generated through exhaustive search of the torsional minima, and the conformers are clustered in a combinatorial fashion. The search begins with a rough positioning and scoring phase that significantly narrows the search space and reduces the number of poses to be further considered. The selected conformations are subjected to standard minimization with the OPLS-AA force field in the receptor binding site. Then, the 10 lowest-energy poses go through a MC procedure in which nearby torsional minima are examined, and orientation of peripheral groups of the ligand is then refined. The minimized poses are rescored using the GlideScore function, which is a more sophisticated version of ChemScore with force field based components and additional term accounting for the solvation. Freisner *et al.* recently reported a new Glide scoring function (Glide XP 4.0) with the goal to recognize diversity in target active sites by rewarding or penalizing interaction types that contribute significantly to gain or loss of binding free energy [57]. Particularly original are the new terms recognizing hydrophobic enclosure, neutral-neutral and charge-charge hydrogen bond binding interactions.

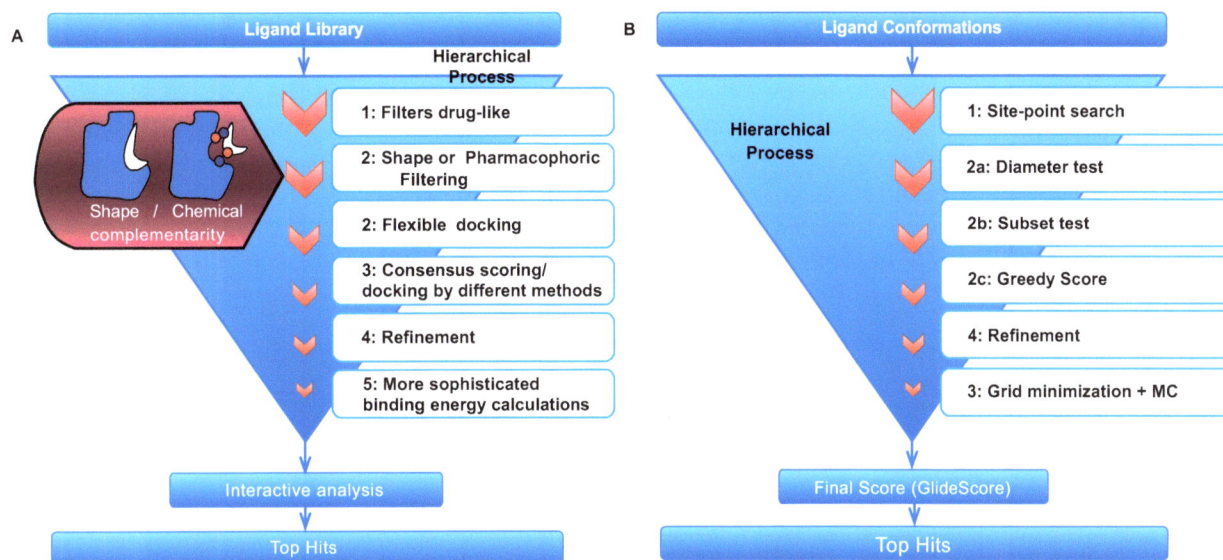

Figure 9: A: An example for a hierarchical SBVS protocol. B: The hierarchical algorithm of Glide [57].

In contrast to consecutively applying different docking programs Wolf and co-authors [143] suggested to develop an integrated system of different modeling schemes and algorithms in order to consider the particular advantages of various docking tools as an alternative of the consensus scoring techniques. They combined the docking algorithm of AutoDock with incorporated parts of the interaction scheme of FlexX in a docking workflow AutoxX. Additionally, a combined visualization may help to identify regions of the protein surface which energetically favor binding. Further, combination of methods can be important, not only to improve the results but also to reduce the computational cost. For example, studies combining FLOG and ICM-dock [144] or FRED-Surflex-DOCK [79] have been reported and successfully applied for identification of potent inhibitor of protein kinase CK2 [138].

Several recent studies reported new promising hits found via hierarchical SBVS (see Table **1**). For instance we have investigated the structures of various domains involved in transient membrane interactions and we found that in several cases a druggable pocket was present at the protein-membrane interface, suggesting that it should possible to use SBVS to identify molecules able to disrupt protein-membrane interactions [139]. We used a multi-step SBVS protocol on the second discoidin (C2) domain of coagulation factor V (FV) binding transiently to the appropriate membrane surface. We started with a collection of over 300,000 drug-like molecules. These molecules were rigidly docked using FRED into two different crystal forms of the FV C2 domain. The top 60,000 molecules after FRED docking forms were selected for flexible docking with Surflex. The top 2000 docked poses were re-docked and investigated with LigandFit to search for consensus poses with Surflex. The final lists of about 1000 selected molecules were tested *in vitro* in functional assays as well as with the Biacore system. Finally, seven drug-like hits were identified, indicating that therapeutic targets that bind transiently to the membrane surface can be investigated cost-effectively and that inhibitors of protein-membrane interactions can be designed. Similar strategy has been employed to find new non-covalently bound inhibitors of the anti-cancer target the dual-specificity phosphatase Cdc25 [145].

Whereas the majority of hit compounds found by SBVS approaches are in the micromolar range of activity, Cozza *et al.* have identified a nanomolar inhibitor of CK2 [138] implicated in a wide variety of important cell functions. Cozza *et al.* [138] used a multi-step protocol proposed by Miteva *et al.* [79] and combined four docking tools and five scoring functions. They screened the ATP binding site of CK2 with their in-house molecular database of 2000 natural-occurring compounds via a rigid-body docking step with FRED, followed by flexible docking by MOE-Dock, Glide, GOLD and the top 5% was selected for consensus scoring. This procedure resulted in new nanomolar potent CK2 inhibitors. In a different manner, ChemBridge Express Pick and the National Cancer Institute Open database with 350000 and 250000 compounds have been screened subsequently with GOLD, then re-ranked by a consensus scoring of five scoring functions, and finally docking with FlexX on the Mycobacterium tuberculosis HisG, an ATP-phosphoribosyl transferase [146]. Two novel hits showing bacteriocidal whole-cell activity have been found that opens up a new avenue of possibilities for drug design for tuberculosis, specifically through inhibitors of HisG.

Another possibility for hierarchical SBVS of a large database is to employ pharmacophoric filters. For instance in [147], a library of 1.6 million compounds has been screened *in silico* against a model of CCR5 by sequential filters. Identifying GPCR agonists by SBVS is a challenging task, because GPCRs are expected to undergo significant conformational changes upon activation. After drug-like and 2D-pharmacophore filters the initial banks has been reduced to 44000 compounds subjected lately to flexible docking Surflex and GOLD yielded a hit list of 59 compounds. Out of these 59 molecules, 10 compounds were found to display micromolar affinities to the CCR5 receptor.

Table 1: Recent success stories with hierarchical SBVS or combined ligand-based and SBVS methods.

Drug target	Disease function	Receptor structure (PDB code)	VS method	Activity of the best found hits
Protein kinase CK2 [138]	cancer	1JWH	Shape filter by rigid docking: FRED; MOE-Dock/ Glide/GOLG	K_i =20 nM
Protein kinase CK1 delta [150]	cancer; neurodegenerative diseases (Alzheimer's and Parkinson's disorders)	Homology model	MOE-Dock, Glide, GOLD, FlexX	IC50 = 0.3 and 0.6 μM
Factor V membrane Interaction [139]	coagulation system	1CZT, 1CZS, 1CZV	Shape filter by rigid docking: FRED; Surflex, LigandFit	IC50=3.5 μM
Dual-specificity phosphatase CDC25 [145]	cancer	1CWT	Shape filter by rigid docking: FRED; Surflex, LigandFit	IC50=13 μM
Ubiquitin C-terminal hydrolase-L3 [151]	ubiquitin-proteasome system; programmed cell death	1XD3	DOCK, GOLD	IC50=100 μM
Farnesoid X receptor [152]	hyperlipidemia; cholelithiasis	1OSH	DOCK 4.0, CScore FlexX	concentration at half-maximal stimulation 1.86 - 6.92 μM
Mycobacterium tuberculosis HisG [146]	tuberculosis	1NH8	GOLD, FlexX	IC50 = 4-10 μM
GPCRs thyrotropin-releasing hormone receptors antagonists [149]	neurodegeneration	Homology model	Receptor-based pharmacophore filtering, FlexE	K_i = 0.29-3.2 μM
GPCRs: CCR5 receptor Agonists [147]	favorable for protection against HIV-1 infection	Homology model	Pharmacophore filtering, Surflex, GOLD	IC50 = 17 μM
Dipeptidyl Peptidase IV [148]	diabetes type 2	1N1M	Pharmacophore filtering, Glide	82% inhibition at 30 μM
Thymidine monophosphatase kinase [153]	tuberculosis	1W2H	3D pharmacophore filtering, FlexX, SiFT	Minimal inhibitory concentration 3.12 μg/ml
Constitutive Androstane Receptor (CAR) [154]	homeostasis; metabolism	1XVP	Pharmacophore filtering (Unity), GOLD	Agonists from 2 to 8 fold CAR activation at 10μM
SARS-3CL^{pro} enzyme [155]	Severe Acute Respiratory Syndrome	1UK4, 2AMD	Pharmacophore filtering (Catalyst), GOLD	IC_{50} = 17.2 μM
Trypanothione Reductase [156]	African sleeping sickness; Chagas disease; leishmaniasis	1AOG	Similarity screening (Molinspiration), AutoDock4, consensus scoring	IC_{50} = 11.5 μM
Yersinia protein kinase A [157]	Plague	Homology model	Support Vector Machine screening, FlexE	IC_{50} = 1.81 μM
HIV-1 integrase [158]	HIV-1	1BL3	Phamacophore (Catalyst), EIIP screening (general model pseudopotential), AutoDock3.0	IC_{50} = 69 μM
Protein tyrosine phosphatase (SHP2) [159]	Noonan syndrome; cancer	Homology	GOLD, SAR (SARNavigator)	K_i = 0.7 μM

Ward *et al.* have also been using hierarchical SBVS with pharmacophoric pre-filtering to search for inhibitors of Dipeptidyl Peptidase IV (DPP-IV) [148]. A single-conformer database of the 40000 selected molecules was then docked into the DPP-IV crystal structure using Glide. Clustering and visual inspection were used to select a final list of 4000 compounds for experimental screening. Finally 51 active compounds were identified. By employing receptor-based pharmacophore filtering and docking with FlexE [149] one million compounds have been screened leading to experimental identification of five structurally diverse antagonists of the thyrotropin-releasing hormone receptors playing an important role in the regulation of thyrotropin (thyroid-stimulating hormone, TSH).

Combining Structure-based and Ligand-based Techniques

Although the purpose of this chapter is not to describe the various ligand-based methods it is interesting to see how they can be combined to SBVS methods. On one hand ligand-based methods tend to be very efficient and low-time consuming for screening large chemical libraries to identify similar molecules to known potent compounds, but it is often said that they cannot offer new compounds with real scaffold hopping and that they remain trapped in the vicinity of the query compound chemical space region, although some techniques are now capable of such leap within the chemical space [160-163]. On the other hand structure-based techniques have no chemical bias toward any kind of privileged substructures, but the computational cost associated to a confident evaluation of the binding energy of the putative ligands has been a major drawback for a long time precluding the use of such techniques on very large compounds collection (> 1 million). The debate whether ligand-based techniques are more efficient than structure-based techniques considering enrichment results is still intense, and convincing arguments can be found to support either opinion. The comparison between screening techniques is not trivial, and depends on a series of conditions that must be carefully considered. The parameters of the techniques are to be set such as not favoring one method over another, because it is known that certain methods have optimized default parameters and other require user-defined settings without which the method might appear to perform poorly. The nature of the data sets used usually greatly influences the quality of the results, thus certain methods may perform well one a given data set and poorly on some others. For example the assessment of ligand-based screening techniques is dramatically influenced by the nature of the ligand query. So it might be important to evaluate the quality of the methods by considering more than one ligand query per target [162, 164]. The same should be true when dealing with structure-based techniques through the assessment of several experimental 3D structures of the receptors tested. Although for a long time ligand- and structure-based methods have been respectfully ignoring each other with few considerations for their respective improvements, it is now well established that instead of comparing them it is more appropriate to combine these two types of approaches when possible [165] *i.e.* when both 3D structures of the target protein (X-ray, NMR, comparative modeling) and known active compounds are available. This can bring great benefit to the identification of new chemical entities (NCE). It is supported by the growing number of success stories on this matter [155-159]. The goal sought when combining ligand-based and structure-based methods is to balance their own limitations and benefit from their strength. Such combination can be done either as a series or in parallel. The parallel combination aims at comparing the selected compounds of both approaches either as a complementary selection (top ranked compounds from each methods) or as a consensual selection (compounds selected by both approaches). The serial combination is designed as a hierarchical procedure with, in general, the fastest and less refined approaches as the first step, and the most computational expensive techniques as the second step (Fig. **10**).

A recent study combined successfully several techniques from ligand-based machine-learning (Support Vector Machine - SVM), homology modeling, ensemble docking and consensus scoring [157]. It aimed at identifying inhibitors of the Yersinia protein kinase A (YpkA) to target plague virulence factors (Fig. **11**). In this study, the authors first designed a kinase focused chemical library using machine-learning techniques on known kinase inhibitors and inactive compounds on kinases. Initially, 364 kinase inhibitors on one hand and 4220 compounds randomly selected from the MDDR database (Elsevier MDL, San Leandro, CA) on the other hand were filtered for ADME/tox properties and used as the learning data set. The SVM model was employed to create a kinase focused library of 200,000 compounds starting from a 1,000,000 compound library. Because no 3D structure was available for the Yersinia protein kinase A, two homology models were generated using MAPK templates both apo- and holo-structures to address the induced fit issue of the ATP binding site upon binding. In order to further sample the binding site conformations, 8 representative conformations were selected from a 2.0 ns MD simulations applied of these two models. An ensemble docking process was applied on these 8 conformations using the docking program FlexE. A scoring procedure consisting of re-ranking with Xscore the top 1000 compounds scored by the FlexE scoring function combined to a visual inspection lead to the selection of 45 compounds, which were tested experimentally against YpkA. Seven of these 45 molecules showed a complete inhibition of YpkA at 225-450 μM with an IC50 ranging from 1.81 to 50 μM.

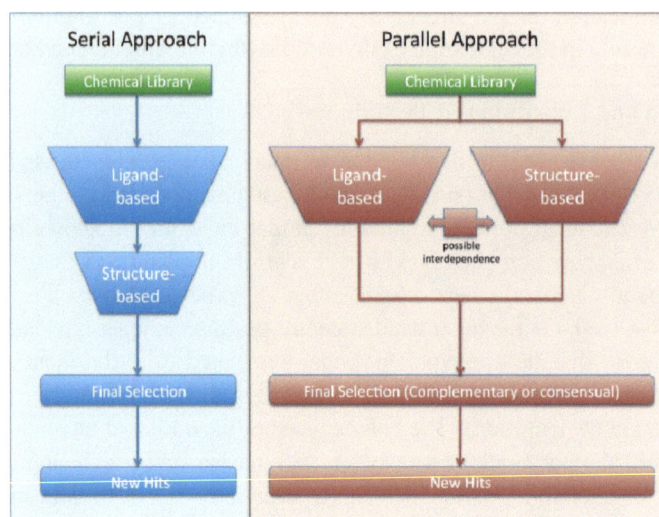

Figure 10: Examples of mix-process of ligand- and structure-based approaches either as a serial (hierarchical) combination, or as a parallel (consecutive) combination with possible interdependence of the processes in the latter case (see example below).

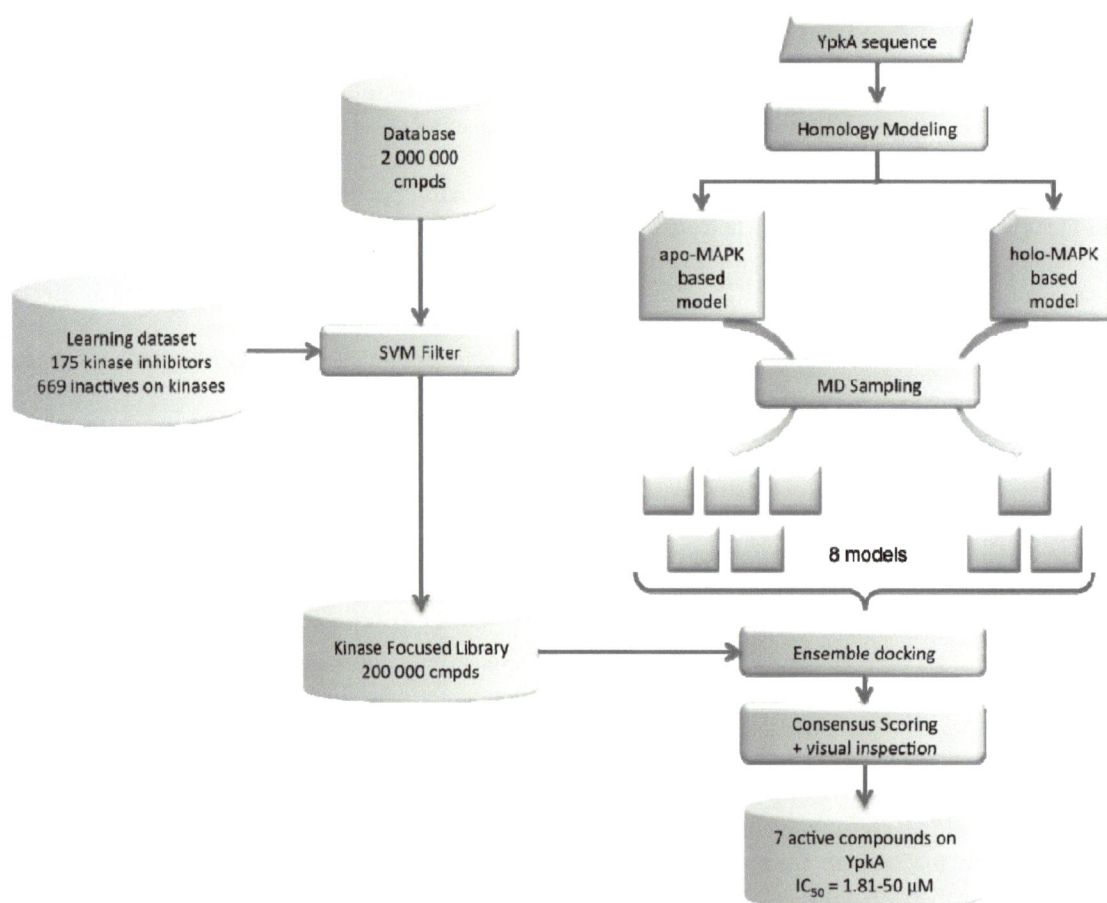

Figure 11: Combining machine-learning, homology modeling, ensemble docking and consensus scoring for to identify inhibitors of the Yersinia protein kinase A [157].

Finally, as an example of a sophisticated parallel combination with interdependence, a recent study compared the performances of a conventional structure-based screening with a series of more sophisticated protocol including multiple target screening (Multiple Active Site Correction - MASC), direct score modification (from homologous proteins), machine-learning score modification (using known active compounds on the same target), and a pure ligand-based method using docking scores as descriptors [166]. This study was carried out on 4 different proteins HIV-1 protease, cyclooxygenase-2, thermolysin and glutathione S-transferase. The structure-based technique including the machine-learning score modification achieved far better enrichment results than conventional docking. The concept of this method is to calculate an optimized score for each a compound on a protein by setting a weighted average of the raw docking scores of that compound. Also performing well in the same study was the ligand-based machine learning methods using molecular descriptors. This ligand-based method used docking scores on many proteins as descriptors and compounds similar to active compounds were selected as candidate hit compounds. The study showed that the two best protocols were already a form of a combination of raw screening techniques, but interestingly the ultimate combination of these two hybrid protocols gave even better enrichment results.

ADVANCES AND CHALLENGES IN SBVS METHODOLOGY

Following the presented methods and success of SBVS one can confirm the enormous potential to find new scaffolds without bias towards known hits or leads. However, there is still a room for improving the efficiency of SBVS. While SBVS is known to give valuable information for selection of new hits among millions of compounds, it is also costly in terms of computational time, *e.g.* the extensive treatment of ligand and receptor flexibility is critical for docking accuracy but increases significantly the computational time. To increase the speed, computations can be carried out on clusters and/or via grid computing technologies but limiting factors are still present, such as the number of commercial software licenses to perform docking simultaneously on multiple processors. Therefore developing new, faster and more efficient docking methods to reduce biocomputing cost has become a priority for many research groups [112, 167-171]. In this context a number of tools and programs essentially free for academic groups are proposed [32] and recently summarized in [35] where links for Structural Bioinformatics/Chemoinformatics tools are provided to assist SBVS (*in silico* ADME/tox prediction tools, compound collections preparation, *in silico* screening, some ligand-based methods, characterization/simulation of 3D targets and homology modeling tools, druggable pocket predictions, active site comparisons, analysis of macromolecular interfaces, protein docking tools to help identifying binding pockets and protein-ligand docking/scoring methods). Successful applications of free tools for identification of new hit molecules are also given in this review [35].

One of the keys for a successful SBVS is the performance of current methods to rank the proceeded chemical database according to ligand binding affinities. Numerous studies benchmarking docking-scoring methods [22, 25, 34, 172-176] have shown that even sometimes one docking program outperforms others, in most of the cases the commonly used docking engines provide the good pose among the generated docking poses. However, distinguishing among the docking poses as well as among the likely candidates is a challenge, and reliable rank-ordering corresponding to the binding affinities remains a weakness of today's scoring algorithms. Indeed, evaluation of affinity has to be very fast in virtual screening while at the same time very complex energy terms have to be assessed. These include pi- stacking, cation-pi interactions, salt-bridges, hydrogen bonds, entropy, hydrophobic contacts while at the same time one should deal with flexibility and water molecules or ions. Clearly, this is not fully feasible at present to compute precisely all these terms. Even, using expensive computational methods like MD or MM-PBSA it is difficult to assure sufficient sampling to obtain good averaging of thermodynamic properties or to achieve a complete induced-fit [41]. A comprehensive analysis of scoring functions advantages and weaknesses can be seen in [33, 37, 44]. A common conclusion is that no single scoring function performs well across different protein targets [33, 58, 177-180]. In this line, consensus scoring [181-183] or consensus docking-consensus scoring can also be used to prevent the approximations inherent to each scoring function [177, 179, 182, 184]. This type of approaches relies on the combination of several scoring functions in order to reciprocally compensate the drawbacks of each function used and promote the identification of true positive hits. The importance of consensus scoring is however still under debate (see for example [79, 185, 186]. One problem might be that a large number of known active molecules should be available in order to select the best combination of several scoring functions. In a recent paper [187] it is suggested to use two scoring functions as the objectives for the binding conformation search as a more effective docking strategy. Some research groups suggest to employ quantum mechanics-based scoring functions [188, 189], fingerprints, like SIFt (structural interaction fingerprint) [153, 190] or hot-spots-guided knowledge-based approaches [191] to ensure that only realistic binders are prioritized. Optimization in scoring can also be achieved by integrated description of hydrogen bonding and dehydration, as for instance in the HYDE scoring function [192].

Another direction to optimize ranking and enrichment is to tailor scoring functions to specific properties of the binding site. Several recent studies provide an analysis of the performance of commonly used scoring functions on binding sites with specific physicochemical properties in terms of size, shape, polarity [79, 175, 193]. These studies demonstrated that the selection of an appropriate scoring function for a binding pocket with particular characteristics is a multifactor and complex task. A common conclusion is that scoring functions performed better on the proteins with well-shaped and closed binding pockets than on those with very open binding sites. It has also been observed that some docking-scoring method do better on pockets of predominant hydrophobic nature while others perform better on the mixed or hydroplilic pockets [175, 184, 193]. Further, in many cases, it will be important to tune the scoring function for given targets assuming one has enough experimental information to start with [36, 74, 79, 194-196].

Recently it has been suggested that post-docking optimization, either after conventional docking-scoring procedures or after hierarchical SBVS protocols may help to further improve both, the docking pose and the scoring, and such the overall efficiency of SBVS experiments. Post-docking optimization prior to re-scoring with force field scoring functions [197] could be applied to avoid some problems with steric clashes generated during the docking. Recent examples of docking poses and enrichment improvements after post-docking energy minimization support this view [94, 197-199]. In the same context, PostDOCK [200] performs a post-processing filtering by biochemical descriptors and machine learning tools to select true binding ligand-protein complexes generated by DOCK4. Other methods for post-processing of docking results have been investigated, such as MASC (multiple active site correction) [74]. Recently, a new open source program AMMOS [201] (http://www.mti.univ-paris-diderot.fr/en/downloads.html) has been proposed which addresses this point by employing energy minimization of pre-generated by docking protein-ligand complexes allowing partial or full atom flexibility from both, the ligand and the receptor sides.

Despite of the above-mentioned advances, there is room for improvements in several important SBVS issues. First, the solvent contribution is crucial for ligand binding [202, 203], and as such for the docking-scoring methodology. Accurate computations in explicit water environment are much more costly than in implicit one. Among the most precise methods are the thermodynamic integration/ free energy perturbation (TI/FEP) which can calculate the differences in binding affinities about 1 kcal.mol^{-1} [33, 204], obviously not feasible for screening of large chemical libraries. In practice, in most of the SBVSs the solvent is completely omitted from the binding energy calculations. One may consider such effects in docking-scoring by incorporating a small number of water molecules explicitly in the binding site. However, placing water might diminish the volume of binding pocket and as such prevent potential ligands to dock successfully. In addition, the presence of water molecules makes the scoring problem more complex – how to score the interactions between the ligand/receptor and the water molecules? Including water molecules as a part of the protein moiety can be useful in cases where tightly bound waters are present in protein structures in complexes with different ligands (for instance a water molecule has been observed within the co-crystal structures of HIV-1 protease with KNI-272 and analogues [205, 206]). On the other hand, often ligand binding can provoke reordering of water molecules, for example some other inhibitors of HIV-1 protease displace this water molecule [207, 208]. The dynamics of conserved water molecules upon ligand binding has been also investigated by MD simulations in the case of Hsp90 [209] where it has been revealed that only four conserved water molecules among many present in the X-ray binding pocket structure are important for new inhibitor design, because their positions are stabilized when the conformation of the protein is changed. Recently, a protocol to identify "relevant" water molecules in protein binding sites using HINT score and a geometric descriptor termed Rank has been described [210]. Clearly, the best would be water molecules that could be displaced during docking experiments. Currently only a few docking methods propose such possibilities. It has previously been shown that AutoDock gained accuracy when water molecules were made displaceable [211]. GOLD [212] can include user-defined waters present in the protein input file while FlexX [213] can place water molecules within the binding site and can search for the ones interacting strongly. These programs score with the water present (on) or not (off) and select the best scoring option. The Glide XP scoring function takes into account the displacement of waters by the ligand from hydrophobic regions of the protein active site. The docking program FITTED [214] also deals with movable water molecules. Along with a genetic algorithm, FITTED incorporates a switching function that effectively turns off the water and allows them to be displaced when needed.

Second, the pH and ionic strength for the binding experiments are well regulated within a narrow pH range (typically pH 7) [41]. The protonation states from both sides, receptor and ligand, can change upon binding [215, 216]. In general, it might be possible to proceed docking with a dynamic prediction of protonation (by pK computations), however, it requires much more complicated procedures than to use automated programs for docking-based screening. One may include small molecules in different protonation forms and fixed protein binding site groups' protonation, still charged distribution in binding site and small molecules should be carefully determined [217, 218]. Along the same line, it is preferable to involve in SBVS experiments multiple tautomer molecules that can change shape, surface, and hydrogen bonds [219, 220]. The importance of treating the ligand protonation states, stereoisomers, and tautomers on SBVS results was recently investigated using docking with GOLD and PLANTS [221] and Surflex, eHiTS, GOLD, and FlexX-Pharm [222].

Third, the entropy estimation is a challenge in current SBVS methods. Most of the current scoring functions take into consideration only terms relating to enthalpy and its correlation to binding affinity. Several methods (simple or detailed physical descriptions) attempt to include the entropy contribution upon ligand binding (see the reviews [42, 223]). For instance a widely applied manner is to compute the ligand configurational entropy (ICM, AutoDock) by estimation of the entropic loss from torsion ligand freezing due to the binding. So far, recent studies [224] demonstrated that the major component of the entropy arises from the vibrational entropy rather than the conformational one. The former can be predicted for instance by time consuming calculations of normal mode analysis [54, 225]. Thus, often SBVS experiments consider only the conformational entropy by estimating the loss of rotational degrees of freedom on ligand upon binding. Therefore, a better treatment of entropic terms should be carried out in order to improve the accurately of binding affinity prediction.

CONCLUSION

Virtual screening methods based on the 3D structure of the protein receptor provide a real opportunity for the identification of new active compounds, without bias towards known hits or leads. In numerous cases, SBVS represents a useful alternative strategy to experimental HTS. The tendency today is to combine both, ligand- based and structure-based approaches. We have introduced in this chapter various docking-scoring methods as well as concepts for further improvements, like optimized algorithms that should better handle receptor flexibility while better scoring functions and binding affinities' predictions are also required.

ACKNOWLEDGEMENTS

Supports from the Inserm institute and University Paris Diderot are greatly appreciated.

REFERENCES

[1] Ertl, P. Cheminformatics analysis of organic substituents: Identification of the most common substituents, calculation of substituent properties, and automatic identification of drug-like bioisosteric groups. *J. Chem. Inf. Comput. Sci.* **2003**, *43*, 374-380.

[2] Davies, J.W.; Glick, M.; Jenkins, J.L. Streamlining lead discovery by aligning *in silico* and high-throughput screening. *Curr. Opin. Chem. Biol.* **2006** *10*, 343-351.

[3] Lyne, P.D. Structure-based virtual screening: An overview. *Drug Discov. Today* **2002**, *7*, 1047-1055.

[4] Shoichet, B.K. Virtual screening of chemical libraries. *Nature* **2004**, *432*, 862-865.

[5] Kubinyi, H. Drug research: Myths, hype and reality. *Nat. Rev. Drug Discov.* **2003**, *2*, 665-668.

[6] Lundqvist, T. The devil is still in the details-driving early drug discovery forward with biophysical experimental methods. *Curr. Opin. Drug Discov. Devel.* **2005**, *8*, 513-519.

[7] Alvarez, J.C. High-throughput docking as a source of novel drug leads. *Curr. Opin. Chem. Biol.* **2004**, *8*, 365-370.

[8] Ekins, S.; Mestres, J.; Testa, B. *In silico* pharmacology for drug discovery: Applications to targets and beyond. *Br. J. Pharmacol.* **2007**, *152*, 21-37.

[9] Clark, D. What has virtual screening ever done for drug discovery? *Expert Opin. Drug Discov.* **2008**, *3*, 841-851.

[10] Miteva, M. Hierarchical structure-based virtual screening for drug design *Biotechnol. Biotechnol. Eq.* **2008**, *22*, 634-638.

[11] Sperandio, O.; Miteva, M.A.; Segers, K.; Nicolaes, G.; Villoutreix, B.O. Screening outside the catalytic site: Inhibition of macromolecular interactions through structure-based virtual ligand screening experiments. *The Open Biochemistry Journal*

2008, *2,* 29-37.

[12] Altman, M.D.; Ali, A.; Reddy, G.S.; Nalam, M.N.; Anjum, S.G.; Cao, H.; Chellappan, S.; Kairys, V.; Fernandes, M.X.; Gilson, M.K.; Schiffer, C.A.; Rana, T.M.; Tidor, B. HIV-1 protease inhibitors from inverse design in the substrate envelope exhibit subnanomolar binding to drug-resistant variants. *J. Am. Chem. Soc.* **2008,** *130,* 6099-6113.

[13] Villoutreix, B.O.; Eudes, R.; Miteva, M.A. Structure-based virtual ligand screening: Recent success stories. *Combinatorial Chemistry & High Throughput Screening* **2009,** *12,* 1000-1016.

[14] Kuntz, I.D. Structure-based strategies for drug design and discovery. *Science* **1992,** *257,* 1078-1082.

[15] Prathipati, P.; Dixit, A.; Saxena, A.K. Computer-aided drug design: Integration of structure-based and ligand-based approaches in drug design. *Curr. Comp. Aided Drug Design* **2007,** *3,* 341-352.

[16] Radestock, S.; Weil, T.; Renner, S. Homology model-based virtual screening for gpcr ligands using docking and target-biased scoring. *J. Chem. Inf. Model.* **2008,** *48,* 1104-1117.

[17] Bajorath, J. Integration of virtual and high-throughput screening. *Nat. Rev. Drug Discov.* **2002,** *1,* 882-894.

[18] Lengauer, T.; Lemmen, C.; Rarey, M.; Zimmermann, M. Novel technologies for virtual screening. *Drug Discov. Today* **2004,** *9,* 27-34.

[19] Stahura, F.L.; Bajorath, J. New methodologies for ligand-based virtual screening. *Curr. Pharm. De.s* **2005,** *11,* 1189-1202.

[20] Auer, J.; Bajorath, J. Molecular similarity concepts and search calculations. *Methods Mol. Biol.* **2008,** *453,* 327-347.

[21] Douguet, D. Ligand-based approaches in virtual screening. *Curr. Comp. Aided Drug Design* **2008,** *4,* 180-190.

[22] Cummings, M.D.; DesJarlais, R.L.; Gibbs, A.C.; Mohan, V.; Jaeger, E.P. Comparison of automated docking programs as virtual screening tools. *J. Med. Chem.* **2005,** *48,* 962-976.

[23] Schneidman-Duhovny, D.; Nussinov, R.; Wolfson, H.J. Predicting molecular interactions *in silico*: II. Protein-protein and protein-drug docking. *Curr. Med. Chem.* **2004,** *11,* 91-107.

[24] Sperandio, O.; Miteva, M.A.; Delfaud, F.; Villoutreix, B.O. Receptor-based computational screening of compound databases: The main docking-scoring engines. *Curr. Protein Pept. Sci.* **2006,** *7,* 369-393.

[25] Zhou, Z.; Felts, A.K.; Friesner, R.A.; Levy, R.M. Comparative performance of several flexible docking programs and scoring functions: Enrichment studies for a diverse set of pharmaceutically relevant targets. *J. Chem. Inf. Model.* **2007,** *47,* 1599-1608.

[26] Rockey, W.M.; Elcock, A.H. Progress toward virtual screening for drug side effects. *Proteins* **2002,** *48,* 664-671.

[27] Evers, A.; Klebe, G. Successful virtual screening for a submicromolar antagonist of the neurokinin-1 receptor based on a ligand-supported homology model. *J. Med. Chem.* **2004,** *47,* 5381-5392.

[28] Kairys, V.; Fernandes, M.X.; Gilson, M.K. Screening drug-like compounds by docking to homology models: A systematic study. *J. Chem. Inf. Model.* **2006,** *46,* 365-379.

[29] Mohan, V.; Gibbs, A.C.; Cummings, M.D.; Jaeger, E.P.; DesJarlais, R.L. Docking: Successes and challenges. *Curr. Pharm. Des.* **2005,** *11,* 323-333.

[30] Kitchen, D.B.; Decornez, H.; Furr, J.R.; Bajorath, J. Docking and scoring in virtual screening for drug discovery: Methods and applications. *Nat. Rev. Drug Discov.* **2004,** *3,* 935-949.

[31] Cavasotto, C.N.; Orry, A.J. Ligand docking and structure-based virtual screening in drug discovery. *Curr. Top. Med. Chem.* **2007,** *7,* 1015-1023.

[32] Geldenhuys, W.J.; Gaasch, K.E.; Watson, M.; Allen, D.D.; Van der Schyf, C.J. Optimizing the use of open-source software applications in drug discovery. *Drug Discov. Today* **2006,** *11,* 127-132.

[33] Leach, A.R.; Shoichet, B.K.; Peishoff, C.E. Prediction of protein-ligand interactions. Docking and scoring: Successes and gaps. *J. Med. Chem.* **2006,** *49,* 5851-5855.

[34] Krovat, E.M.; Steindl, T.; Langer, T. Recent advances in docking and scoring. *Current Computer-Aided Drug Design* **2005,** *1,* 93-102.

[35] Villoutreix, B.O.; Renault, N.; Lagorce, D.; Sperandio, O.; Montes, M.; Miteva, M.A. Free resources to assist structure-based virtual ligand screening experiments. *Curr. Protein Pept. Sci.* **2007,** *8,* 381-411.

[36] Kroemer, R.T. Structure-based drug design: Docking and scoring *Curr. Prot. Peptide Sci.* **2007,** *8,* 312-328.

[37] Moitessier, N.; Englebienne, P.; Lee, D.; Lawandi, J.; Corbeil, C.R. Towards the development of universal, fast and highly accurate docking/scoring methods: A long way to go. *Br. J. Pharmacol.* **2008,** *153,* S7-26.

[38] Miteva, M.A.; Sperandio, O.; Villoutreix, B.O. Virtual ligand screening for structure-based drug design: Approaches and progress. *Bioautomation* **2007,** *7,* 104-121.

[39] Taylor, R.D.; Jewsbury, P.J.; Essex, J.W. A review of protein-small molecule docking methods. *J. Comput. Aided Mol. Des.* **2002,** *16,* 151-166.

[40] Alonso, H.; Bliznyuk, A.A.; Gready, J.E. Combining docking and molecular dynamic simulations in drug design. *Med. Res. Rev.* **2006,** *26,* 531-568.

[41] Brooijmans, N.; Kuntz, I.D. Molecular recognition and docking algorithms. *Annu. Rev. Biophys. Biomol. Struct.* **2003,** *32,* 335-373.

[42] Mobley, D.L.; Dill, K.A. Binding of small-molecule ligands to proteins: "What you see" Is not always "What you get". *Structure* **2009,** *17,* 489-498.

[43] Aqvist, J.; Medina, C.; Samuelsson, J.E. A new method for predicting binding affinity in computer-aided drug design. *Protein Eng.* **1994,** *7,* 385-391.

[44] Jain, A.N. Scoring functions for protein-ligand docking. *Curr. Protein Pept. Sci* **2006,** *7,* 407-420.

[45] Meng, E.C.; Shoichet, B.K.; Kuntz, I.D. Automated docking with grid-based energy evaluation. *J. Comput. Chem.* **1992,** *13,* 505-524.

[46] Goodsell, D.S.; Olson, A.J. Automated docking of substrates to proteins by simulated annealing. *Proteins* **1990,** *8,* 195-202.

[47] Shoichet, B.K.; Leach, A.R.; Kuntz, I.D. Ligand solvation in molecular docking. *Proteins* **1999,** *34,* 4-16.

[48] Zou, X.; Sun, Y.; Kuntz, I.D. Inclusion of solvation in ligand binding free energy calculations using the generalized-born model *J. Am. Chem. Soc.* **1999,** *121,* 8033-8043.

[49] Wang, J.; Morin, P.; Wang, W.; Kollman, P.A. Use of MM-PBSA in reproducing the binding free energies to HIV-1 RT of TIBO derivatives and predicting the binding mode to HIV-1 RT of efavirenz by docking and MM-PBSA. *J. Am. Chem. Soc.* **2001,** *123,* 5221-5230.

[50] Yoon, S.; Welsh, W.J. Identification of a minimal subset of receptor conformations for improved multiple conformation docking and two-step scoring *J. Chem. Inf. Comput. Sci.* **2004,** *44,* 88-96.

[51] Lyne, P.D.; Lamb, M.L.; Saeh, J.C. Accurate prediction of the relative potencies of members of a series of kinase inhibitors using molecular docking and MM-GBSA scoring. *J. Med. Chem.* **2006,** *49,* 4805-4808.

[52] Ferrari, A.M.; Degliesposti, G.; Sgobba, M.; Rastelli, G. Validation of an automated procedure for the prediction of relative free energies of binding on a set of aldose reductase inhibitors. *Bioorg. Med. Chem.* **2007,** *15,* 7865-7877.

[53] Graves, A.P.; Shivakumar, D.M.; Boyce, S.E.; Jacobson, M.P.; Case, D.A.; Shoichet, B.K. Rescoring docking hit lists for model cavity sites: Predictions and experimental testing. *J. Mol. Biol.* **2008,** *377,* 914-934.

[54] Wang, J.; Kang, X.; Kuntz, I.D.; Kollman, P.A. Hierarchical database screenings for HIV-1 reverse transcriptase using a pharmacophore model, rigid docking, solvation docking, and MM-PB/SA. *J. Med. Chem.* **2005,** *48,* 2432-2444.

[55] Michel, J.; Verdonk, M.L.; Essex, J.W. Protein-ligand binding affinity predictions by implicit solvent simulations: A tool for lead optimization? *J. Med. Chem.* **2006,** *49,* 7427-7439.

[56] Steinbrecher, T.; Case, D.A.; Labahn, A. A multistep approach to structure-based drug design: Studying ligand binding at the human neutrophil elastase. *J. Med. Chem.* **2006,** *49,* 1837-1844.

[57] Friesner, R.A.; Murphy, R.B.; Repasky, M.P.; Frye, L.L.; Greenwood, J.R.; Halgren, T.A.; Sanschagrin, P.C.; Mainz, D.T. Extra precision glide: Docking and scoring incorporating a model of hydrophobic enclosure for protein-ligand complexes. *J. Med. Chem.* **2006,** *49,* 6177-6196.

[58] Kulharia, M.; Goody, R.S.; Jackson, R.M. Information theory-based scoring function for the structure-based prediction of protein-ligand binding affinity. *J. Chem. Inf. Model.* **2008,** *48,* 1990-1998.

[59] Bohm, H.J. The development of a simple empirical scoring function to estimate the binding constant for a protein-ligand complex of known three-dimensional structure. *J. Comput. Aided Mol. Des.* **1994,** *8,* 243-256.

[60] Weng, Z.; Vajda, S.; Delisi, C. Prediction of protein complexes using empirical free energy functions. *Protein Sci.* **1996,** *5,* 614-626.

[61] Schulz-Gasch, T.; Stah, M. Scoring functions for protein-ligand interactions: A critical perspective. *Drug Discovery Today: Technologies* **2004,** *1,* 231-239.

[62] de Azevedo, W.F., Jr.; Dias, R. Evaluation of ligand-binding affinity using polynomial empirical scoring functions. *Bioorg. Med. Chem.* **2008,** *16,* 9378-9382.

[63] Raub, S.; Steffen, A.; Kamper, A.; Marian, C.M. AIscore chemically diverse empirical scoring function employing quantum chemical binding energies of hydrogen-bonded complexes. *J. Chem. Inf. Model.* **2008,** *48,* 1492-1510.

[64] Wang, B.; Westerhoff, L.M.; Merz, K.M., Jr. A critical assessment of the performance of protein-ligand scoring functions based on NMR chemical shift perturbations. *J. Med. Chem.* **2007,** *50,* 5128-5134.

[65] Wallqvist, A.; Jernigan, R.L.; Covell, D.G. A preference-based free-energy parameterization of enzyme-inhibitor binding. Applications to HIV-1-protease inhibitor design. *Protein Sci.* **1995,** *4,* 1881-1903.

[66] Huang, S.Y.; Zou, X. An iterative knowledge-based scoring function to predict protein-ligand interactions: Ii. Validation of the scoring function. *J. Comput. Chem.* **2006,** *27,* 1876-1882.

[67] Dominy, B.N.; Shakhnovich, E.I. Native atom types for knowledge-based potentials: Application to binding energy prediction. *J. Med. Chem.* **2004,** *47,* 4538-4558.

[68] Baroni, M.; Cruciani, G.; Sciabola, S.; Perruccio, F.; Mason, J.S. A common reference framework for analyzing/comparing proteins and ligands. Fingerprints for ligands and proteins(FLAP): Theory and application. *J. Chem. Inf. Model.* **2007,** *47,* 279-294.

[69] Marcou, G.; Rognan, D. Optimizing fragment and scaffold docking by use of molecular interaction fingerprints. *J. Chem. Inf. Model.* **2007,** *47,* 195-207.

[70] Seifert, M.H. Targeted scoring functions for virtual screening. *Drug Discov. Today* **2009,** *14,* 562-569.

[71] Seifert, M.H. Optimizing the signal-to-noise ratio of scoring functions for protein-ligand docking. *J. Chem. Inf. Model.* **2008,** *48,* 602-612.

[72] Kuntz, I.D.; Blaney, J.M.; Oatley, S.J.; Langridge, R.; Ferrin, T.E. A geometric approach to macromolecule-ligand interactions. *J. Mol. Biol.* **1982,** *161,* 269-288.

[73] McGann, M.R.; Almond, H.R.; Nicholls, A.; Grant, J.A.; Brown, F.K. Gaussian docking functions. *Biopolymers* **2003,** *68,* 76-90.

[74] Vigers, G.P.; Rizzi, J.P. Multiple active site corrections for docking and virtual screening. *J. Med. Chem.* **2004,** *47,* 80-89.

[75] Miller, M.D.; Kearsley, S.K.; Underwood, D.J.; Sheridan, R.P. Flog: A system to select 'quasi-flexible' ligands complementary to a receptor of known three-dimensional structure. *J. Comput. Aided Mol. Des.* **1994,** *8,* 153-174.

[76] Lawrence, M.C.; Davis, P.C. Clix: A search algorithm for finding novel ligands capable of binding proteins of known three-dimensional structure. *Proteins* **1992,** *12,* 31-41.

[77] Pang, Y.P.; Perola, E.; Xu, K.; Prendergast, F.G. Eudoc: A computer program for identification of drug interaction sites in macromolecules and drug leads from chemical databases. *J. Comput. Chem.* **2001,** *22,* 1750-1771.

[78] Mozziconacci, J.C.; Arnoult, E.; Bernard, P.; Do, Q.T.; Marot, C.; Morin-Allory, L. Optimization and validation of a docking-scoring protocol; application to virtual screening for COX-2 inhibitors. *J. Med. Chem.* **2005,** *48,* 1055-1068.

[79] Miteva, M.A.; Lee, W.H.; Montes, M.O.; Villoutreix, B.O. Fast structure-based virtual ligand screening combining FRED, DOCK, and Surflex. *J. Med. Chem.* **2005,** *48,* 6012-6022.

[80] Lorber, D.M.; Shoichet, B.K. Flexible ligand docking using conformational ensembles. *Protein Sci.* **1998,** *7,* 938-950.

[81] Lorber, D.M.; Shoichet, B.K. Hierarchical docking of databases of multiple ligand conformations. *Curr. Top. Med. Chem.* **2005,** *5,* 739-749.

[82] Sauton, N.; Lagorce, D.; Villoutreix, B.O.; Miteva, M.A. MS-DOCK: Accurate multiple conformation generator and rigid docking protocol for multi-step virtual ligand screening. *BMC Bioinformatics* **2008,** *9,* 184.

[83] Moustakas, D.T.; Lang, P.T.; Pegg, S.; Pettersen, E.; Kuntz, I.D.; Brooijmans, N.; Rizzo, R.C. Development and validation of a modular, extensible docking program: DOCK 5. *J. Comput. Aided. Mol. Des.* **2006,** *20,* 601-619.

[84] Kirchmair, J.; Wolber, G.; Laggner, C.; Langer, T. Comparative performance assessment of the conformational model generators omega and catalyst: A large-scale survey on the retrieval of protein-bound ligand conformations. *J. Chem. Inf. Model.* **2006,** *46,* 1848-1861.

[85] Sperandio, O.; Souaille, M.; Delfaud, F.; Miteva, M.A.; Villoutreix, B.O. MED-3DMC: A new tool to generate 3D conformation ensembles of small molecules with a Monte Carlo sampling of the conformational space. *Eur. J. Med. Chem.* **2009,** *44,* 1405-1409.

[86] Leite, T.B.; Gomes, D.; Miteva, M.A.; Chomilier, J.; Villoutreix, B.O.; Tuffery, P. Frog: A free online drug 3D conformation generator. *Nucleic Acids Res.* **2007,** *35,* W568-572.

[87] Vainio, M.J.; Johnson, M.S. Generating conformer ensembles using a multiobjective genetic algorithm. *J. Chem. Inf. Model.* **2007,** *47,* 2462-2474.

[88] Lagorce, D.; Pencheva, T.; Villoutreix, B.O.; Miteva, M.A. DG-AMMOS: A new tool to generate 3D conformation of small molecules using Distance Geometry and Automated Molecular Mechanics Optimization for *in silico* Screening. *BMC Chem. Biol.* **2009,** *9,* 6.

[89] Makino, S.; Kuntz, I.D. Automated flexible ligand docking method and its application for database search *J. Comput. Chem.* **1997,** *18,* 1812-1825.

[90] Grant, J.A.; Pickup, B.T.; Nicholls, A. A smooth permittivity function for Poisson-Boltzmann solvation methods. *J. Comp. Chem.* **2001,** *22,* 608-640.

[91] Fradera, X.; Knegtel, R.M.; Mestres, J. Similarity-driven flexible ligand docking. *Proteins* **2000,** *40,* 623-636.

[92] Joseph-McCarthy, D.; Thomas, B.E. t.; Belmarsh, M.; Moustakas, D.; Alvarez, J.C. Pharmacophore-based molecular docking to account for ligand flexibility. *Proteins* **2003,** *51,* 172-188.

[93] Jain, A.N. Surflex: Fully automatic flexible molecular docking using a molecular similarity-based search engine. *J. Med. Chem.* **2003,** *46,* 499-511.

[94] Jain, A.N. Surflex-dock 2.1: Robust performance from ligand energetic modeling, ring flexibility, and knowledge-based search. *J. Comput. Aided Mol. Des.* **2007,** *21,* 281-306.

[95]   Welch, W.; Ruppert, J.; Jain, A.N. Hammerhead: Fast, fully automated docking of flexible ligands to protein binding sites. *Chem. Biol.* **1996**, *3*, 449-462.

[96]   Rarey, M.; Kramer, B.; Lengauer, T.; Klebe, G. A fast flexible docking method using an incremental construction algorithm. *J. Mol. Biol.* **1996**, *261*, 470-489.

[97]   Bohm, H.J. The computer program ludi: A new method for the de novo design of enzyme inhibitors. *J. Comput. Aided Mol. Des.* **1992**, *6*, 61-78.

[98]   Klebe, G.; Mietzner, T. A fast and efficient method to generate biologically relevant conformations. *J. Comput. Aided Mol. Des.* **1994**, *8*, 583-606.

[99]   Zsoldos, Z.; Reid, D.; Simon, A.; Sadjad, B. S.; Johnson, A.P. Ehits: An innovative approach to the docking and scoring function problems. *Curr. Protein Pept. Sci.* **2006**, *7*, 421-435.

[100]   Venkatachalam, C.M.; Jiang, X.; Oldfield, T.; Waldman, M. Ligandfit: A novel method for the shape-directed rapid docking of ligands to protein active sites. *J. Mol. Graph. Model.* **2003**, *21*, 289-307.

[101]   Krammer, A.; Kirchhoff, P.D.; Jiang, X.; Venkatachalam, C.M.; Waldman, M. Ligscore: A novel scoring function for predicting binding affinities. *J. Mol. Graph. Model.* **2005**, *23*, 395-407.

[102]   Verkhivker, G.M.; Rejto, P.A.; Bouzida, D.; Arthurs, S.; Colson, A.B.; Freer, S.T.; Gehlhaar, D.K.; Larson, V.; Luty, B.A.; Marrone, T.; Rose, P.W. Towards understanding the mechanisms of molecular recognition by computer simulations of ligand-protein interactions. *J. Mol. Recognit.* **1999**, *12*, 371-389.

[103]   Liu, M.; Wang, S. Mcdock: A Monte Carlo simulation approach to the molecular docking problem. *J. Comput. Aided Mol. Des.* **1999**, *13*, 435-451.

[104]   Hart, T.N.; Ness, S.R.; Read, R.J. Critical evaluation of the research docking program for the CASP2 challenge. *Proteins* **1997**, *Suppl 1*, 205-209.

[105]   McMartin, C.; Bohacek, R.S. Qxp: Powerful, rapid computer algorithms for structure-based drug design. *J. Comput. Aided Mol. Des.* **1997**, *11*, 333-344.

[106]   Todorov, N.P.; Alberts, I.L.; Esch, I.J.; Dean, P.M. QUASI: A novel method for simultaneous superposition of multiple flexible ligands and virtual screening using partial similarity. *J. Chem. Inf. Model.* **2007**, *47*, 1007-1020.

[107]   Tietze, S.; Apostolakis, J. GlamDock: Development and validation of a new docking tool on several thousand protein-ligand complexes. *J. Chem. Inf. Model.* **2007**, *47*, 1657-1672.

[108]   Jones, G.; Willett, P.; Glen, R.C.; Leach, A.R.; Taylor, R. Development and validation of a genetic algorithm for flexible docking. *J. Mol. Biol.* **1997**, *267*, 727-748.

[109]   Taylor, J.S.; Burnett, R.M. Darwin: A program for docking flexible molecules. *Proteins* **2000**, *41*, 173-191.

[110]   Cecchini, M.; Kolb, P.; Majeux, N.; Caflisch, A. Automated docking of highly flexible ligands by genetic algorithms: A critical assessment *J. Comput. Chem.* **2004**, *25*, 412-422.

[111]   Morris, G.; Goodsell, D.S.; Halliday, R.S.; Huey, R.; Hart, W.E.; Belew, R. K.; Olson, A.J. Automated docking using a lamarckian genetic algorithm and an empirical binding free energy function. *J. Comput. Chem.* **1999**, *19*, 1639-1662.

[112]   Vaque, M.; Arola, A.; Aliagas, C.; Pujadas, G. BDT: An easy-to-use front-end application for automation of massive docking tasks and complex docking strategies with AutoDock. *Bioinformatics* **2006**, *22*, 1803-1804.

[113]   Zhang, S.; Kumar, K.; Jiang, X.; Wallqvist, A.; Reifman, J. DOVIS: An implementation for high-throughput virtual screening using AutoDock. *BMC Bioinformatics* **2008**, *9*, 126.

[114]   Trott, O.; Olson, A.J. AutoDock Vina: Improving the speed and accuracy of docking with a new scoring function, efficient optimization, and multithreading. *J. Comput. Chem.* **2010**, *31,* 455-461.

[115]   Namasivayam, V.; Günther, R. PSO@AUTODOCK: A fast flexible molecular docking program based on swarm intelligence. *Chem. Biol. Drug Des.* **2007**, *70*, 475 - 484.

[116]   Stroganov, O.V.; Novikov, F.N.; Stroylov, V.S.; Kulkov, V.; Chilov, G.G. Lead Finder: An approach to improve accuracy of protein-ligand docking, binding energy estimation, and virtual screening. *J. Chem. Inf. Model.* **2008**, *48*, 2371-2385.

[117]   Baxter, C.A.; Murray, C.W.; Clark, D.E.; Westhead, D.R.; Eldridge, M.D. Flexible docking using tabu search and an empirical estimate of binding affinity. *Proteins* **1998**, *33*, 367-382.

[118]   Pei, J.; Wang, Q.; Liu, Z.; Li, Q.; Yang, K.; Lai, L. PSI-DOCK: Towards highly efficient and accurate flexible ligand docking. *Proteins* **2006**, *62*, 934-946.

[119]   Gorelik, B.; Goldblum, A. High quality binding modes in docking ligands to proteins. *Proteins* **2008**, *71*, 1373-1386.

[120]   Korb, O.; Stutzle, T.; Exner, T.E. PLANTS: Application of ant colony optimization to structure-based drug design. **2006;** *ANTS LNCS 4150*, pp. 247-258.

[121]   Erickson, J.A.; Jalaie, M.; Robertson, D.H.; Lewis, R.A.; Vieth, M. Lessons in molecular recognition: The effects of ligand and protein flexibility on molecular docking accuracy. *J. Med. Chem.* **2004**, *47*, 45-55.

[122]   Sivanesan, D.; Rajnarayanan, R.V.; Doherty, J.; Pattabiraman, N. In-silico screening using flexible ligand binding pockets:

A molecular dynamics-based approach. *J. Comput. Aided Mol. Des.* **2005**, *19*, 213-228.

[123] Bolstad, E.S.; Anderson, A.C. In pursuit of virtual lead optimization: Pruning ensembles of receptor structures for increased efficiency and accuracy during docking. *Proteins* **2009**, *75*, 62-74.

[124] Wu, G.; Vieth, M. Sdocker: A method utilizing existing x-ray structures to improve docking accuracy. *J. Med. Chem.* **2004**, *47*, 3142-3148.

[125] Kuhn, B.; Gerber, P.; Schulz-Gasch, T.; Stahl, M. Validation and use of the MM-PBSA approach for drug discovery. *J. Med. Chem.* **2005**, *48*, 4040-4048.

[126] Khandelwal, A.; Lukacova, V.; Comez, D.; Kroll, D.M.; Raha, S.; Balaz, S. A combination of docking, QM/MM methods, and MD simulation for binding affinity estimation of metalloprotein ligands. *J. Med. Chem.* **2005**, *48*, 5437-5447.

[127] Ferrara, P.; Curioni, A.; Vangrevelinghe, E.; Meyer, T.; Mordasini, T.; Andreoni, W.; Acklin, P.; Jacoby, E. New scoring functions for virtual screening from molecular dynamics simulations with a quantum-refined force-field (QRFF-MD). Application to cyclin-dependent kinase 2. *J. Chem. Inf. Model.* **2006**, *46*, 254-263.

[128] Teague, S.J. Implications of protein flexibility for drug discovery *Nat. Rev. Drug Discov.* **2003**, *2*, 527-541.

[129] Cavasotto, C.N.; Abagyan, R.A. Protein flexibility in ligand docking and virtual screening to protein kinases. *J. Mol. Biol.* **2004**, *337*, 209-225.

[130] Cozzini, P.; Kellogg, G.E.; Spyrakis, F.; Abraham, D.J.; Costantino, G.; Emerson, A.; Fanelli, F.; Gohlke, H.; Kuhn, L.A.; Morris, G.M.; Orozco, M.; Pertinhez, T.A.; Rizzi, M.; Sotriffer, C.A. Target flexibility: An emerging consideration in drug discovery and design. *J. Med. Chem.* **2008**, *51*, 6237-6255.

[131] Hornak, V.; Simmerling, C. Targeting structural flexibility in HIV-1 protease inhibitor binding. *Drug Discov. Today* **2007**, *12*, 132-138.

[132] Cochran, A.G. Protein-protein interfaces: Mimics and inhibitors. *Curr. Opin. Chem. Biol.* **2001**, *5*, 654-659.

[133] Toogood, P.L. Inhibition of protein-protein association by small molecules: Approaches and progress. *J. Med. Chem.* **2002**, *45*, 1543-1558.

[134] Barril, X.; Morley, S.D. Unveiling the full potential of flexible receptor docking using multiple crystallographic structures. *J. Med. Chem.* **2005**, *48*, 4432-4443.

[135] Polgar, T.; Keseru, G.M. Ensemble docking into flexible active sites. Critical evaluation of FlexE against JNK-3 and beta-secretase. *J. Chem. Inf. Model.* **2006**, *46*, 1795-1805.

[136] Totrov, M.; Abagyan, R. Flexible ligand docking to multiple receptor conformations: A practical alternative. *Curr. Opin. Struct. Biol.* **2008**, *18*, 178-184.

[137] Floriano, W.B.; Vaidehi, N.; Zamanakos, G.; Goddard, W.A., 3rd. HierVLS hierarchical docking protocol for virtual ligand screening of large-molecule databases. *J. Med. Chem.* **2004**, *47*, 56-71.

[138] Cozza, G.; Bonvini, P.; Zorzi, E.; Poletto, G.; Pagano, M.A.; Sarno, S.; Donella-Deana, A.; Zagotto, G.; Rosolen, A.; Pinna, L.A.; Meggio, F.; Moro, S. Identification of ellagic acid as potent inhibitor of protein kinase CK2: A successful example of a virtual screening application. *J. Med. Chem.* **2006**, *49*, 2363-2366.

[139] Segers, K.; Sperandio, O.; Sack, M.; Fischer, R.; Miteva, M.A.; Rosing, J.; Nicolaes, G.A.; Villoutreix, B.O. Design of protein-membrane interaction inhibitors by virtual ligand screening, proof of concept with the C2 domain of Factor V. *Proc. Natl. Acad. Sci. USA* **2007**, *104*, 12697-12702.

[140] Beautrait, A.; Leroux, V.; Chavent, M.; Ghemtio, L.; Devignes, M. D.; Smail-Tabbone, M.; Cai, W.; Shao, X.; Moreau, G.; Bladon, P.; Yao, J.; Maigret, B. Multiple-step virtual screening using VSM-G: Overview and validation of fast geometrical matching enrichment. *J. Mol. Model.* **2008**, *14*, 135-148.

[141] Cho, A.E.; Wendel, J.A.; Vaidehi, N.; Kekenes-Huskey, P.M.; Floriano, W. B.; Maiti, P.K.; Goddard, W.A., 3rd. The MPSim-Dock hierarchical docking algorithm: Application to the eight trypsin inhibitor cocrystals. *J. Comput. Chem.* **2005**, *26*, 48-71.

[142] Halgren, T.A.; Murphy, R.B.; Friesner, R.A.; Beard, H.S.; Frye, L.L.; Pollard, W.T.; Banks, J.L. Glide: A new approach for rapid, accurate docking and scoring. 2. Enrichment factors in database screening. *J. Med. Chem.* **2004**, *47*, 1750-1759.

[143] Wolf, A.; Zimmermann, M.; Hofmann-Apitius, M. Alternative to consensus scoring-a new approach toward the qualitative combination of docking algorithms. *J. Chem. Inf. Model.* **2007**, *47*, 1036-1044.

[144] Maiorov, V.; Sheridan, R.P. Enhanced virtual screening by combined use of two docking methods: Getting the most on a limited budget. *J. Chem. Inf. Model.* **2005**, *45*, 1017-1023.

[145] Montes, M.; Braud, E.; Miteva, M.A.; Goddard, M.L.; Mondesert, O.; Kolb, S.; Brun, M.P.; Ducommun, B.; Garbay, C.; Villoutreix, B.O. Receptor-based virtual ligand screening for the identification of novel Cdc25 phosphatase inhibitors. *J. Chem. Inf. Model.* **2008**, *48*, 157-165.

[146] Cho, Y.; Ioerger, T.R.; Sacchettini, J.C. Discovery of novel nitrobenzothiazole unhibitors for Mycobacterium tuberculosis ATP phosphoribosyl transferase (HisG) through virtual screening. *J. Med. Chem.* **2008**, *51*, 5984-5992

[147] Kellenberger, E.; Springael, J.Y.; Parmentier, M.; Hachet-Haas, M.; Galzi, J.L.; Rognan, D. Identification of nonpeptide CCR5 receptor agonists by structure-based virtual screening. *J. Med. Chem.* **2007,** *50*, 1294-1303.

[148] Ward, R.A.; Perkins, T.D.; Stafford, J. Structure-based virtual screening for low molecular weight chemical starting points for dipeptidyl peptidase iv inhibitors. *J. Med. Chem.* **2005,** *48*, 6991-6996.

[149] Engel, S.; Skoumbourdis, A.P.; Childress, J.; Neumann, S.; Deschamps, J.R.; Thomas, C.J.; Colson, A.O.; Costanzi, S.; Gershengorn, M.C. A virtual screen for diverse ligands: Discovery of selective G protein-coupled receptor antagonists. *J. Am. Chem. Soc.,* **2008,** *130*, 5115-5123.

[150] Cozza, G.; Gianoncelli, A.; Montopoli, M.; Caparrotta, L.; Venerando, A.; Meggio, F.; Pinna, L. A.; Zagotto, G.; Moro, S. Identification of novel protein kinase CK1 delta (CK1delta) inhibitors through structure-based virtual screening. *Bioorg. Med. Chem. Lett.* **2008,** *18*, 5672-5675.

[151] Hirayama, K.; Aoki, S.; Nishikawa, K.; Matsumoto, T.; Wada, K. Identification of novel chemical inhibitors for ubiquitin C-terminal hydrolase-L3 by virtual screening. *Bioorg. Med. Chem.* **2007,** *15*, 6810-6818.

[152] Deng, G.; Li, W.; Shen, J.; Jiang, H.; Chen, K.; Liu, H. Pyrazolidine-3,5-dione derivatives as potent non-steroidal agonists of farnesoid X receptor: Virtual screening, synthesis, and biological evaluation. *Bioorg. Med. Chem. Lett.* **2008,** *18*, 5497-5502.

[153] Kumar, A.; Chaturvedi, V.; Bhatnagar, S.; Sinha, S.; Siddiqi, M. I. Knowledge based identification of potent antitubercular compounds using structure based virtual screening and structure interaction fingerprints. *J. Chem. Inf. Model.* **2009,** *49*, 35-42.

[154] Küblbeck, J.; Jyrkkärinne, J.; Poso, A.; Turpeinen, M.; Sippl, W.; Honkakoski, P.; Windshügel, B. Discovery of substituted sulfonamides and thiazolidin-4-one derivatives as agonists of human constitutive androstane receptor. *Biochem. Pharmacol.* **2008,** *76*, 1288–1297

[155] Mukherjee, P.; Desai, P.; Ross, L.; White, E.L.; Avery, M.A. Structure-based virtual screening against SARS-3CL(pro) to identify novel non-peptidic hits. *Bioorg. Med. Chem.* **2008,** *16*, 4138-4149.

[156] Perez-Pineiro, R.; Burgos, A.; Jones, D.C.; Andrew, L.C.; Rodriguez, H.; Suarez, M.; Fairlamb, A.H.; Wishart, D.S. Development of a novel virtual screening cascade protocol to identify potential trypanothione reductase inhibitors. *J. Med. Chem.* **2009,** *52*, 1670-1680.

[157] Hu, X.; Prehna, G.; Stebbins, C.E. Targeting plague virulence factors: A combined machine learning method and multiple conformational virtual screening for the discovery of Yersinia Protein Kinase A inhibitors. *J. Med. Chem.* **2007,** *50*, 3980-3983.

[158] Tintori, C.; Manetti, F.; Veljkovic, N.; Perovic, V.; Vercammen, J.; Hayes, S.; Massa, S.; Witvrouw, M.; Debyser, Z.; Veljkovic, V.; Botta, M. Novel virtual screening protocol based on the combined use of molecular modeling and electron-ion interaction potential techniques to design HIV-1 integrase inhibitors. *J. Chem. Inf. Model.* **2007,** *47*, 1536-1544.

[159] Hellmuth, K.; Grosskopf, S.; Lum, C.T.; Würtele, M.; Röder, N.; von Kries, J.P.; Rosario, M.; Rademann, J.; Birchmeier, W. Specific inhibitors of the protein tyrosine phosphatase Shp2 identified by high-throughput docking. *Proc. Natl. Acad. Sci. U S A* **2008,** *105*, 7275-7280.

[160] Rush, T.S., 3rd; Grant, J.A.; Mosyak, L.; Nicholls, A. A shape-based 3-D scaffold hopping method and its application to a bacterial protein-protein interaction. *J. Med. Chem.* **2005,** *48*, 1489-1495.

[161] Quintus, F.; Sperandio, O.; Grynberg, J.; Petitjean, M.; Tuffery, P. Ligand scaffold hopping combining 3D maximal substructure search and molecular similarity. *BMC Bioinformatics* **2009,** *10*, 245.

[162] Sperandio, O.; Andrieu, O.; Miteva, M.A.; Vo, M.Q.; Souaille, M.; Delfaud, F.; Villoutreix, B.O. Med-sumolig: A new ligand-based screening tool for efficient scaffold hopping. *J. Chem. Inf. Model.* **2007,** *47*, 1097-1110.

[163] Kearsley, S.K.; Sallamack, S.; Fluder, E.M.; Andose, J.D.; Mosley, R.T.; Sheridan, R.P. Chemical similarity using physiochemical property descriptors. *J. Chem. Inf. Model.* **1996,** *36*, 118-127.

[164] Sperandio, O.; Petitjean, M.; Tuffery, P. Wwligcsrre: A 3D ligand-based server for hit identification and optimization. *Nucleic Acids Res.* **2009,** *37*, W504-509.

[165] Muegge, I. Synergies of virtual screening approaches. *Mini Rev. Med. Chem.* **2008,** *8*, 927-933.

[166] Omagari, K.; Mitomo, D.; Kubota, S.; Nakamura, H.; Fukunishi, Y. A method to enhance the hit ratio by a combination of structure-based drug screening and ligand-based screening. *Advances and Applications in Bioinformatics and Chemistry* **2008,** *1*, 19-28.

[167] Li, H.; Li, C.; Gui, C.; Luo, X.; Chen, K.; Shen, J.; Wang, X.; Jiang, H. Gasdock: A new approach for rapid flexible docking based on an improved multi-population genetic algorithm. *Bioorg. Med. Chem. Lett.* **2004,** *14*, 4671-4676.

[168] Park, H.; Lee, J.; Lee, S. Critical assessment of the automated AutoDock as a new docking tool for virtual screening. *Proteins* **2006,** *65*, 549-554.

[169] Schellhammer, I.; Rarey, M. FlexX-Scan: Fast, structure-based virtual screening. *Proteins* **2004,** *57*, 504-517.

[170] Yoon, S.; Smellie, A.; Hartsough, D.; Filikov, A. Surrogate docking: Structure-based virtual screening at high throughput speed. *J. Comput. Aided Mol. Des.* **2005**, *19*, 483-497.

[171] Favia, A.D.; Nobeli, I.; Glaser, F.; Thornton, J.M. Molecular docking for substrate identification: The short-chain dehydrogenases/reductases. *J. Mol. Biol.* **2008**, *375*, 855-874.

[172] Stahl, M.; Rarey, M. Detailed analysis of scoring functions for virtual screening. *J. Med. Chem.* **2001**, *44*, 1035-1042.

[173] Ferrara, P.; Gohlke, H.; Price, D.J.; Klebe, G.; Brooks, C.L., 3rd. Assessing scoring functions for protein-ligand interactions. *J. Med. Chem.* **2004**, *47*, 3032-3047.

[174] Seifert, M.H. Assessing the discriminatory power of scoring functions for virtual screening. *J. Chem. Inf. Model.* **2006**, *46*, 1456-1465.

[175] Cheng, T.; Li, X.; Li, Y.; Liu, Z.; Wang, R. Comparative assessment of scoring functions on a diverse test set. *J. Chem. Inf. Model.* **2009**, *49*, 1079-1093.

[176] Montes, M.; Miteva, M.A.; Villoutreix, B.O. Structure-based virtual ligand screening with LigandFit: Pose prediction and enrichment of compound collections. *Proteins* **2007**, *68*, 712-725.

[177] Wang, R.; Lu, Y.; Wang, S. Comparative evaluation of 11 scoring functions for molecular docking. *J. Med. Chem.* **2003**, *46*, 2287-2303.

[178] Warren, G.L.; Andrews, C.W.; Capelli, A.M.; Clarke, B.; LaLonde, J.; Lambert, M.H.; Lindvall, M.; Nevins, N.; Semus, S.F.; Senger, S.; Tedesco, G.; Wall, I.D.; Woolven, J.M.; Peishoff, C.E.; Head, M.S. A critical assessment of docking programs and scoring functions. *J. Med. Chem.* **2006**, *49*, 5912-5931.

[179] Teramoto, R.; Fukunishi, H. Consensus scoring with feature selection for structure-based virtual screening. *J. Chem. Inf. Model.* **2008**, *48*, 288-295.

[180] Spyrakis, F.; Amadasi, A.; Fornabaio, M.; Abraham, D.J.; Mozzarelli, A.; Kellogg, G.E.; Cozzini, P. The consequences of scoring docked ligand conformations using free energy correlations. *Eur. J. Med. Chem.* **2007**, *42*, 921-33.

[181] Feher, M. Consensus scoring for protein-ligand interactions. *Drug Discov. Today* **2006**, *11*, 421-428.

[182] Clark, R.D.; Strizhev, A.; Leonard, J.M.; Blake, J.F.; Matthew, J.B. Consensus scoring for ligand/protein interactions. *J. Mol. Graph. Model.* **2002**, *20*, 281-295.

[183] Betzi, S.; Suhre, K.; Chetrit, B.; Guerlesquin, F.; Morelli, X. Gfscore: A general nonlinear consensus scoring function for high-throughput docking. *J. Chem. Inf. Model.* **2006**, *46*, 1704-1712.

[184] Kontoyianni, M.; McClellan, L.M.; Sokol, G.S. Evaluation of docking performance: Comparative data on docking algorithms. *J. Med. Chem.* **2004**, *47*, 558-565.

[185] Verdonk, M.L.; Berdini, V.; Hartshorn, M.J.; Mooij, W.T.; Murray, C.W.; Taylor, R.D.; Watson, P. Virtual screening using protein-ligand docking: Avoiding artificial enrichment. *J. Chem. Inf. Comput. Sci.* **2004**, *44*, 793-806.

[186] Hevener, K.E.; Zhao, W.; Ball, D.M.; Babaoglu, K.; Qi, J.; White, S.W.; Lee, R.E. Validation of molecular docking programs for virtual screening against dihydropteroate synthase. *J. Chem. Inf. Model.* **2009**, *49*, 444-460.

[187] Li, H.; Zhang, H.; Zheng, M.; Luo, J.; Kang, L.; Liu, X.; Wang, X.; Jiang, H. An effective docking strategy for virtual screening based on multi-objective optimization algorithm. *BMC Bioinformatics* **2009**, *10*, 58.

[188] Raha, K.; Merz, K.M., Jr. Large-scale validation of a quantum mechanics based scoring function: Predicting the binding affinity and the binding mode of a diverse set of protein-ligand complexes. *J. Med. Chem.* **2005**, *48*, 4558-4575.

[189] Fischer, B.; Fukuzawa, K.; Wenzel, W. Receptor-specific scoring functions derived from quantum chemical models improve affinity estimates for in-silico drug discovery. *Proteins* **2008**, *70*, 1264-1273.

[190] Deng, Z.; Chuaqui, C.; Singh, J. Structural interaction fingerprint (SIFt): A novel method for analyzing three-dimensional protein-ligand binding interactions. *J. Med. Chem.* **2004**, *47*, 337-344.

[191] Barillari, C.; Marcou, G.; Rognan, D. Hot-spots-guided receptor-based pharmacophores (HS-Pharm): A knowledge-based approach to identify ligand-anchoring atoms in protein cavities and prioritize structure-based pharmacophores. *J. Chem. Inf. Model.* **2008**, *48*, 1396-1410.

[192] Reulecke, I.; Lange, G.; Albrecht, J.; Klein, R.; Rarey, M. Towards an integrated description of hydrogen bonding and dehydration: Decreasing false positives in virtual screening with the HYDE scoring function. *ChemMedChem* **2008**, *3*, 885-897.

[193] Pencheva, T.; Samna Soumana, O.; Pajeva, I.; Miteva, M.A. Post-docking virtual screening of diverse binding pockets: Comparative study using DOCK, AMMOS, X-score and FRED scoring functions. *Eur. J. Med. Chem.* **2010**, doi:10.1016/j.ejmech.2009.12.025.

[194] Schulz-Gasch, T.; Stahl, M. Binding site characteristics in structure-based virtual screening: Evaluation of current docking tools. *J. Mol. Model.* **2003**, *9*, 47-57.

[195] Timmers, L.F.; Caceres, R.A.; Vivan, A.L.; Gava, L.M.; Dias, R.; Ducati, R.G.; Basso, L.A.; Santos, D.S.; de Azevedo, W.F.J. Structural studies of human purine nucleoside phosphorylase: Towards a new specific empirical scoring function.

Arch. Biochem. Biophys. **2008**, *479*, 28-38.

[196] Catana, C.; Stouten, P.F. Novel, customizable scoring functions, parameterized using N-PLS, for structure-based drug discovery. *J. Chem. Inf. Model.* **2007**, *47*, 85-91.

[197] Huang, N.; Kalyanaraman, C.; Irwin, J.J.; Jacobson, M.P. Physics-based scoring of protein-ligand complexes: Enrichment of known inhibitors in large-scale virtual screening. *J. Chem. Inf. Model.* **2006**, *46*, 243-253.

[198] Perola, E.; Walters, W.P.; Charifson, P.S. A detailed comparison of current docking and scoring methods on systems of pharmaceutical relevance. *Proteins* **2004**, *56*, 235-249.

[199] Koska, J.; Spassov, V.Z.; Maynard, A.J.; Yan, L.; Austin, N.; Flook, P.K.; Venkatachalam, C.M. Fully automated molecular mechanics based induced fit protein-ligand docking method. *J. Che.m Inf. Model.* **2008**, *48*, 1965-1973.

[200] Springer, C.; Adalsteinsson, H.; Young, M.M.; Kegelmeyer, P.W.; Roe, D.C. PostDOCK: A structural, empirical approach to scoring protein ligand complexes. *J. Med. Chem.* **2005**, *48*, 6821-6831.

[201] Pencheva, T.; Lagorce, D.; Pajeva, I.; Villoutreix, B.O.; Miteva, M.A. AMMOS: Automated molecular mechanics optimization tool for *in silico* screening. *BMC Bioinformatics* **2008**, *9*, 438.

[202] Ladbury, J.E. Just add water! The effect of water on the specificity of protein-ligand binding sites and its potential application to drug design. *Chem. Biol.* **1996**, *3*, 973-980.

[203] Barillari, C.; Taylor, J.; Viner, R.; Essex, J. W. Classification of water molecules in protein binding sites. *J. Am. Chem. Soc.* **2007**, *129*, 2577-2587.

[204] Pearlman, D.A. Evaluating the Molecular Mechanics Poisson-Boltzmann Surface Area free energy method using a congeneric series of ligands to p38 MAP kinase. *J. Med. Chem.* **2005**, *48*, 7796-7807.

[205] Wang, Y.X.; Freedberg, D.I.; Wingfield, P.T.; Stahl, S.J.; Kaufman, J.D.; Kiso, Y.; Bhat, T.N.; Erickson, J.W.; Torchia, D.A. Bound water molecules at the interface between the HIV-1 protease and a potent inhibitor, KNI-272, determined by NMR. *J. Am. Chem. Soc.* **1996**, *118*, 12287–12290.

[206] Louis, J.M.; Dyda, F.; Nashed, N.T.; Kimmel, A.R.; Davies, D.R. Hydrophilic peptides derived from the transframe region of Gag-Pol inhibit the HIV-1 protease. *Biochemistry* **1998**, *37*, 2105-2110.

[207] Lam, P.Y.; Jadhav, P.K.; Eyermann, C.J.; Hodge, C.N.; Ru, Y.; Bacheler, L.T.; Meek, J.L.; Otto, M.J.; Rayner, M.M.; Wong, Y.N.; *et al.* Rational design of potent, bioavailable, nonpeptide cyclic ureas as HIV protease inhibitors. *Science* **1994**, *263*, 380-384.

[208] Grzesiek, S.; Bax, A.; Nicholson, L.K.; Yamazaki, T.; Wingfield, P.; Stahl, S.J.; Eyermann, C.J.; Torchia, D.A.; Nicholas Hodge, C.; Lam, P.Y.S.; Jadhav, P.K.; Chang, C.H. NMR evidence for the displacement of a conserved interior water molecule in HIV protease by a non-peptide cyclic urea-based inhibitor. *J. Am. Chem. Soc.* **1994**, *116*, 1581-1582.

[209] Yan, A.; Grant, G.H.; Richards, W.G. Dynamics of conserved waters in human Hsp90: Implications for drug design. *J. R. So.c Interface.* **2008**, *5 Suppl 3*, S199-205.

[210] Amadasi, A.; Surface, J.A.; Spyrakis, F.; Cozzini, P.; Mozzarelli, A.; Kellogg, G.E. Robust classification of "Relevant" Water molecules in putative protein binding sites. *J. Med. Chem.* **2008**, *51*, 1063-1067.

[211] Moitessier, N.; Westhof, E.; Hanessian, S. Docking of aminoglycosides to hydrated and flexible RNA. *J. Med. Chem.* **2006**, *49*, 1023-1033.

[212] Verdonk, M.L.; Chessari, G.; Cole, J.C.; Hartshorn, M.J.; Murray, C.W.; Nissink, J.W.; Taylor, R.D.; Taylor, R. Modeling water molecules in protein-ligand docking using GOLD. *J. Med. Chem.* **2005**, *48*, 6504-6515.

[213] Rarey, M.; Kramer, B.; Lengauer, T. The particle concept: Placing discrete water molecules during protein-ligand docking predictions. *Proteins* **1999**, *34*, 17-28.

[214] Corbeil, C.R.; Moitessier, N. Docking ligands into flexible and solvated macromolecules. 3. Impact of input ligand conformation, protein flexibility, and water molecules on the accuracy of docking programs. *J. Chem. Inf. Model.* **2009**, *49*, 997-1009.

[215] Czodrowski, P.; Sotriffer, C.A.; Klebe, G. Atypical protonation states in the active site of HIV-1 protease: A computational study. *J. Chem. Inf. Model.* **2007**, *47*, 1590-1598.

[216] Steuber, H.; Czodrowski, P.; Sotriffer, C.A.; Klebe, G. Tracing changes in protonation: A prerequisite to factorize thermodynamic data of inhibitor binding to aldose reductase. *J. Mol. Biol.* **2007**, *373*, 1305-1320.

[217] Mitra, R.; Shyam, R.; Mitra, I.; Miteva, M.A.; Alexov, E. Protonation states of proteins and small molecules: Implications to ligand-receptor interactions. *Current Computer-Aided Drug Design* **2008**, *4*, 169-179.

[218] Lee, A.C.; Crippen, G.M. Predicting pka. *J. Chem. Inf. Model.* **2009**, *49*, 2013–2033.

[219] Pospisil, P.; Ballmer, P.; Scapozza, L.; Folkers, G. Tautomerism in computer-aided drug design. *J. Recept. Signal. Transduct. Res.* **2003**, *23*, 361-371.

[220] Milletti, F.; Storchi, L.; Sforna, G.; Cross, S.; Cruciani, G. Tautomer enumeration and stability prediction for virtual screening on large chemical databases. *J. Chem. Inf. Model.* **2009**, *49*, 68-75.

[221]   ten Brink, T.; Exner, T.E. Influence of protonation, tautomeric, and stereoisomeric states on protein-ligand docking results. *J. Chem. Inf. Model.* **2009**, *49*, 1535-1546.

[222]   Polgar, T.; Magyar, C.; Simon, I.; Keseru, G.M. Impact of ligand protonation on virtual screening against beta-secretase (BACE1). *J. Chem. Inf. Model.* **2007**, *47*, 2366-2373.

[223]   Rajamani, R.; Good, A.C. Ranking poses in structure-based lead discovery and optimization: Current trends in scoring function development. *Curr. Opin. Drug Discov. Devel.* **2007**, *10*, 308-315.

[224]   Chang, C.E.; Chen, W.; Gilson, M.K. Ligand configurational entropy and protein binding. *Proc. Natl. Acad. Sci. U S A* **2007**, *104*, 1534-1539.

[225]   Zoete, V.; Meuwly, M.; Karplus, M. Study of the insulin dimerization: Binding free energy calculations and per-residue free energy decomposition. *Proteins* **2005**, *61*, 79-93.

In Silico Lead Discovery, 2011, 47-59

CHAPTER 3

3D Similarity Search for Lead Compound Identification

Frédéric Guyon and Pierre Tufféry*

MTi, INSERM UMR-S973, Université Paris Diderot - Paris 7, 75205, Paris, France; RPBS, Université Paris Diderot - Paris 7, 75205, Paris, France; Fax: +331 57 27 83 72, E-mail: pierre.tuffery@univ-paris-diderot.fr

Abstract: The search for new lead compounds when no sufficient structural information on the receptor binding site, neither on the binding mechanism, are available is challenging, but it can be proceeded if biological activity data are provided for at least one active compound towards the target under consideration. In the recent years, traditional approaches with respect to this situation such as QSAR and pharmacophore techniques have been supplemented by 3D similarity search techniques to mine large banks of compounds, searching for small compounds similar to the bioactive ones. We introduce some of the concepts related to the latter techniques, and we discuss briefly foreseen developments.

INTRODUCTION

The search for new lead compounds when no sufficient structural information on the receptor binding site, neither on the binding mechanism, are available is challenging, but it can be proceeded if biological activity data are provided for at least one active compound towards the target under consideration. The structure of the biological target might be unknown, or no information about the drug-target interaction could be obtained. Techniques addressing this situation are usually termed as Ligand-based virtual screening (LBVS) approaches.

To address such cases, one usually assumes that "similar" compounds are likely to exhibit similar biological activities [1]. It is important to understand that, in essence, the concept of compound "similarity" encompasses many aspects including physico-chemical considerations, ADME/tox properties, structural properties, etc. For that reason, it has naturally found an echo in different audiences. Although early works on compound common feature extraction emerged in the 1960s, the question of assessing compound similarity has become a central concern over the last decades. Among the various approaches addressing this question, three major approaches can be identified: QSAR (Quantitative Structure Activity Relationship), pharmacophore and similarity search approaches (Fig. **1**). These paradigms have obvious overlaps. For instance, QSAR techniques are able to produce pharmacophore models, as well similarity search techniques share common goals with practical application of pharmacophore for compound libraries' screening. However, the communities supporting each paradigm generally have different backgrounds so that the distinction remains largely accepted.

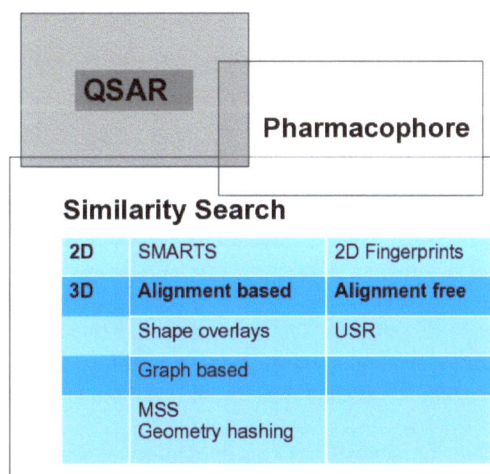

Figure 1: Overview of different approaches for Ligand-based virtual screening. The similarity search techniques are detailed.

When a sufficient number of active compounds are known and significant amount of structure activity relationship data have been validated, one can apply QSAR [2] or pharmacophore techniques (see for instance [3]). Although there are obvious pitfalls in such approaches (for instance, one should validate that all compounds' acting is relied on the same molecular mechanism, and that they do not affect unrelated molecular processes), they have been the subject of many developments over the last decades, and their efficiency is largely proven (*e.g.* [4]). QSAR techniques aim at deriving molecular descriptors that are statistically discriminative of a particular activity of the compounds of interest starting from the available 2D or 3D structural data. The obtained statistical model can then be used to design new active molecules. Pharmacophore based techniques also rely on the knowledge of the biological activity of multiple hits to identify key features. According to the IUPAC Glossary of Terms Used in Medicinal Chemistry [5], "a pharmacophore is the ensemble of steric and electronic features that is necessary to ensure the optimal supramolecular interactions with a specific biological target structure and to trigger (or to block) its biological response. A pharmacophore does not represent a real molecule or a real association of functional groups, but a purely abstract concept that accounts for the common molecular interaction capacities of a group of compounds towards their target structure. The pharmacophore can be considered as the largest common denominator shared by a set of active molecules." Practical applications of this concept imply two steps: (i) pharmacophore identification, and (ii) compound search based on the created pharmacophore. Although this concept has traditionally been associated to the 2D topological description of compounds and search techniques such as fingerprints, the necessity of 3D molecular descriptors has been described since the early 1990s [6, 7]. Recent developments (*e.g.* [8-10]) have moved this concept up to 3D analysis of series of bioactives and new developments on that direction are to be expected.

The third large category represents similarity search techniques. They address the question of the direct quantification of the similarity between two compounds. In essence, they can be applied as soon as one bioactive compound is known. Whereas early approaches mostly encompass 2D-similarity-search in which fingerprints approaches are a major component (*e.g.* [11]), 3D-similarity search has emerged in the last two decades as a major new component of this field of research (*e.g.* [12, 13]). 2D- and 3D-based approaches attempt to quantify compound similarity based on the sharing of chemical groups, the shape of compounds and their chemistry. Overall, these approaches rely on different concepts and include shape overlay (*e.g.* [13]), graph decomposition of compounds (*e.g.* [14, 15]), pruned brute force atom coordinate superimposition (*e.g.* [16, 17]), or non superimposition techniques comparing sets of compound shape descriptors (*e.g.* [18]).

Although in this chapter we focus on 3D similarity search, some important aspects common to the mentioned three paradigms should be emphasized. Firstly, practical libraries' screening to identify new lead compounds is confronted with the difficult balance between search specificity and search plasticity. Search specificity can be related to receptor selectivity. Candidate compounds should be able to biologically interact with the target, and this could result in restraining the search to compounds too similar to the bioactive compounds. Search plasticity is related to the necessary alleviation to any scaffold dependency, to propose new relevant scaffolds divergent if possible from the known bioactive compounds. As pointed out in [19, 18], this latter aspect is particularly important "to break out the protected "Patent space" around drugs or to hop for molecules with different scaffolds when lead compounds have undesirable features such as intractable chemistry or poor pharmacological properties". These two goals, apparently in contradiction, result in the necessary combination of different kind of information to measure the chemical compounds' similarity. For 3D similarity search approaches, one should combine both geometry and chemistry properties evaluation to mine chemical compounds.

Secondly, *in silico* lead search usually implies screening of several millions of compounds. On date of November 2009, the size of the PubChem [20] collection is over 25,000 000 compound conformers, the size of the Zinc database [21] is over 8,000 000 compounds. Although it is possible to filter huge libraries to obtain focused libraries for some particular applications using ADME/tox (*e.g.* [22, 23]) or contextual property filters, the size of the libraries to screen usually remains over several hundred thousand compounds. In order to be efficient, all techniques are confronted with speed issues. This is particularly sensitive for the 3D similarity search or 3D pharmacophore techniques, which in essence attempt to perform systematic similarity assessment over compound collections. Obviously, efforts are focused on development of strategies permitting to perform screening, testing of various hypotheses, and experiment combinations, in a reasonable amount of time.

Finally, another important point is that the goal of LBVS applications is to propose candidate molecules to further experimental essays, using HTS techniques for instance [24, 25] that are very costly and much longer than *in silico* experiments. Thus, an essential task of employing LVBS approaches is to propose for further experimental validation a limited set of candidates. Therefore, it is important that LBVS are able to properly rank the potential actives and that the true active compounds can be retrieved in the top of the proceeded compounds' collection, which is often denoted as "early enrichment".

In the following, we focus on the elementary components underlying 3D search techniques, including alignment based and alignment-free techniques and library screening strategies.

APPROACHES FOR 3D ALIGNMENT OF COMPOUNDS

The 3D search for compound alignment is usually considered as a highly non trivial problem. Unlike for protein structural alignment, one major difficulty comes from the lack of a priori knowledge of the neighbors, that should allow the use of the classical dynamic programming alignment techniques for instance. Evaluating similarity through an alignment technique amounts to search for maximal feature sets common for the compounds to align. Non surprisingly, the fuzziness of the concept of similarity has resulted in many developments based on various techniques, such as the quantum mechanics, protein modeling, but also such as more distant fields as graph theory or image recognition techniques. In the following section, we present three categories of approaches. The first one is the shape overlay. Its bases are issued from the quantum mechanics, where Carbo *et al.* [26] introduced a similarity index to compare the molecular electron density, and a representation of atomic orbitals with Gaussian functions [27]. In these methods the volume drives the superimposition. The second one exploits a graph decomposition of the compound and applies techniques such as clique search. The third one considers atomic positions, and has usually been introduced by the protein structure community.

It is important to note that depending on the community driving the development of the methods, the underlying hypotheses, as well as the true goals reached, might differ. For instance, the objective of the shape based methods described in the next section is to maximize the volume overlap of the compounds or parts of compounds. For approaches as Maximal Substructure Search or Geometry hashing techniques, the objective is to maximize the number of atoms paired. While the difference could appear minute, both approaches may lead to different results. Basically, volume overlap techniques do not compute and do not provide alignments (or pairing) of atoms but only a similarity index between the shapes is calculated. Therefore, for the same pair of compounds, maximal volume overlap and maximal atom pairing could result in different relative orientations of the compounds. However, applied to large collections of compounds, it is likely that such differences would have minor impact.

Finally, alignment approaches can consider various degrees of flexibility for the compounds. Rigid approaches that consider one conformation per conformer and perform a rigid superimposition not affecting the compound geometry are conformation sensitive and as such they require a relevant compound conformation. This can be avoided by taking multiple conformers per compound, treating each in a rigid manner (Fig. **2**). This semi-flexible strategy is presently dominant for 3D screening of large libraries. Full flexibility of at least one compound during the superimposition has been considered by only few approaches so far, but remains of limited use since it is usually more computationally intensive.

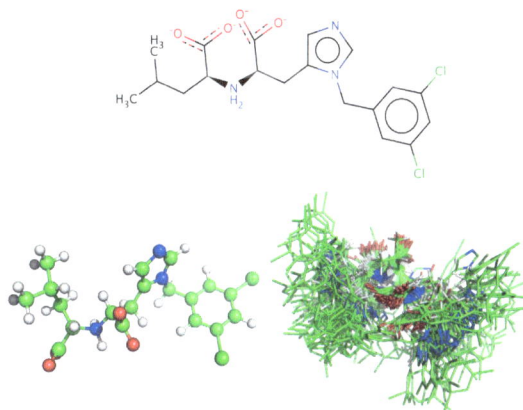

Figure 2: Different representations of a compound. Top: The 2D representation of the compound depicts its topology. Bottom left: A 3D conformation of the same compound. Bottom right: Semi-flexible approaches consider an ensemble of representative conformers per compound.

Shape Based Overlays

Gaussian Shape Representation

Gaussian representation allows a simple evaluation of molecules' volumes and volume intersection. Gaussian description of molecular shape was first described by Grant *et al.* [28]. It made an improvement over previous hard-sphere model where a molecule is represented by a set of atom-centered spheres of different radii:

$$r_i(x) = b_i \exp(-a_i \|x - c_i\|^2)$$

(1)

where *a's* and *b's* are constants depending on atom types and *c's* are the atom center coordinates. The shape function giving the total volume is:

$$r_{Vol} = 1 - \prod_{i=1}^{N}(1 - r_i)$$

(2)

and is developed to give:

$$r_{Vol} = \sum_{i} r_i - \sum_{i,j>i} r_i r_j + \sum_{i,j>i,k>j} r_i r_j r_k - \dots + r_1 r_2 \dots r_N$$

(3)

Using the fact that a product of Gaussian functions is a finite sum of Gaussian functions, integrals giving volumes and derivatives of volumes with respect to inter-atom distances can easily be computed. Based on this statement Grant *et al.* [29] proposed a fast method to compute the maximal volume overlap. The maximization of the volume over six parameters describing rigid body motion (a quaternion for rotation and three coordinates for translation) is performed by a classical gradient methods optimization. To reduce the amount of computation, only a few neighbors are considered for each atoms. This score results in a precise measure of shape similarity, which is implemented in the ROCS software [13].

3D Fingerprint

In 2005, Haigh *et al.* introduced shape-fingerprint methods to further reduce the computational times of optimal volume overlap [30]. Let's consider two molecular shapes *A* and *B*. Gaussian representation allows for simple evaluation of volumes of *A*, *B* and the volume intersection. This is used to define the Shape-Tanimoto (ST) score :

$$ST = \frac{V_{AB}}{V_A + V_B - V_{AB}}$$

(4)

A set of reference shapes is selected from the molecule database by an agglomerative clustering. This process clusters together shapes with a Shape-Tanimoto score greater than a threshold parameter called Design-Tanimoto. A reference shape corresponds to each cluster. Therefore, all pairs of reference shapes have a ST similarity greater than the Design-Tanimoto threshold. A shape fingerprint is a vector of bits, each bit corresponding to a reference shape. A bit is set to "1" when the described shape is sufficiently close to the reference in the sense of the ST score.

The shape-fingerprint score is then given by:

$$SF = \frac{N_{AB}}{N_A + N_B - N_{AB}}$$

(5)

with N 's are the number of equal bits. This idea leads to a fast technique, which nevertheless may be insufficiently discriminative allowing dissimilar molecules to exhibit high scores. Physico-chemical fingerprint score can be added to the ST score to improve superimposition of groups with similar properties (hydrogen bond donor, hydrogen bond acceptor, hydrophobic patch, cation, anion and ring). This color score added to this shape score is called the combo score.

Graph Based Techniques

There is a lot of information in the literature on the graph based representation for searching similarities between chemical compounds based on their structural formula. Molecules can be represented by reduced graphs which summarize their chemical structure [31]. However, such graphs only represent molecule connectivities and can not differentiate conformers. More recent methods consider graphs based on distance matrices and convey information on the 3D structure. Nodes are labeled by atoms types (or atom properties) and edges by inter-atomic distances [32, 14, 15].

The search for common chemical structure is formulated as a common subgraph or maximal common subgraph problem that is a search for isomorphisms between graphs. These problems are known to be NP-hard (no polynomial time algorithms known to solve this problem) and many algorithms exist to approximately solve them. Most optimal algorithms (that are able to find the best solution) are based on maximal clique detection in a composite graph constructed from the two graphs to be compared [33, 34]. For example, RASCAL [35] searches for maximal common edge subgraphs more efficient for chemical graphs than former vertices induced searches. This approach has been extended to take 3D topology into account by [36] and [32].

A second class of algorithm are based on a backtracking procedure [37] and [38]. When dealing with planar graphs, the Ullman algorithm runs in linear time, and therefore is an algorithm of choice for similar chemical substructure detection [38]. This algorithm has been extended by Krissinel [14] and is considerably more efficient than classical maximum clique detection algorithms.

In a recent method due to Vainio *et al.* [15], molecules are represented as graphs whose vertices are labeled with electrostatic potential. Then the method searches for maximal common subgraph. Vertices are geometrical locations added around each heavy atom following simple geometric rules. These locations are chosen in order to regularly span the space around the molecule without collision with atoms. Then at each vertices the electrostatic potential is evaluated and a vector of shape descriptors is computed. This vector corresponds to histograms of signed distances of all atoms from the graph vertices. The CSI algorithm of Krissinel is used to search for maximal common subgraphs and subgraphs values (vertice values) are compared during the search in order to reject non similar substructures. Superimposition is then performed and is scored by an overlay optimization of Gaussian shape densities (ROCS like).

Other approaches address more explicitly the question of the 3D alignment of compounds. Among those, we briefly discuss the geometry matching based techniques and one Maximal Substructure Search approach that, in principle, are not limited in the chemical vicinity of one seed compound.

Maximal Substructure Search

The CSR algorithm introduced in the late 90s [39] is a general approach for search for the maximal 3D motif common (or Maximal SubStructure - MSS) to two sets of coordinates. It iteratively and stochastically searches for the largest set of atom pairings between two clouds of atom coordinates - no a priori pairings or a priori rules such as knowledge of the neighbors are required. Atomic natures, bonds and connectivity information are ignored. Briefly, each iteration starts from a random initial superposition, and iterative pairings of atoms are performed until no new pairing occurs. One particular feature of this approach is the use of an intrinsic definition of the stop criterion. Pairing is based on distance sort of the N1*N2 inter-atomic distances between the N1 atoms of the molecule 1 and the N2 atoms of the molecule 2. The array of the N1*N2 distances is sorted by increasing values. The first atom-pair, corresponding to the smallest distances, is always included in the common motif. Next pairs are included until a member of a pair already included in the common motif appears. The latter pair is not included in the motif and the pairing terminates. Then the complete sets of coordinates are best superimposed from the current pairings and the whole process distance sorting/best fit is iterated until no new pair is accepted. This whole process is performed for a series of random starting points and CSR returns the largest motif identified. Thus, the approach attempts to maximize the number of pairings. CSR was shown to efficiently retrieve similarities in large sets of coordinates. However, limitations are present since similarities between biological molecules must also consider that atomic properties of the pairs are compatible.

The LigCSRre evolution of the algorithm [16] embedded several additional particular features. Firstly, LigCSRre accepts a regular expression formalism that allows to define for each atom which pairings are possible based on some physico-chemical properties, similarly to the Escan approach [40]. This results in smaller search space and increases the search efficiency: LigCSRre usually requires much less iterations than CSR. Secondly, LigCSRre extends the set of pairings at search convergence. The extended pair collection embeds the MSS identified by the CSR algorithm enlarged by atom pairs that are distant by less than a user specified tolerance. This can result in more relevant similarity search since it is possible that CSR stops to enlarge its MSS for an atom already paired, but hiding subsequent pairs.

Geometry Hashing

The geometry hashing algorithm was firstly introduced in the late 80s [41, 42] for model-based recognition in computer vision. Its underlying basic idea is to transform the objects to match in a redundant representation invariant by rigid transformation, and to take advantage of these invariant encodings to search for similarity between objects. Its transposition to the sequence independent structural comparison of proteins and the detection of three dimensional structural motifs in proteins as soon of the early 90s (*e.g.* [43-45] has also found a concrete application for drug alignment [46, 47]. However, unlike the original geometry hashing algorithm, there is no use in these applications of hash tables. These approaches are also called Geometry Matching. One recent approach exploiting the concepts of geometry hashing is LigMatch [17]. For each compound, a series of triplets of atoms are generated, to serve as the basis of comparison with other compounds. For compound alignment, all possible triplets for each compound are compared, in terms of inter-atomic distances and atom types. Compatible triplets are considered as matched. These matches are then used to superimpose the complete set of atoms of the two compounds and each superimposition is scored. The score is calculated as the number of coincident atoms in the mapping. Among all the possible matches between two compounds, only the largest score is retained. Important for this approach are the definition of the triplets of atoms, depending on inter-atomic distances and atomic types. Another related approach based on triplets but coupled to graph techniques is MedSUMOLig [12], a derivation for small compound of the Sumo approach [48] originally developed for proteins.

Coloring Atoms and Shapes to Combine Physico-chemical Properties

Whereas all these search engines are based on some pure geometric criterion to optimize (volume, RMS deviation,...), one should keep in mind that there is a need to enhance the physico-chemical relevance of the geometric 3D similarity search. Indeed, for all approaches taking into account the physico-chemical properties associated with compound overlap, a significant gain in the results has been observed (Fig. **3**). For shape based techniques, the ROCS software for database screening adds to a geometric similarity index (shape Tanimoto) a color score which takes into accounts similar physico-chemical properties among overlapped groups (hydrogen bond donor, hydrogen bond acceptor, hydrophobic group, cation, anion and ring). Ranking hits on this combined score, improves greatly the screening performances [13].

The addition of physico-chemical properties may facilitate the resolution of the difficult optimization problem of purely geometrical overlapping of molecules. The field-based alignment method FieldScreen [49, 50] is a such example. It computes the extrema of four types of three-dimensional molecular potential fields (van der Waals, electrostatic, hydrophobic). The extrema called field points are rigidly overlayed. However, such field-based alignment is a difficult global optimization problem full of local minima. This problem is solved by considering the field points as a graph with nodes labeled with their field type and edges labeled with distances between pairs of field points. A clique detection algorithm retrieves all common subgraphs which are used to compute alignment of field points. These alignments give initial values for maximizing the volume overlay. The best overlay is selected and is very close to the global minima. In the LigMatch geometry hashing approach, implicit physico-chemical compatibility is based on the fact that matches can only occur for triplet of equivalent atoms, where all atoms of the same element are treated as equivalent. In the LigCSRre approach, a more sophisticated tunable strategy to define atom physico-chemical property equivalence classes is implemented. It makes use of the UNIX regular expressions to define equivalence classes. Those expressions are user-defined, which makes this program totally customizable, offering enough plasticity to construct complex rules such that compound specific expertise from the chemists could be used, although so far only the use of generic set of rules has been reported.

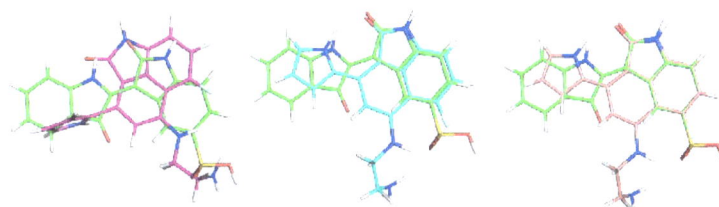

Figure 3: Two active compounds of the Cyclin Dependant Kinase 2 (CDK2) are superimposed using ROCS (left), ROCScff (middle), and LigCSRre (right). Important groups for their binding to the well-known "hinge" region of the CDK2 protein include the N-H and C=O on top. The uncolored ROCS superposition appears much less relevant than those obtained using ROCScff or LigCSRre, that take into account the physico-chemical properties of the atoms. One also denotes the small variations in the alignment between ROCScff and LigCSRre that illustrate how the alignment is dependant on each approach.

Flexible Alignment

All above alignment-based or alignment-free methods consider shapes as rigid body and may lead to comparison errors as varying shapes of the same molecules has to be considered. Note that the context of libraries' screening is very different from the context of compound alignment for pharmacophore extraction, for which numerous approaches for flexible multiple alignment of compounds have been described (*e.g.* [51]). In the context of large library screening some approaches also consider flexibility. Among them are FLEXS [52], fFlash [53], Surflex-sim [54], or Fang *et al.*' [55].

Most methods which handle flexibility can be characterized by their similarity criteria, the 3D superposition approach and the optimization algorithm used to perform the superposition. In general, they try to optimize a scoring function combining chemical similarity, given by the number of common chemical features of atoms or group of atoms, the superposition based volume overlay (rigid or Gaussian, and/or electrostatic fields) and the internal deformation energy. Optimization variables are sets of angles of rotation around flexible bonds. In FLEXS for instance, the molecules are partitioned into fragments and each fragment can adopt a discrete set of conformations using a discrete set of torsional angles for each rotatable bond. The discrete optimization process firstly places an initial fragment using a rigid superposition Fourier method. Then it incrementally adds fragments selecting the best scored fragment given already among placed ones. The score function simultaneously accounts for geometrical (steric overlap) and chemical similarities. In fFlash, the optimization process is performed by a graph clique search. Then an explicit 3D superposition taking into account the angles of torsions of the molecules is applied to all the retrieved cliques.

In a very recent approach developed by Shin *et al.* [56], the superposition is driven by a sum of Gaussian functions combining volume and electrostatic field. This score is penalized by an internal energy term that approximately constrains the energy of deformation to not overpass a user given limit. The optimization is performed by a Quasi-Newton method.

Overall however, these methods are usually not considered as well suited for large scale *in silico* screening due to computational time limitations. Interestingly, new fast approaches, such as developed by Yi Fang *et al.* [55], use deformation invariant shapes descriptors instead of considering explicitly a model of flexibility with torsion angles. It describes a shape by a set of local diameters. To each selected points on a Connolly surface, corresponds its local diameter that is the distance to the opposite side of the molecule (going inward following the normal to the surface). Histograms of local diameters are compared to provide a shape similarity score. This alignment-free method is a promising approach for large scale *in silico* screening.

Alignment-free Approaches

All the alignment-free methods describe molecules via a set of descriptors independent on rigid movements that can be eigenvalues (BCUT indices, [57, 58], moments of charges, mass distribution [59], inter-atomic distances [18] or histograms comparison methods [60-63]). Any adequate similarity scores between such matrix can be used to perform a fast 3D comparison without molecules alignment or superimposition.

Early non-superposition methods considered all atom triplets of the molecules and encoded this information into bit strings and histograms for fast comparison, [60-62]. These methods are fast but are less effective than superposition methods. In [63], a ligand is described by a ray tracing technique. The shape is represented by distribution (histograms) of ray segment length (1D description) and distribution of electrostatic potential along the ray path at reflection point at the molecule surface. The histograms of two molecules are compared by a distance.

Recently, [18] developed a non-superposition method named USR (Ultrafast Shape Recognition). A molecular shape is represented by 4 sets of atomic distances: distances of all the atoms of the molecule to four specific locations: the molecular centroid (ctd), the closest atoms to centroid (cst), the farthest atom to ctd (fct) and farthest atoms to fct (ftf). The last three locations are well separated points and the four set of distances are mathematically sufficient to reconstruct all the atomic positions up to a rigid body displacement. The four sets of distances are then reduced to the first three centered moments of their distribution that is the mean, the variance and the skewness. Hence, the molecule shape M is represented by only 12 parameters. A simple score between molecule A and B is given by:

$$S = \frac{1}{1 + \frac{1}{12} \sum_i |M_i(A) - M_i(B)|}$$

$$(6)$$

Despite the information compression, USR achieves similar performance as ROCS and Shape signature of Zauhar on identifying good matches (molecules that share a common biological activity) while being thousand of times faster.

SCREENING STRATEGIES FOR LARGE COMPOUND COLLECTIONS

Speed Issues

The screening of large compound collections, usually each represented by multiple conformers, relies on scoring of the quality of superimpositions, which highly vary for each approach (see previous section). Depending on bank size, additional strategies have been setup to enhance the search efficiency. The basic ideas underlying efficient search against large collections are to (i) favor pre-computation when possible (ii) favor hierarchical strategies when possible. Original Geometric hashing for instance embedded such pre-computation during the calculation of the hash table, taking advantage of this to gain in the query phase. Alignment recycling [64] is another very recent hierarchical approach to smartly reduce search computation (Fig. **4**). It extends the 3D shape-fingerprint of Haigh [30] with precise but slower ROCS similarity. In this approach, the optimal shape overlap of the query to the database conformers is approximated by evaluating the best move passing through reference shapes. That is the rigid movement (translation and rotation) giving the best overlap is approximated by the chaining of a movement overlaying the query to a reference shape (as selected in [30]) with overlapping of the reference to the database conformers. Only overlapping between references and database conformers of sufficient similarity need to be retained and are pre-calculated. A Shape Tanimoto score over 0.73 (minimum dissimilarity between reference shapes) gives a good balance between speed and granularity (or density of reference shape database covering). This technique shows a considerable improvement of speed over optimal shape overlay. On average, the CPU time needed to perform one query against a fifteen million conformers' database is divided by 100.

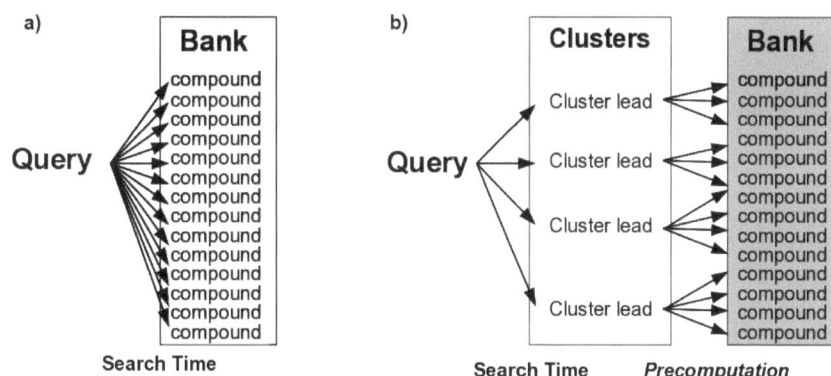

Figure 4: Alignment recycling strategy. The brute force strategy (left) consists in aligning the query and all compounds of the bank. Alignment recycling reduces the search-time by pre-computing the alignment between the representatives and the compounds associated with them. At search-time, only the alignments of the query and representatives are performed, and pre-calculated values are used to derive alignments against each compound of the bank, resulting in important gain in search speed.

Limiting Conformers Per Compound

Except for the few approaches that consider compound flexibility, both alignment based and alignment-free approaches are highly dependent on the compound 3D conformation. This is usually compensated by using a large "enough" number of conformations per compound. Whereas for Structure Based Virtual Screening it can be important that a conformation close the bio-active one is present, and that the conformations sample rather uniformly the space, the 3D similarity search in Ligand Based Virtual Screening context is more sensitive to the fact that negative match between two compounds should not result from the absence of conformers matching with scores sufficiently close to the best possible score for the two compounds.

Whereas the number of conformers of a compound is, in a first approximation, a function of its number of rotatable bonds, the question of the maximal number of conformers to be considered is under discussion in the literature [65, 12, 16, 17]. MacGaughey *et al.* [65] have shown that ROCScff performances are not deeply affected by increasing the number of conformations per compound up to 100. Quintus *et al.* [16] have also shown on some cases that the performance of LigCSRre is not considerably changed when the maximal number of conformers increased from 50 to 100. Validating LigMatch, Kinnings and Jackson [17] also reported a study with used a maximal number of conformers of 100. Various tools allow for such multiconformer generation, such as OMEGA [66] or Frog [67].

Combining Searches from Several Active Compounds

Similarity based techniques are confronted with the limitation that the search is based on one compound, whereas QSAR or pharmacophore techniques aim at extracting representative features from a collection of bioactive compounds. Indeed, large variations in the enrichments have been observed depending on the template compound [12, 16, 68], highlighting the need to consider the diversity of the information available. To our knowledge, no 3D similarity search technique combining several compounds during the search has been proposed so far other than in the pharmacophore context (*e.g.* [69, 8, 10]). Possible alternatives to overcome this limitation are (i) to carefully select from the collection of the known bioactive compounds one representative [17] or (ii) to consider the combination of the screenings performed using each compound separately [16], in the same spirit that the group fusion techniques used in 2D search approaches [70]. The fact that full ligand structures are used rather than pharmacophoric hypothesis makes the search more dependent on the chemical properties of each query, and might inevitably bring noise to the results. Nonetheless, this approach has the advantage to avoid focusing on what co-actives must have in common for activity (pharmacophore) but rather on what might bring specificity to protein binding. Thus, by combining several query searches a higher number of hits and therefore a higher number of therapeutic tracks will be opened. Indeed, the combination of the results from independent screenings involving several compounds chemically diverse has been reported to improve the quality of the compound enrichment [12,

16]. It is particularly interesting that early enrichment when multiple template is used seems to benefit from such combination. Undoubtedly, the benefit of cumulating the available information of several actives is to retrieve complementary novel chemical entities rather than focusing on consensual chemical features.

HOW FAR ARE WE ?

The developments of these recent years have led to important evolutions for *in silico* lead candidate identification. One major advance is undoubtedly the emergence of new search engines and strategies that make possible the search for candidates in a few seconds to a few hours only, whereas early approaches would have required much more time. Noticeably, today screening engines are mostly based on the semi-flexible model. The question of the development of engines considering the flexible model is opened in the context of large bank screening. Obviously, one question is related to the quantification of the gain expected from the flexible model compared to the semi-flexible one, in particular with respect to the goal of new scaffold hopping that should be performed. The size of the banks to screen could become a critical parameter to that respect. Although estimates of the chemical space are on the order of over 10^{60} [71], it is today unlikely that screening experiments will screen banks of more than a few million compounds. Indeed, it is likely that the better the focus libraries, the most efficient the screening.

The efficiency of the search engines must not conceal one last bottleneck related to the assessment of the similarity. Indeed, the balance between the search for too similar only, and thus a risk for poor applicability due to patents or delivery aspects, and the hop for new scaffolds remains a challenge. Whereas efforts to derive new scoring functions have to be pursued, in particular to derive context specific functions, depending on the foreseen interaction mechanism of the compound, the cellular, sub-cellular or molecular environment, it seems that far or unrelated scaffolds to the starting compound lie by essence at the limit of the scope of the similarity search. Certainly, one direction to dig is the integration within the search of the human expertise of medicinal chemists of each particular pathology or each particular mechanism. Some tunability should be possible for instance to weight particular aspects, or prune some branches of the bank mining. Interestingly, many of the current approaches are not based on the same underlying hypotheses or the same heuristics to drive the search. Clearly, one promising perspective lies in the combination of the different approaches to address the search using as many point of views as possible and extract compounds common to all or specific of one particular method, under the supervision of experts of medicinal chemistry. As for the end user, it thus seems desirable that the approaches described here be available either as standalone programs or via the Web, in order to be easily compared, combined, adapted, modified and further developed. Indeed many of these approaches are available either as commercial packages (*e.g.* [11, 13, 12],...), some are freely available to the community either as standalone programs or via the Web [72, 73, 20].

ACKNOWLEDGEMENTS

Supports from the Inserm institute and University Paris Diderot are greatly appreciated.

REFERENCES

[1]   Stockwell, B.R. Chemical genetics: ligand-based discovery of gene function. *Nat. Rev. Genet.* **2000**, *1*, 116–25.

[2]   Esposito, E.X.; Hopfinger, A.J.; Madura, J.D. Methods for applying the quantitative structure-activity relationship paradigm. *Methods Mol. Biol.* **2004**, *275*, 131–214.

[3]   Khedkar, S.A.; Malde, A.K.; Coutinho, E.C.; Srivastava, S. Pharmacophore modeling in drug discovery and development: an overview. *Med. Chem.* **2007**, *3*, 187–97.

[4]   Ekins, S.; Mestres, J.; Testa, B. *In silico pharmacology for drug discovery: applications to targets and beyond.* **2007**, *152*, 21-37.

[5]   IUPAC, Glossary of Terms used in Medicinal Chemistry, **1998.** http://www.chem.qmul.ac.uk/iupac/medchem/ix.html.

[6]   Martin, Y.C.; Bures, M.G.; Willett, P. Searching databases of three dimensional structures. In *Reviews in Computational Chemistry*; Lipkowski, K., Boyd, D.B., Eds.; VHC Publishers: New York, 1990; Vol. 1, pp 213–256.

[7]   Clark, D.E.; Willett, P.; Kenny, P.W. Pharmacophoric pattern matching in files of three-dimensional chemical

structures: use of bounded distance matrices for the representation and searching of conformationally flexible molecules. *J. Mol. Graph.* **1992**, *10*, 194–204.

[8]　Hessler, G.; Zimmermann, M.; Matter, H.; Evers, A.; Naumann, T.; Lengauer, T.; Rarey, M. Multiple-ligand-based virtual screening: methods and applications of the MTree approach. *J. Med. Chem.* **2005**, *48*, 6575–84.

[9]　Ebalunode, J.O.; Ouyang, Z.; Liang, J.; Zheng, W. Novel approach to structure-based pharmacophore search using computational geometry and shape matching techniques. *J. Chem. Inf. Model.* **2008**, *48*, 889–901.

[10]　Dror, O.; Schneidman-Duhovny, D.; Inbar, Y.; Nussinov, R.; Wolfson, H.J. Novel approach for efficient pharmacophore-based virtual screening: method and applications. *J. Chem. Inf. Model.* **2009**, *49*, 2333–43.

[11]　Girke, T.; Cheng, L.-C.; Raikhel, N. ChemMine. A compound mining database for chemical genomics. *Plant. Physiol.* **2005**, *138*, 573–7.

[12]　Sperandio, O.; Andrieu, O.; Miteva, M.A.; Vo, M.-Q.; Souaille, M.; Delfaud, F.; Villoutreix, B.O. MED-SuMoLig: a new ligand-based screening tool for efficient scaffold hopping. *J. Chem. Inf. Model.* **2007**, *47*, 1097–110.

[13]　OpenEyes, *ROCS version 2.2*, 2006.

[14]　Krissinel, E.; Henrick, K. Secondary-structure matching (SSM), a new tool for fast protein structure alignment in three dimensions. *Acta Crystallogr. D Biol. Crystallogr.* **2004**, *60*, 2256–68.

[15]　Vainio, M.J.; Puranen, J.S.; Johnson, M.S. ShaEP: molecular overlay based on shape and electrostatic potential. *J. Chem. Inf. Model.* **2009**, *49*, 492-502.

[16]　Quintus, F.; Sperandio, O.; Grynberg, J.; Petitjean, M.; Tuffery, P. Ligand scaffold hopping combining 3D maximal substructure search and molecular similarity. *BMC Bioinformatics* **2009**, *10*, 245.

[17]　Kinnings, S.L.; Jackson, R.M. LigMatch: a multiple structure-based ligand matching method for 3D virtual screening. *J. Chem. Inf. Model.* **2009**, *49*, 2056–66.

[18]　Ballester, P.J.; Richards, W.G. Ultrafast shape recognition to search compound databases for similar molecular shapes. *J. Comput. Chem.* **2007**, *28*, 1711–23.

[19]　Jenkins, J.L.; Glick, M.; Davies, J.W. A 3D similarity method for scaffold hopping from known drugs or natural ligands to new chemotypes. *J. Med. Chem.* **2004**, *47*, 6144–59.

[20]　Wang, Y.; Xiao, J.; Suzek, T. O.; Zhang, J.; Wang, J.; Bryant, S.H. PubChem: a public information system for analyzing bioactivities of small molecules. *Nucleic Acids Res.* **2009**, *37*, W623–33.

[21]　Irwin, J.J.; Shoichet, B.K. ZINC–a free database of commercially available compounds for virtual screening. *J. Chem. Inf. Model.* **2005**, *45*, 177–82.

[22]　Miteva, M.A.; Violas, S.; Montes, M.; Gomez, D.; Tuffery, P.; Villoutreix, B.O. FAF-Drugs: free ADME/tox filtering of compound collections. *Nucleic Acids Res.* **2006**, *34*, W738–44.

[23]　Lagorce, D.; Sperandio, O.; Galons, H.; Miteva, M.A.; Villoutreix, B.O. FAF-Drugs2: free ADME/tox filtering tool to assist drug discovery and chemical biology projects. *BMC Bioinformatics* **2008**, *9*, 396.

[24]　Sittampalam, G.S.; Kahl, S.D.; Janzen, W.P. High-throughput screening: advances in assay technologies. *Curr. Opin. Chem. Biol.* **1997**, *1*, 384–91.

[25]　Sun, H. Pharmacophore-based virtual screening. *Curr. Med. Chem.* **2008**, *15*, 1018–24.

[26]　Carbo, R.; Leyda, L.; Arnau, M. How Similar is a Molecule to Another? An Electron Density Measure of Similarity Between two Molecular Structures. *Int. J. Quantum. Chem.* **1980**, *17*, 1185–1189.

[27]　Boys, S.F. Electronic Wave Functions. I. A General Method of Calculation for the Stationary States of Any Molecular System. *Proc. R. Soc. Lond. A* **1950**, *200*, 542–554.

[28]　Grant, J.A.; Pickup, B.T. A Gaussian Description of Molecular Shape. *J. Phys. Chem.* **1995**, *99*, 3503–3510.

[29]　Grant, J.A.; Gallardo, M.A.; Pickup, B.T. A Fast Method of Molecular Shape Comparison: A Simple Application of a Gaussian Description of Molecular Shape. *Journal of Computational Chemistry* **1996**, *17*, 1653–1666.

[30]　Haigh, J.A.; Pickup, B.T.; Grant, J.A.; Nicholls, A. Small molecule shape-fingerprints. *J. Chem. Inf. Model.* **2005**, *45*, 673–84.

[31]　Gillet, V.J.; Willett, P.; Bradshaw, J. Similarity searching using reduced graphs. *J. Chem. Inf. Comput. Sci.* **2003**, *43*, 338–45.

[32]　Marialke, J.; Korner, R.; Tietze, S.; Apostolakis, J. Graph-Based Molecular Alignment (GMA). *J. Chem. Inf. Model.* **2007**, *47*, 591–601.

[33]　Levi, G.A note on the derivation of maximal common subgraphs of two directed or undirected graphs. *Calcolo* **1972**, *9*, 341–354.

[34]　Bron, C.; Kerbosch, J. Algorithm 457-finding all cliques of an undirected graph. *Communications of the ACM*

1973, *16*, 575–577.

[35] Raymond, J.W.; Gardiner, E.J.; Willett, P. RASCAL: Calculation of Graph Similarity Using Maximum Common Edge Subgraphs. *The Computer Journal* **2002**, *45*, 631–644.

[36] Stahl, M.; Mauser, H.; Tsui, M.; Taylor, N.R. A robust clustering method for chemical structures. *J. Med. Chem.* **2005**, *48*, 4358–66.

[37] McGregor, J. Backtrack search algorithms and the maximal common subgraph problem. *Software Practice and Experience* **1982**, *12*, 23–34.

[38] Ullman, J.R. An algorithm for subgraph isomorphism. *JACM* **1976**, *23*, 31–42.

[39] Petitjean, M. Interactive Maximal Common 3D Substructure Searching with the Combined SDM/RMS Algorithm. *Comp. Chem.* **1998**, *22*, 463–465.

[40] Escalier, V.; Pothier, J.; Soldano, H.; Viari, A. Pairwise and multiple identification of three-dimensional common substructures in proteins. *J. Comput. Biol.* **1998**, *5*, 41–56.

[41] Lamdan, Y.; Wolfson, H.J. Geometric hashing: a general and efficient model-based recognition scheme. *Proceedings of the Second ICCV* **1988**, 3–26.

[42] Wolfson, H.J. Model-based recognition by geometric hashing. *Proceedings of the 1st European Conference on Computer Vision (ECCV 90)*, New York, 1990; pp 526–536.

[43] Fischer, D.; Bachar, O.; Nussinov, R.; Wolfson, H. An efficient automated computer vision based technique for detection of three dimensional structural motifs in proteins. *J. Biomol. Struct. Dyn.* **1992**, *9*, 769–89.

[44] Bachar, O.; Fischer, D.; Nussinov, R.; Wolfson, H. A computer vision based technique for 3-D sequence-independent structural comparison of proteins. *Protein Eng* **1993**, *6*, 279–88.

[45] Pennec, X.; Ayache, N. A geometric algorithm to find small but highly similar 3D substructures in proteins. *Bioinformatics* **1998**, *14*, 516–22.

[46] Lemmen, C.; Lengauer, T. Time-efficient flexible superposition of medium-sized molecules. *J. Comput. Aided. Mol. Des.* **1997**, *11*, 357–68.

[47] Putta, S.; Lemmen, C.; Beroza, P.; Greene, J. A novel shape-feature based approach to virtual library screening. *J. Chem. Inf. Comput. Sci.* **2002**, *42*, 1230–40.

[48] Jambon, M.; Imberty, A.; Deléage, G.; Geourjon, C. A new bioinformatic approach to detect common 3D sites in protein structures. *Proteins* **2003**, *52*, 137–45.

[49] Cheeseright, T.; Mackey, M.; Rose, S.; Vinter, A. Molecular field extrema as descriptors of biological activity: definition and validation. *J. Chem. Inf. Model* **2006**, *46*, 665–76.

[50] Cheeseright, T.J.; Mackey, M.D.; Melville, J.L.; Vinter, J.G. FieldScreen: virtual screening using molecular fields. Application to the DUD data set. *J. Chem. Inf. Model.* **2008**, *48*, 2108–17.

[51] von Korff, M.; Freyss, J.; Sander, T. Flexophore, a new versatile 3D pharmacophore descriptor that considers molecular flexibility. *J. Chem. Inf. Model.* **2008**, *48*, 797–810.

[52] Lemmen, C.; Lengauer, T.; Klebe, G. FLEXS: a method for fast flexible ligand superposition. *J. Med. Chem.* **1998**, *41*, 4502–20.

[53] Krämer, A.; Horn, H.W.; Rice, J.E. Fast 3D molecular superposition and similarity search in databases of flexible molecules. *J. Comput. Aided Mol. Des.* **2003**, *17*, 13–38.

[54] Jain, A.N. Ligand-based structural hypotheses for virtual screening. *J. Med. Chem.* **2004**, *47*, 947–61.

[55] Fang, Y.; Liu, Y.-S.; Ramani, K. Three dimensional shape comparison of flexible proteins using the local-diameter descriptor. *BMC Struct. Biol.* **2009**, *9*, 29.

[56] Shin, W.; Hyun, S.A.; Chae, C.H.; Chon, J.K. Flexible alignment of small molecules using the penalty method. *J. Chem. Inf. Model.* **2009**, *49*, 1879–88.

[57] Burden, F.R. Molecular Identification Number for Substructure Searches. *J. Chem. Inf. Comput. Sci.* **1989**, *29*, 225–227.

[58] Pearlman, R.S.; Smith, K.M. Metric Validation and the Receptor-Relevant Subspace Concept. *J. Chem. Inf. Comput. Sci.* **1999**, *39*, 28–35.

[59] Silverman, B.D.; Platt, D.E. Comparative Molecular Moment Analysis (CoMMA): 3D-QSAR without Molecular Superposition. *J. Med. Chem.* **1996**, *39*, 2129–2140.

[60] Bemis, G.W.; Kuntz, I.D. A fast and efficient method for 2D and 3D molecular shape description. *J. Comput. Aided Mol. Des.* **1992**, *6*, 607–28.

[61] Nilakantan, R.; Bauman, N.; Venkataraghavan, R. New method for rapid characterization of molecular shapes: applications in drug design. *J. Chem. Inf. Comput. Sci.* **1993**, *33*, 79–85.

[62] Good, A.C.; Ewing, T.J.; Gschwend, D.A.; Kuntz, I.D. New molecular shape descriptors: application in

database screening. *J. Comput. Aided Mol. Des.* **1995**, *9*, 1–12.

[63] Zauhar, R.J.; Moyna, G.; Tian, L.; Li, Z.; Welsh, W.J. Shape signatures: a new approach to computer-aided ligand- and receptor-based drug design. *J. Med. Chem.* **2003**, *46*, 5674–90.

[64] Fontaine, F.; Bolton, E.; Borodina, Y.; Bryant, S.H. Fast 3D shape screening of large chemical databases through alignment-recycling. *Chem. Cent. J.* **2007**, *1*, 12.

[65] McGaughey, G.B.; Sheridan, R.P.; Bayly, C.I.; Culberson, J.C.; Kreatsoulas, C.; Lindsley, S.; Maiorov, V.; Truchon, J.-F.; Cornell, W.D. Comparison of topological, shape, and docking methods in virtual screening. *J. Chem. Inf. Model.* **2007**, *47*, 1504–19.

[66] OpenEyes, *OMEGA version 2.2.1*, **2007**.

[67] Leite, T.B.; Gomes, D.; Miteva, M.A.; Chomilier, J.; Villoutreix, B.O.; Tufféry, P. Frog: a FRee Online druG 3D conformation generator., *Nucleic Acids Res.* **2007**, *35*, W568-72.

[68] Kirchmair, J.; Distinto, S.; Markt, P.; Schuster, D.; Spitzer, G.M.; Liedl, K.R.; Wolber, G. How to optimize shape-based virtual screening: choosing the right query and including chemical information. *J. Chem. Inf. Model.* **2009**, *49*, 678-92.

[69] Mestres, J.; Rohrer, D.C.; Maggiora, G.M. A molecular field-based similarity approach to pharmacophoric pattern recognition. *J. Mol. Graph. Model.* **1997**, *15*, 114–21, 103–6.

[70] Willett, P. Searching techniques for databases of two- and three-dimensional chemical structures. *J. Med. Chem.* **2005**, *48*, 4183–99.

[71] Dobson, C.M. Chemical space and biology. *Nature* **2004**, *432*, 824-8.

[72] Bauer, R.A.; Bourne, P.E.; Formella, A.; Frömmel, C.; Gille, C.; Goede, A.; Guerler, A.; Hoppe, A.; Knapp, E.-W.; Pöschel, T.; Wittig, B.; Ziegler, V.; Preissner, R. Superimpose: a 3D structural superposition server. *Nucleic Acids Res.* **2008**, *36*, W47-54.

[73] Sperandio, O.; Petitjean, M.; Tuffery, P. wwLigCSRre: a 3D ligand-based server for hit identification and optimization. *Nucleic Acids Res.* **2009**, *37*, W504-9.

CHAPTER 4

Fragment-Based Methods for Lead Discovery

Will R. Pitt[1,2], Alicia P. Higueruelo[2] and Nikolay P. Todorov[3,*]

[1]*UCB Celltech, branch of UCB Pharma S.A., Slough, UK;* [2]*Department of Biochemistry, University of Cambridge, Cambridge, UK and* [3]*Molscape Therapeutics, Cambridge, UK; E-mail: nikolay.todorov@molscape.com*

Abstract: Fragment-based drug discovery (FBDD) has received much attention in recent years by the pharmaceutical industry and academia. Here published FBDD methods are reviewed with a particular focus on *in silico* methods and theoretical approaches. Included are techniques useful at each stage of the process, from fragment library design, to virtual fragment screening and on to fragment growth and lead generation. The strengths and weaknesses of FBDD approaches are presented and the benefits of the complementary use of experimental and computational methods are discussed.

Key Words: fragment-based drug discovery, *de novo* drug design, library design, binding energy, scoring function.

INTRODUCTION

The discovery and development of new drugs is a highly-regulated, costly and time-consuming process taking over 15 years, about $800M, and the testing of more than fifteen thousand chemical compounds [1]. Given this huge cost, it is highly attractive to seek novel approaches that would afford a more efficient development. The completion of the sequencing of the Human Genome and the rapidly growing number of protein crystal structures provides increased opportunities for structure-based design. A range of recent methods, known collectively as Fragment-Based Drug Discovery (FBDD), can be applied even when High-Throughput Screening (HTS) fails to identify initial hits for starting a development program [2].

FBDD methods involve sensitive measurements to detect weak binders and to provide information about fragment binding modes. FBDD offers great promise for the exploration of new areas of chemical space, increased efficiency of the drug design process and has already provided a number of compounds in clinical trials [3]. Hann *et al.* proposed a simple model examining the relationship between the probability of discovering lead molecules from screening and their molecular complexity [4], they rationalized the difficulties encountered by HTS methods and how FBDD could help to solve them.

Computational methods are widely used in conjunction the experimental FBDD to select and prioritize fragment libraries suitable for screening. In addition, computational FBDD methods are considered in their own right as a complementary approach and an analogue to experimental FBDD that can be applied even in the absence of structural protein information. Computational FBDD has been developed in parallel as part of virtual screening and especially *de novo* design methodology. Within the context of the pharmaceutical industry value chain, computational FBDD is a relatively recent addition in the early research stage of drug discovery (Fig. **1**). The importance of computational technologies is highlighted in a Price Waterhouse Coopers Report, "*In silico* technologies will enable drug manufacturers to accelerate the selection process, reduce the cost of preclinical and clinical studies and increase their overall chance of success. We estimate that they could collectively save at least $200m and two to three years per drug." [5].

This chapter reviews current *in silico* FBDD technologies. First, the strengths and weaknesses of FBDD approaches are presented and the benefits from the complementary use of experimental and computational methods are discussed. The theoretical studies and analyses, which have done so much to reveal the potential power of the approach, are described. Methods for fragment-library design and filtering are presented, followed by experimental and *in silico* methods for fragment screening. Prioritization of fragment hits is discussed next, where the concept of ligand efficiency is introduced and methods for computational estimation of binding energy are reviewed. Three general strategies for developing fragments into leads, namely fragment growing, linking and merging are presented and illustrated with validated examples. The concept of modular and emergent type of fragment binding is highlighted, as it has important implications for rational design strategies. Finally, an attempt is made to extrapolate and trace the path of future developments in *in silico* FBDD, viewed in conjunction with parallel developments in experimental technology.

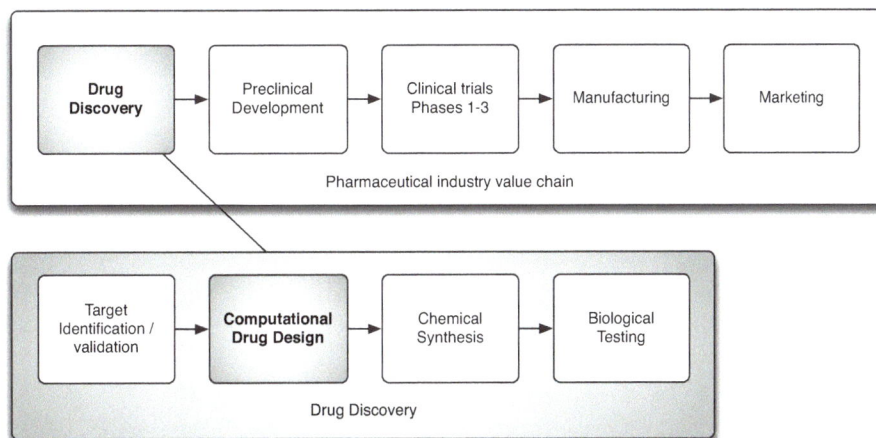

Figure 1: Stages in the biopharmaceutical value chain with focus on drug discovery.

THE PROS AND CONS OF FBDD

The theoretical advantages of testing very small compounds are many folds and there has been widespread uptake of the FBDD approach. By focussing on low molecular weight molecules, the chemical space to be explored is significantly smaller and more manageable. Fink and Reymond estimated that the number of molecules with less than 11 atoms is 26.4M [6]. It has been recognised that, because the size of the chemical universe grows exponentially with the number of atoms, chemical space can be sampled much more effectively with very small molecules. In fact, it has been claimed that using fragments instead of traditional screening libraries, reduces the size of the chemical universe by at least ten orders of magnitude. In addition, less complex molecules will have more chances to match a section of the active site as demonstrated by Hann *et al.* [4] (see below).

Another consideration is the combinatorial explosion of the possible arrangements of these low molecular weight pieces into bigger, more complex molecules. In a stepwise fashion, fragment approaches take advantage of the more efficient exploration of all the possible molecules, by anchoring building blocks one at the time (Fig. **2**). For instance, instead of screening all 25 products of a combinatorial binary library of 5 monomers for each site, one can first screen the 5 monomers for site A and then proceed only with the one that shows complementarity with this sub-site. Only 10 products are synthesised in this scenario. When it comes to growing fragments into larger, more potent molecules, it is possible to retain this sampling efficiency if additional fragments are added in an iterative fashion. For example, if one can link two fragments that bind to separate sites in a neutral fashion, the resulting binding affinity is the product of the 3 component parts (both fragments and linker). This would lead to a leap in competitive inhibitor potency.

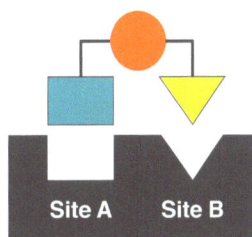

Figure 2: Fragment-based drug discovery.

These considerations lead to the conclusion that employing a fragment-based ligand design has the potential to produce compounds with greatly enhanced binding efficiency [7]. In addition, if after carrying out a fragment screen no useful hits are discovered, the target can safely be discarded at this early stage, allowing resource to be focussed on more tractable targets [8].

There are also practical advantages to fragment screening. Firstly, because fragment hit rates are expected to be higher than traditional HTS campaigns, the screening libraries do not need to be so large. Typically fragment libraries of general utility can be as small as a few thousand compounds or less. By contrast, the HTS decks of pharmaceutical companies are usually larger than several hundred thousand compounds. This represents a huge cost saving not only in terms of purchase price but also in maintenance and informatics overheads. Secondly, the biophysical techniques normally employed for fragment screening can be deployed much more quickly than biochemical assays because they require little or no prior knowledge of protein function. As a result, fragment screening is relatively cheap to carry out and is accessible to small companies and academic labs.

Inevitably, there are practical problems to be overcome before all the theoretical benefits of fragment-based drug discovery can be fully realised. Fragment libraries and many methods for screening them are available, and indeed hit rates are probably higher than for traditional HTS campaigns. However, they are perhaps one order of magnitude higher, not ten! This difference between theory and practice can be explained in a number of ways. Firstly, there are few if any head to head comparisons of HTS compared to fragment screens. HTS successes are highest where they are done against homologs of previous targets for which many known actives populate the screening deck. Secondly, HTS decks will almost always contain relatively small and simple examples, which are close to being fragments. Thirdly, screening compounds will contain low complexity substructures that can bind leaving other portions extending into solvent. Finally, at high micromolar or millimolar concentrations many fragments are at or near saturation point leading to an increased false negative rate.

Once fragment hits are identified there is an urgency to increase potency because of the lowly starting point. If a crystal structure of a hit fragment bound to its target can be generated then rapid progress can be made [9]. However, such crystal structures are not always easily obtained in time. Unlike HTS screens, fragments screens usually detect binding and do not provide any kind of functional read-out. Further assays are needed for this purpose but, almost by definition, these cannot be used until the original fragments have been developed into tighter binders. As a result, there is a risk that time and effort are wasted on inactive molecules. Indeed, even when a crystal structure is available, there is the danger that fragment hits represent a false dawn, leading to project failure in the fragment to hit phase of a medicinal chemistry program. Here FBDD computational techniques that can more reliably predict the maximum affinity or druggability of a given binding site could prove invaluable.

Fragment linking is also not so simple in practice. Neighbouring fragment binding sites are not always available. When they are, designing optimal linkers is a difficult task because the short distances and constrained terminal vectors, coupled with synthetic feasibility considerations, severely limit the available chemical space. Computational ligand design software can highlight this problem or provide solutions where possible. Fragment growth and merging approaches are more usually employed than fragment linking. These are the ways of proven medicinal chemistry and structure-based drug design. See below for more discussion on fragment growth strategies reported in the literature.

As fragment-based drug discovery is still a relatively new paradigm, progress is being made quite rapidly. However, there are still some uncertainties about its utility in different situations. For instance, there is the thought that fragments may not be strong enough binders to induce conformational changes (or select rare conformations) in proteins. Therefore they may prove ineffective tools for the detection of highly prized allosteric binding sites. It also remains unclear whether, as the use of fragment screening in combination with structure-based drug design (SBDD) becomes more widely used, patent congestion will increase. This is possible because more efficient searching of chemical space could lead to different teams converging on similar solutions to binding the same active site. This risk can be reduced by the generation of proprietary fragment screening decks. As with HTS screening decks, computational tools can be used to help design such a library, for example by highlighting areas of fragment space not covered by commercial catalogues (see below).

There are also several other general issues that are especially pertinent to fragment screening. The first is solubility. Fragments need to be highly soluble because of the high concentrations used in screening. The second is the greater need for structural information which in turn puts more pressure on protein production and purification facilities.

Thirdly some types of proteins will be more difficult to target e.g. membrane proteins [10]. However, each drug discovery approach has its limitations and logistical difficulties. FBDD is another approach to add to the growing armoury of methodologies available to drug discovers. Below we describe some of the aspects that should be considered when deciding when, where and how to use FBDD.

Complexity

Widely popular in the 90's, the methods of HTS of compounds stored in the in-house collections of pharmaceutical companies, often prove insufficient and costly for the identification of initial hits to start lead-development programs. In a seminal paper Hann and colleagues [4] showed how molecular complexity works against the chances of finding a hit in a biological assay. This work has been referenced since as the theoretical background which identifies the limitations of HTS and supports the fragment-based approach.

The authors elaborated a simple model to describe the binding event as a match of all ligand features with the features of the surface of the receptor. Ligand and receptor features are conceptually reduced to +/- localized recognition points where a + ligand needs to match a − in the receptor (Fig. **3A**). In this way one can calculate the probability of a binding event for a randomly chosen ligand of a particular size (accounted as number of features) within a given active site described by a finite number of points or features. This probability is computed by enumeration of all possible configurations of ligand and active site and considering the binding event as a complete match of all the ligand recognition points with those of the receptor (Fig. **3B**). For a given active site, this probability can be plotted against the number of ligand features (as a measure of complexity). Although this model is simple and does not take into account the full flexibility or structural reorganisation, it clearly shows how the chances of finding a matching molecule decrease as the complexity of the molecule increases.

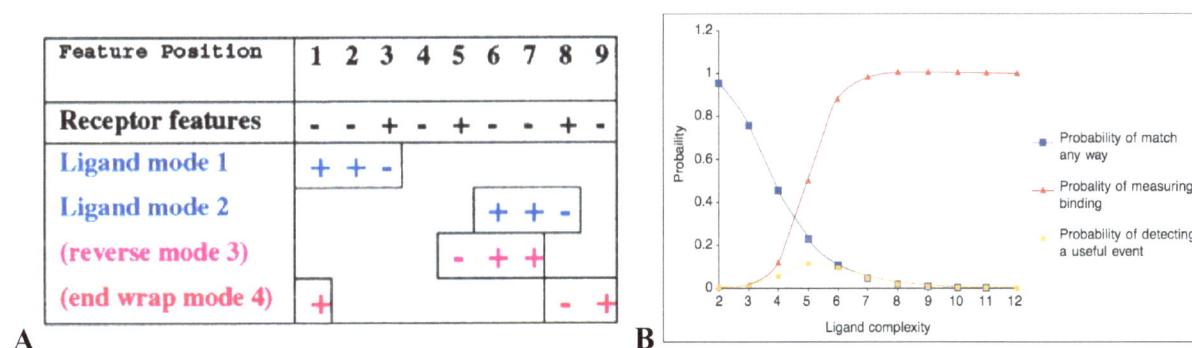

Figure 3: Hann's complexity model. A. Description of the model in terms of receptor and ligand features. B. Probability of a match (blue), probability of measuring binding (red), probability of detecting useful event (yellow) as a function of ligand complexity.

Having very simple molecules reduces the likelihood of actually achieving measurable binding events. In addition low complexity ligands can have multiple binding modes. Although a small number of features are easier to match in the active site, they might not give sufficient affinity for binding to be experimentally detected. Therefore, there is an "optimal" complexity that balances the chances of having a perfect fit between ligand and receptor, and enough interactions to reach a detectable binding (Fig. **3B**).

However, complexity is more a relative concept than a calculable property. Although one can estimate complexity in many different ways (e.g. molecular weight, fingerprints), the precise optimal values that would maximise the chances of having a binding ligand and the ability of measuring it, would depend on the system being studied and the assays employed.

Chemical Space

Chemical space is boundless, infinite. With this premise, it seems hopeless to search for a drug molecule unless a "drug" section of this space can be defined. However, it is difficult to set the boundaries of this drug region; and even if one could delimit and map it, it would be still too large to exhaustively screen it. The size of chemical space varies with the constraint set in the hypothesis used to calculate it [11], but what is clear is that it is immense: one of the most cited estimates is 10^{60} molecules with less than 30 atoms [12].

As already mentioned, the size of the chemical space increases exponentially with the number of atoms per molecule. As more and more atoms are added there are more and more ways of combining them. Therefore, focussing in low molecular weight molecules, the chemical space to be explored is significantly smaller and more manageable [13]. In addition, less complex molecules will have more chances to match a section of the active site as demonstrated by Hann *et al.* [4] (see above).

Efforts have been made towards the exhaustive enumeration of parts of the fragment space, for example the chemical space constituted by molecules up to 11 atoms [6-14]. More relevant to fragments, a complete enumeration of heteroaromatic ring systems has been recently published, named VEHICLe [11]. There are 23,895 unique heteroaromatic (5, 6, 5-5, 6-6 and 5-6) neutral ring systems formed by C, N, O, S and H. This comprehensive enumeration gives a robust and consistent reference system to map known regions of the chemical space and identify unexplored areas. For instance, from all these possible ring systems, only 1,701 have been reported in chemical databases and literature. However, only a few hundred of these are commonly used (*privileged structures*), but authors predict that around 3000 could be synthetically accessible.

Fragmentation Methods

Here the distinction is made between chemical fragments and computer generated fragments. A fragment in a chemical sense is a small molecule from which to build a bigger more potent molecule. A fragment in the computational sense is the result of an artificial deconstruction procedure, and does not have to correspond to an existing compound. This fragmentation procedure dissects the molecule into scaffolds according to predefined rules and identifies all fragments that comprise the chemical compound.

This scaffold identification or fragmentation can be done computationally using two different methods: substructure and building blocks [15]. Fragmentation using substructure methods are also known as subgraph mining techniques. These procedures break the molecules into all possible fragments of all possible sizes. This exhaustive enumeration leads to a huge number of fragments per molecule, but can reveal patterns otherwise difficult to perceive. For example, a common application of this method is to find the MCSS (maximum common substructure) between two or more molecules. The second fragmentation method breaks molecules into building blocks, i.e. into chemically meaningful fragments rather than abstract subgraphs. The rules that define chemical meaningful building blocks are used to break the molecules.

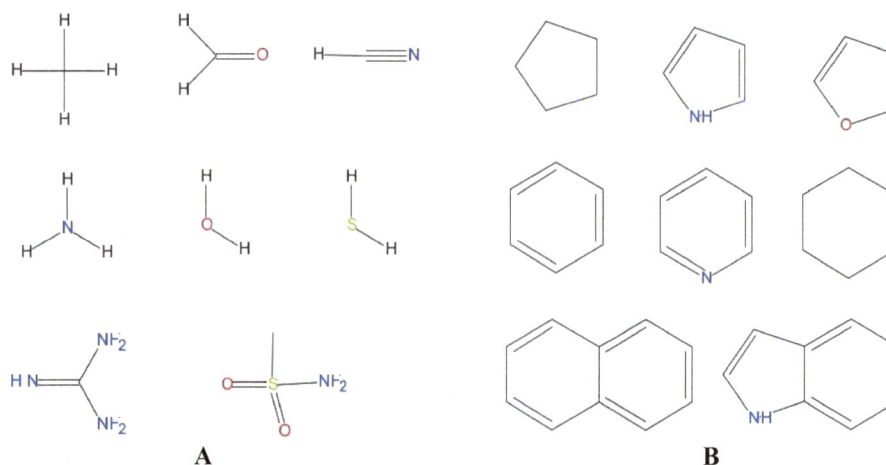

Figure 4: Fragments: A. acyclic; B. ring fragments.

Probably the most applied method in this second category is the procedure described by Bemis and Murcko in 1996 [16]. The authors broke molecules into unique non-overlapping fragments that describe the molecule in hierarchical fashion. In this way, compounds are broken down into frameworks and side chains. Frameworks are composed of rings (cycles and cycles sharing an edge) and linkers (atoms connecting rings). Side chains are composed of atoms not belonging to a ring or a linker. These frameworks can be abstract graphs representing the skeleton of the molecule, or they can incorporate atom types. The power of this fragmentation procedure is the uniqueness and non-overlapping character of the fragments derived. Analyses done with different molecular collections are easy to interpret and generate a useful reference set. They can be used for compound deck enrichment to either enrich towards or to fill the gaps in scaffold architectures. Bemis and Murcko Frameworks (or subtle modifications of their method) have been used as a tool to catalogue molecular composition and as a reference filter, for example in SCONP (Structural Classification of Natural Products) [17]. In addition, common frameworks found in drugs were the main components of the SHAPES library used by Vertex [18].

In these ways, fragment sets are derived by fragmentation analysis of molecules from any compound collection. These could be either a generally available database such as the NCI (National Cancer Institute) database, or commercial databases such as ACD (Available Chemical Directory, Symyx) or WDI (World Drug Index, Derwent), or the private collection of a pharmaceutical company. Fragments are sorted by frequency and the trivial ones removed. This is done both by hand and by computational techniques. Different databases have different emphases and contain different chemical classes. To some extent this will be reflected in the fragments that are derived, with a consequential effect on the ligands that are generated from those fragments. For example, target-based fragments can be derived from the fragmentation of a database of known ligands for the target of interest. Such a database will be much smaller than a general database, but the chemical classes of the structures it contains will be highly relevant. Experiments based on this idea have been performed for kinase and GPCR targets with encouraging results [19].

RECAP [20] is an interesting computational fragmentation methodology which splits the input molecules according to known synthetic transformations. This approach has the major advantage that fragments thus derived can be recombined in a synthetically feasible manner. This procedure has proven popular despite the limited types of molecules the method is able to suggest. It appears that this can be a quite powerful tool for tackling the synthetic feasibility problem, if the right selection of reaction transforms is used [21].

Privileged Structures

The term *privileged structures* refers to chemical substructures or scaffolds that are recurrently present in bioactive molecules for a particular target or a family target [22]. Optimal decoration of these scaffolds should yield specificity and required pharmacokinetic properties. In the context of fragment library design, this concept is remarkably useful. Indeed, many studies have analysed bioactive molecules (drug like but also natural products) in order to identify scaffolds readily recognised by proteins. These scaffolds are then used for general library design and compound collections analysis.

Privileged structures or masterkeys [23] have been used with good results in well known target families like kinases and GPCRs [24] where the wealth of ligand information allows to distinguish favoured substructures for specific target families.

Drug-likeness and Lead-likeness

If one wants to develop a drug to fight a particular disease, it would be best to search for starting points in the appropriate drug section of available ingredients. In addition, drugs that target brain molecules have a different profile to drugs targeting receptors outside the brain, for instance, and this difference should also be taken into account.

Several studies have tried to analyse, classify and navigate the chemical space [25]. In particular, attempts have been made to map the drug, lead and fragment-like areas. Probably the most referenced and the one that set up this "drug-like" concept was the analysis of Lipinski [26] of the World Drug Index. He found some simple rules (to calculate and remember) to flag compounds with a low chance of becoming an orally bioavailable drug. However, the well-known "Rule of five" (Ro5) may have limited in scientist minds the frontiers of the drug-like region of chemical space. However, drug-likeness is a multifaceted concept. Only half of the FDA approved drugs are for oral administration and comply with Lipinski's rules [27]. The importance of the Ro5, however, is that it created a picture of the existence of more interesting regions within the chemical space for drug discovery.

Historically, the process of developing hits into drug candidates has often lead to an increase in molecular weight [28]. Hits are usually weak binders; in order to improve their affinity medicinal chemists will add and optimise functional groups. The concept of "lead-likeness" is based in an original study at AstraZeneca [29] where the authors demonstrated the increase in molecular weight and lipophilicity of compounds optimized from leads to drug candidates. The odds of finding a fully developed drug in the early stages of discovery are very small. It seems more appropriate to focus in the lead-likeness region of the chemical space when one is searching for starting points, rather than the drug-like one. The qualities required for a molecule to be a drug are certainly not encapsulated by a few easy to calculate descriptors. However, they can be used in a quick way to lead the scientist towards a more favourable region of the chemical space. Fragment-based approaches take this concept one step further. They can be seen as the building blocks of leads. In the next section we summarise the current boundaries of what makes a molecule "fragment-like".

FRAGMENT LIBRARY DESIGN

What is a Fragment?

Here we summarise some published guidelines for fragment library design and selection. It is worth emphasising the word guideline. Compound selection and library design, both for fragments or bigger molecules, is a dynamic process. It receives input from the specifics of the project that the library is designed for, like target family, assays used, financial budget, and screening capacities as well as feedback input from the previous designs and screens [30].

Physico-chemical descriptors allow fast filtering of large numbers of compounds *in silico*; perhaps the most frequently used filters in practice are molecular weight and number of heavy atoms. The general philosophy favours the use of Lipinski's like properties [26], (MW, logP, number of hydrogen bond acceptors and donors, rotatable bonds and polar surface area) to rapidly filter down big catalogues (Table **1**). Further refined selection is usually done with more complicated calculations or costly measures, which include solubility estimations or 3D shape, for example.

Table 1: Desired molecular properties of fragments.

Property	Common Ranges
MW	150-300/350 [*]
# heavy atoms	≤ 22 10-20
logP	≤ 3 ≤ 2.2
HBD	≤ 3
HBA	≤ 4 ≤ 8
HBD + HBA + # of Rings	4-7
# of Rot Bonds	≤ 3 ≤ 5 ≤ 6
PSA	≤ 60 ≤ 70
Chemical Eye	Final selection by chemists. Medicinal chemist expertise

[*]Lower limit to minimise multiple binding modes.

Increase upper limit to allow certain chemotypes

Scientists at Astex [31] have proposed the "Rule of 3" to select fragments, an analogue of the Lipinski Rule of 5. The beauty of this rule is that it is easy to remember and calculate. As with Lipinski's rule, it leads the researches into a defined region of the chemical space. However, it is also worth mentioning the existence of successful fragments not compliant with the rule [32], and the open-mindedness that it is best to adopt when using these sorts of guidelines.

Chemical Filters

Besides the property profile of a fragment library, there are more considerations to include in its design. Here, we have collated the filters routinely used to make up fragment libraries. Filtering, clustering and synthetic analysis tools are used both in the initial selection of fragments and in the selection of designed ligands in order to meet the numerous requirements that would make a compound acceptable from a medicinal chemistry perspective.

Solubility in the assay media is probably the most important characteristic for a successful fragment to be soluble in the media of the assay. Fragments are weak binders and therefore have to be screened at high concentration (0.2-1.0mM) [33]. Still, accurately calculating solubility from the chemical structure alone is not a solved problem. Predictions are useful to cut down number of compounds, but the general practice is to experimentally measure water solubility of the final fragment library prior to screening. Normal values used to filter compounds, based on their calculated water solubility, are in the region of 1-2mM [30, 34].

Substructure filters distil much of what has been learnt from the HTS results and historical medicinal chemistry data. Some of this knowledge can be conveniently summarised in specific chemical moieties or substructures. These substructures can be used to not only filter out compounds with undesirable properties in an efficient way, but also to filter in molecules with a pattern that a target might recognise. These universal principles are also applied to fragment libraries, perhaps with the exception of more challenging targets. For new or difficult targets, sometimes researchers are looking for probe compounds, which would provide invaluable information, at the expense of lowered requirements. Usually, substructure searching and filtering is done with SMARTS, which allows powerful, fast and efficient queries.

It is now common practice to filter out molecules that contain known toxicophores or reactive groups that bind covalently with the protein or with other fragments in the same cocktail; only fragments containing H, C, N, O, F, Cl or S are kept [30]. Other substructure filters are the condition of the assay used [33] e.g. the fragments must contain SH for tethering [35], ^{19}F labelled molecules for NMR competition binding or enrich towards aromatic bromine for crystallography and follow up chemistry. It is to avoid having fluorophores if the assay is based in fluorescence.

A fragment hit will be further developed into a lead compound. The presence of synthetically amenable chemical handles, such as halogen substituted aromatic rings, is used regularly to select compounds for fragment libraries [30-36]. Alternatively, the existence of readily available analogues, is another criteria used to select compounds for a fragment library. In this way, the designed library will have close analogues ready to screen and to define valuable early Structure-Activity Relationships (SAR) from the initial hits.

Another consideration to include in fragment library design is the accessibility of the molecules [32]. Some vendors include in their catalogues "virtual" molecules, i.e. molecules that can be synthesised upon order but are not in stock. In addition, it is common that some compounds are out of stock. Brewer *et al.* [34] have quoted an order dropout rate of around 10%. It is good practice to increase the final number of molecules requested by about this amount. In addition, because of the high concentration needed for the assays, and the importance of having a stock of compound for further testing and modification, one can also filter out compounds that are available only in a low quantity (for example solid powder of less than 100mg) [32]. Furthermore, the purity of the physical sample should be always checked on arrival to avoid confusing assay results. All of these factors, as well as the cost of the molecules, past experience of the reliability and delivery time, lead researchers to prioritise vendors. It is more efficient to order compounds from a smaller number of suppliers.

In crystallography and NMR screening, fragments are tested in cocktails. In order to aid structural elucidation by X-ray crystallography, these cocktails of molecules are best when their 3D shape is as different as possible from each other [32]. In addition, variability in the shape of the compounds selected assures conformational diversity of the library.

The size of the final library is a very important factor in the design of the selection. This number will be mainly determined by the available budget and resource, in terms compound logistics, storage and assay capabilities. Published fragment (MW< 350) libraries vary from a few hundreds of molecules to 20,000 [33]. However, around a thousand of diverse small fragments should be enough to deliver hits against most targets [30].

Diversity and Clustering

Clustering and comparison with known ligands is used to help to evaluate how thorough the coverage of known chemical classes is, and to classify novel scaffolds. Scaffold Hunter [37] is an interactive computer-based tool for the navigation of complex structural relationships associated with bioactivity; the method correlates the structures in a hierarchical tree-like arrangement, and annotates them with bioactivity; branching from structurally complex to simple scaffolds allows identification of new ligand types [37-38].

Even using all the filters described so far, it is likely that the remaining list of molecules is still too large. For example, in a recent review Brewer *et al.* [34] found 65,000 rule-of-three [31] compliant commercially available compounds. Fragment screening technologies are not yet high throughput. If one has a well-behaved target along with a robust assay that can handle HCS (High Concentration Screening), screening of libraries up to 20,000 fragments is possible. However, a lower number of compounds must be selected if the capacity of the experimental screening or project budget is low throughput. Normally, this last refinement in the selection is diversity-based.

The choice of molecular descriptors will define the chemical space, and the algorithms to calculate distances or to classify compounds in that particular space will determine the output of the final "diverse" subset. A huge arsenal of molecular descriptors is available, from simple counts of molecular properties to complex 3D fingerprints [28]. The preference for one or another usually depends upon performance and access to the required software; Vernalis group reports good coverage of diverse scaffolds from 2D-3 point phamacophore fingerprints [30], for example. The same applies to algorithms used to select a diverse subset: clustering, machine learning algorithms, scaffold classification, cell-based diversity algorithms to cite but a few [39]. The designed structures could also be compared to binding modes of known inhibitors by means of similarity measures or based on the interaction pattern with key groups on the receptor i.e. interaction fingerprints.

FRAGMENT SCREENING

Experimental Methods

It is beyond the scope of this book to cover experimental techniques in detail. However, for a full appreciation of the computational techniques discussed in this chapter, some background knowledge on the experimental techniques currently being used to screen and characterise binding modes is essential. Here we provide selected references for this reason. The pioneers of fragment screening were the group at Abbott Laboratories who developed the SAR by NMR (Structure Activity Relationships by Nuclear Magnetic Resonance spectroscopy) approach [40]. Various NMR techniques are still employed for fragment screening, see for example [41] for a review. If the protein can be isotopically labelled then it is possible to obtain information about the binding site of a fragment in solution and not just its affinity can be obtained. Astex Therapeutics are strong advocates of the use of X-ray crystallography for fragment screening and enlargement design strategies, see for example Hartshorn *et al.* [42]. Soaking fragments into crystal structures of the *apo* form of a macromolecular target is a very commonly used approach, usually using fragment hits discovered by another technique. Screening of large numbers of fragments is carried out using a variety of techniques including NMR and optical biosenor techniques (e.g. see Pröll *et al.* [43]) such a Surface Plasmon Resonance (SPR) (e.g. see Perspicace *et al.* [44]). The increase in melting point of proteins on ligand binding is detected using thermal shift assays (see Lo *et al.* [45]) and this approach can be used for high throughput fragment screening. Biochemical assays have the advantage of providing a functional read out, for instance enzyme inhibition (see Barker *et al.* [46]). Isothermal Titration Calorimetry (ITC) is commonly used to confirm binding first measured by a more high throughput technique, and to obtain accurate binding constants. Tethering [47], a technique developed by James Wells and Sunesis Pharmaceuticals is unique in that fragments binders are detected when they form a covalent bond to the protein target. Details and pros and cons of each method can be found in Chung *et al.* [48].

Computational Fragment Binding Site Identification and Druggability Assessment

In days before "fragment" had a special meaning for chemists involved in drug discovery, Mattos and Ringe were experimenting with organic solvents as protein ligands. These molecules are too small to be fragments according to our guidelines (see above) but for the purposes of discussion we shall treat them as such. In their pioneering work [49] using porcine elastase, later extended by English *et al.* [50] using thermolysin, they soaked different organic solvents into protein crystals to probe for possible binding sites. They called this approach Multiple Solvent Crystal Structures (MSCS) (but this has also been done using NMR [51]). What they found was that solvents, such as isopropanol, clustered in sites where known inhibitors were known to bind and very few bound elsewhere on the protein. The technique was found to be useful for the identification of ligand hotspots. However, there are not many structures published using this approach, possibly because crystals may in some cases be disrupted by high concentrations of non-aqueous media or perhaps because computational techniques provide a quicker and cheaper alternative.

Partly due to the success of fragment screening, there has recently been renewed interest in computational approaches to organic solvent soaking, with applications for the prediction of hotspots and protein binding site druggability. Some of these approaches are reviewed below. One of the earliest, and perhaps most elegant, techniques that has been applied for this purpose is called Multiple Copy Simultaneous Search (MCSS) [52]. This technique involves the simultaneous minimisation *in vacuo* with the CHARMM forcefield [53] of 100's-1000's randomly positioned organic solvent fragments. These probes are subjected to the full protein forcefield but not to that of other probe molecules. Protein and probes can be rigid or treated flexibly. Numerous methods have been published for the evaluation of the resulting probe minima. These aspects and the use of the methodology to fragment-based drug design were reviewed recently by Schubert and Stultz [54]. Another well established program is GRID [55] (Molecular Discovery). This software calculates interaction energies for probes at grid points on the protein surface. An alternative approach is provided by SUPERSTAR [56]. This program calculates propensity maps for probe interactions on the surface of proteins derived from statistical analyses of the geometry of non-covalent interactions found in the CSD [57] and the PDB [58]. English *et al.* used MCSS and GRID to try and predict the experimentally determined organic solvent positions for thermolysin [59] but were disappointed by the results. The authors speculate that the electrostatics were over estimated due to the calculation being done *in vacuo* with no entropic and explicit solvent (water) considerations. Others have tried to do better. Locus pharmaceuticals Inc. [60] use a grand canonical Monte Carlo simulation in which rigid ligands interact with rigid protein structures. At first the protein is completely immersed in the solvent, which is then moved about and gradually removed, using a Monte Carlo selection process, until only the lowest energy examples remain. They use proprietary software and a large compute cluster for this purpose. Computational solvent mapping (CS-Map) e.g. Landon *et al.* [61], is similar to MCSS except that probes are allowed to interact. They use 14 probe fragments (all neutral presumably due to problems of self repulsion) scattered in and around cavities, which are then minimised in the presence of rigid protein and clustered. Sites are scored based on the number and diversity of probe clusters. Results show that experimentally derived ligand hotspots often co-inside with high ranking CS-Map sites. The CS-Map software is now owned by FORMA Therapeutics Inc. but a similar method could be scripted in a modelling package such as SYBYL (Tripos International), Maestro (Schrödinger Inc.) or Moe (Chemical Computing Group) but again a large compute cluster would be needed.

Some recent methods treat proteins as flexible entities or as an ensemble of conformers. For example, SILCS (Site Identification by Ligand Competitive Saturation) [62]. Multiple CHARMM Molecular Dynamics (MD) simulations (10 x 5 ns) of protein in high concentration (1M) solutions of propane and benzene (with a repulsive term added to forcefield to prevent aggregation) yield "fragmaps". The authors suggest these could be used as docking grids and to highlight ligand hotspots. In a another attempt to take protein flexibility into account, Ekonomiuk *et al.* [63] docked benezene and two other target specific fragments into 100 MD snapshots. The snapshot that could best accommodate all 3 fragments was chosen for docking and virtual screening using their fragment based docking software SEED/FFLD.

Binding site prediction is a related problem to protein druggability assessment. In an interesting paper Seco *et al.* [64] apply computational organic solvent soaking for this purpose. They carried out an MD simulation of five proteins of pharmacological interest in an isopropanol/water mixture. Solvent atomic interaction populations were used to colour code the protein surface and to calculate the Gibb's free energy of binding (ΔG) by assuming a Boltzmann distribution. The solvent population at grid points was used to estimate the maximum affinity of a drug-like molecule by judicious summing of ΔG contributions within volumes about the size of a drug-like molecule.

Hajduk *et al.* [65] found that fragments tend to bind in binding sites occupied by high affinity ligands and that NMR derived fragment hit rates correlate with the ability to identify them. They went further to characterise the properties of confirmed drug-like binding pockets compared to pockets where few or no fragments were known to bind. Apolar surface area was a key descriptor but shape and surface roughness were also found to be important.

Docking of Fragments

Since fragments are very simple in nature, there have been doubts raised about the ability of docking programs to accurately predict their binding modes. Firstly, fragments may be more likely have multiple binding modes, which would complicate the interpretation of the results. Secondly, because fragments form relatively few interactions with binding partners, the scoring functions may struggle to distinguish the correct binding mode from other similarly scoring wrong answers.

Taylor *et al.* [66] used docking with FlexX (BioSolveIT) combined with chemical descriptor similarity, followed by visual inspection, to select compounds for testing using NMR. The target was the v-Src SH2 domain. Of the 9 compounds tested, 2 bound with similar affinity to the phosphotyrosine fragment standard.

Teotico *et al.* [67] report very positive results for a fragment docking and binding mode prediction targeted at AmpC beta-lactamase using DOCK [68]. Of the ~140,000 fragments extracted from ZINC [69] that were docked and scored, 48 were tested and 23 were active. Not only was this hit rate incredibly high (48%) but when the structures of 8 of the hits bound in the active site where solved by X-ray crystallography only 2 were very different from the corresponding predicted binding modes. This is a very encouraging result for those thinking about carrying out a fragment-based virtual screen. However, this was a favorable case in some ways in that all hits were negatively charged and hand selected from the top 500 scoring fragments by scientists with previous experience of virtual screening against this target.

Astex carried out an evaluation of the utility of docking and virtual screening of fragments with GOLD [70-71]. They found that performance was target dependent but on the whole the results for fragments were on a par with those of drug-like molecules. Fragments have very few rotatable bonds so that conformational sampling is not the problem it is with drug-like molecules. This advantage may balance out the problems mentioned above.

We now discuss in more detail the selection, scoring and free energy prediction of fragment poses in more detail.

PRIORITIZATION OF FRAGMENT HITS

Promiscuous Fragments and Aggregation

One of the frequently asked question regarding fragment-based screening is why not use standard biochemical screening methods at higher concentrations in order to compensate low affinity of the fragments; in this manner expensive biophysical techniques are avoided. The main problem with this approach is the high rate of false-positive hits that have to be additionally validated. Many false positives in early drug discovery are due to non-specific inhibition by colloid-like aggregates of organic molecules. Effects on aggregation predominate, although antagonism is also observed [72]. Aggregate concentration, structure, stoichiometry and affinity for enzymes have been investigated using flow cytometry. Aggregates are formed at high concentration above a critical value (5 - 30 fM) where, aggregate count increases linearly with added material, while the particles disperse when diluted below the critical value. The stoichiometry of binding is about 10 000 enzyme molecules per particle; aggregates are densely packed particles and the enzyme can be comfortably accommodated on the surface of the aggregate [73]. Especially for screens of unbiased libraries, false positives can dominate hit lists. The results suggest that the physical behaviour of organic molecules, not their reactivity, accounts for most screening artefacts [74].

Feng and Shoichet describe simple experiments that can offer confirmation of aggregate-based inhibition [72]. The authors monitor inhibition of β-lactamase in the absence and presence of detergent; inhibition that is attenuated in the presence of detergent is characteristic of an aggregate-based mechanism. Based on these results computational predictive models could be developed and used as filters in order to reduce the rate of false positives while screening. Roche *et al.* [75] proposed one such method for identification of potential "frequent hitters". A scoring scheme was elaborated from substructure analysis, multivariate linear and nonlinear statistical methods applied to several sets of 1D and 2D molecular descriptors. The method was applied to database filtering, yielding between 8% and 35%

potential frequent hitters. Such a filter is a valuable tool for the prioritization of compounds from large databases, for compound purchase and biological testing, and for building new virtual libraries.

By utilizing X-ray structural information for promiscuous inhibitors, Aronov and Murcko proposed a five-point pharmacophore for kinase-specific frequent hitters, demonstrated its ability to discriminate between frequent hitters and selective ligands, and suggest a strategy for selective inhibitor design [76].

Ligand Efficiency

The concept of ligand efficiency (LE) [77] is useful for quantifying the potential of a fragment for lead development. Ligand efficiency is defined as the free energy if binding (in kcal/mol) divided by the number of non-hydrogen atoms. Several related concepts have been proposed, including the binding efficiency index (BEI) defined as pKi, pKd or pIC50 divided by the molecular weight and multiplied by 1000 [78] and lipophilic efficiency [27] defined as LipE = -log(IC50) –clogP. Bembenek *et al.* reviewed ligand efficiency measures and their use in fragment-based drug discovery [79].

The maximal affinity gain per atom has been observed to increase linearly and then flatten out above 20 heavy atoms [7]. The arguments put forward to explain the plateau are based on the observation that the interaction surface area per atom decreases as the molecules get larger and that larger ligands have to adapt to more binding-site constraints.

Hajduk [80] indicated that ligand efficiency remains constant as the fragment is developed into a higher molecular weight lead compounds. He observed that each atom adds approximately 0.3 kcal/mol and that each 64 Da of molecular mass increases potency 10 times. This observation has important implications for the design and development of fragments into leads. As the ligand is grown, it has to reach certain activity milestones at each stage in order to progress further. Such requirement essentially prunes the search over possible extensions and contains the combinatorial problem of searching over all possible molecules of given size.

Computational Scoring

In relation to FBDD, it has been demonstrated that fast scoring functions are capable of providing reasonable enrichment values not only in a set of lead compounds, but also they are equally successful at ranking smaller fragments [67]. Scoring functions need to be robust in that they rank solutions highly with favourable steric and electrostatic interactions and rank ligands lower that are not complementary to a particular target, or are in an unfavourable configuration. Commonly used empirical scoring functions include ChemScore [81], ScreenScore [82] and PMF [83]. Although not all popular scoring functions have yet been validated for use in computational fragment screening, initial results are promising [67].

A number of studies have also attempted to compare the predictive abilities of combinations of several different scoring functions. Charifson *et al.* showed that by combining scoring functions into a consensus score results in an enhancement in the ability to discriminate between active and inactive compounds [84] and improves hit-rates in virtual screening [85]. To our knowledge such investigation has not been performed yet in conjunction with fragment-based screening. Potentially, using this tool, discrimination and ranking of fragments could be improved even further.

Due to the reduced number of fragments that need to be tested, compared to leads, more rigorous scoring methods could also be applied. Methods such as MM-PBSA (Molecular Mechanics Poisson-Boltzmann Surface Area) [86] and Linear Interaction Energy method [87] incorporate a better description of the physics of ligand binding and could be beneficial. Variations allow the use of either explicit or implicit water simulations [88]. A solute entropy term is sometimes included to yield a total free energy; binding free energies could then be calculated from the difference between the free energy of the complex and that of the protein and ligand components. The solvated complex is minimized and the binding energy of a single structure is determined, or an ensemble of structures is acquired by conducting dynamics simulations and the binding energy is averaged over the ensemble. Kuhn *et al.* [89] find that MM-PBSA applied to a single minimized complex is as successful as the averaged free energy approach applied to an ensemble in terms of enrichments and ranking in virtual screening. It is also possible to use these methods on protein models when the experimental 3D structure of the target is not available [90].

BINDING MODES AND FREE ENERGY

The effect has been examined of considering multiple docking solutions on the success rate of obtaining the crystallographic binding mode. By selection of a small set of representative binding modes, the experimentally observed binding mode can be predicted with a higher probability after a ligand-protein docking simulation [91]. Ruvinsky extended further the concept and included in a rigorous manner the entropy of relative and overall molecular motions in the calculations of protein-ligand binding [92].

Clark *et al.* [93] report a fragment-based method for computing protein-ligand binding free energies by systematic sampling followed by *de novo* assembly of fragments into molecules and computation of binding free energies with statistical mechanics. The rigorous sampling provides independence from the choice of initial binding pose and assembling fragments enables evaluation of binding of a large number of molecule poses with relatively little computation. The method allows a full sampling of possible conformations and avoids the "conformational focusing" problem associated with free energy methods that sample only limited conformational and orientation changes from a starting pose. The direct computation of the entropy loss upon assembling fragments into molecules is an innovation for fragment-based methods.

ENERGY LANDSCAPES

It has been suggested that the energy landscape that is characteristic of ligand-protein binding is funnel-shaped and, in that respect, resembles the energy landscape of protein folding. Native ligands and certain so-called "anchor fragments" exhibit minimally frustrated pathways to the global minimum leading to a stable binding mode, whilst non-binding compounds will have a frustrated energy landscape leading to multiple modes [94-95]. There is a considerable energy gap between the global energy state of the bound ligand and the average protein-ligand configuration (Fig. **5A**). Native ligands and anchor fragments fulfil both thermodynamic stability and kinetic accessibility criteria due to the funnel-shaped binding energy landscape and fast convergence to the global energy minimum (Fig. **5B** and **C**).

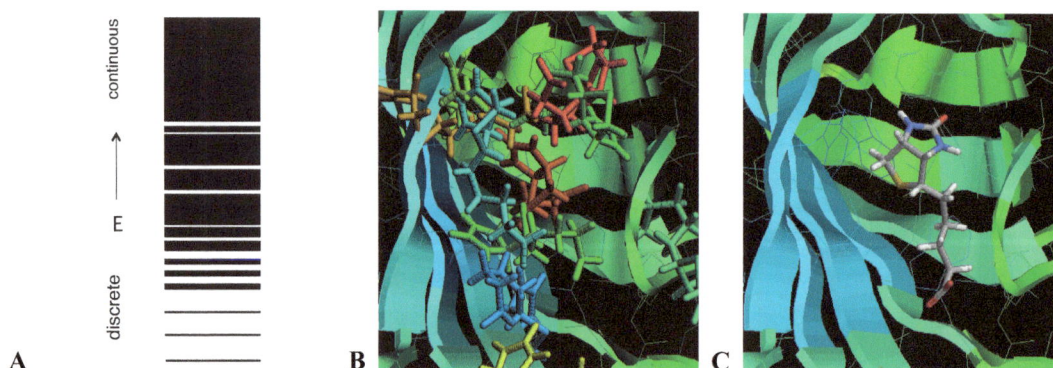

Figure 5: Protein-ligand energy landscape: A. energy levels; B. Random initial positions of ligand; C. Fast convergence of ligand to global energy minimum.

LIGAND-BASED SCORING

In the absence of 3D protein structure ligand-based design could be employed [96]. Computational FBDD offer much potential in this type of scenario. Estimation of binding free energy is less accurate and the methods rely on matching descriptors of a quantitative structure-activity relationship model, pharmacophores of superimposed ligands [97] or pseudoreceptor modelling [98]. Three-dimensional superpositions and pseudoreceptors are validated using an external set of test compounds; they should be able to semi-quantitatively predict the binding affinities of related molecules modelling [99]. Directional force field, ligand desolvation energy and the change of both ligand-internal energy and ligand entropy upon receptor binding could be used [100]. Ligand–based models could be used both for virtual screening and FBDD *de novo* design [101].

Additivity of Binding Energy

Dill [102] examined the connection between the binding energies of two isolated fragments in relation to the binding energy of a compound containing them and stressed the importance of entropic contributions. Even earlier Jencks [103] observed the usefulness of the concept of additivity. The Gibbs free energy changes for the binding of a ligand to a protein is a sum of the intrinsic binding energies of the component parts of the ligand and a connection Gibbs energy, due largely to changes in translational and rotational entropy. Later, Murray and Verdonk [104] succeeded in estimating the barrier to binding, due to the loss of rigid-body entropy, at 15-20 kJ/mol. This large barrier explains the rarity of the simultaneous binding of multiple fragments to non-overlapping adjacent sites in enzymes.

Andrews [105] provided additive functional group contributions to drug-receptor interactions. Later, Ciulli *et al.* [106] used a fragment-based approach to probe "hot spots" at the cofactor-binding site of a model dehydrogenase, *Escherichia coli* ketopantoate reductase. NADPH was broken down into smaller fragments and the biophysical characterization of their binding using NMR spectroscopy, ITC, and inhibition studies. A combination of ITC and site-directed mutagenesis locate the fragments at separate hot spots on opposite ends of the cofactor-binding site, and observe additive binding free energies of fragments.

FROM FRAGMENTS TO LEADS

Once active fragments have been found and validated, they could be extended further either with the help of *in silico* reagent screening, or by direct synthesis and testing. De Kloe *et al.*, 2009 review 23 FBDD examples of successful development of screened fragments into higher molecular weight lead compounds [3].

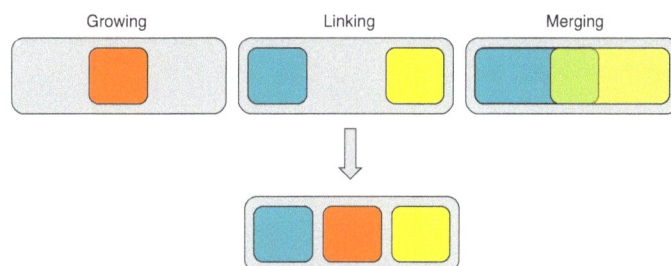

Figure 6: Lead generation strategies: growing, linking and merging.

Computational FBDD ligand design has matured in the last 15-20 years as part of *de novo* design methods and recent developments have been reviewed (see [107-112]).

Three main strategies have been developed for lead generation from fragments: fragment growing, linking and merging. All strategies should take into account the complexities of fragment interaction and possibilities for modular and emergent binding. FBDD provides several examples of fragments successfully grown into lead compounds, some of which are already in clinical trials [3].

Modular and Emergent Binding

Hajduk distinguished "emergent binding" as opposed to "modular binding" when the fragment does not move significantly when incorporated as part of a larger ligand and speculated that emergent binding could be quite challenging for drug design [113]. Babaoglu *et al.* illustrated both modular and emergent binding [114] in an elegant experiment where a beta-lactamase inhibitor was divided into four low-affinity fragments and crystal structures obtained with each fragment. Only one of the four fragments kept its binding mode as seen in the original compound. The other three fragments bound in different regions of the binding site, one region was even created by a conformational rearrangement of the protein. This is a clear example of the complexity and issues involved in moving from smaller, less active, to larger, more active molecules.

The above examples represent the spectrum of changes in fragment orientation upon optimization and illustrate the importance of having the correct pose throughout the fragment optimization process. Since it is generally not known *a priori* whether a fragment will stay in the same place as substituents are added or modifications are made, it is important to take care in determining the binding mode throughout the process, be it by experimental or computational approaches. Furthermore, predicting the single lowest-energy binding mode for a fragment may not be sufficient for many design problems. This highlights one of the key strengths of computational docking programs, which are capable of returning a list of viable poses. Analyzing multiple potential binding modes can provide insights into orientations of the fragment that may arise during the optimization process. While experimental techniques can sometimes observe multiple binding modes when the energies of the states are sufficiently close, computational docking techniques can usually return as many poses as the user requests.

Fragment Growing

Fragment-growing strategies have often been discussed in computational FBDD. In this strategy, fragments are added one-by-one while keeping the ligand fully connected. This strategy allows the application of empirical rules during the building process in order to ensure that the final structure is in a low energy state.

Besong *et al.* reported the FBDD *de novo* design of a novel inhibitor of D-ala-D-ala ligase from *E.coli*. The molecular design program SPROUT was used with the X-ray crystal structures of the bacterial enzymes DdlB and VanA to produce a novel enzyme-selective inhibitor template. Following short and efficient synthesis and in keeping with the design predictions, the resulting inhibitor showed useful levels of enzyme-selective inhibition [115].

The main challenge in the fragment-growing approach is to achieve a strong favourable interaction with the protein site. The choice of starting position and the choice of atoms for placement of successive fragments could be crucial. It is not necessary that at each step the new fragment be added at the best possible position and orientation, since that could prevent an overall stronger interaction after the addition of the next fragments. Schemes have been proposed in which a number of alternatives with scores higher than that of the best possible placement are retained at each step and one of the possibilities is then selected [116-117].

Evolutionary methods also offer a lot of flexibility in the implementation. Small changes either in the chemistry of the structure, or in its position are introduced in order to obtain a new structure. Repetition of these transitions results in a gradual change of the ligand structure. The suitability of this updated structure as a solution of the ligand design problem is evaluated and its score is compared to the previously examined structures; this makes it possible to use feedback to direct the transitions. Once different types of transition and a scoring function have been implemented, standard global optimization methods could be used. Each structure is viewed as a proposal for a complete solution of the design problem and its suitability is evaluated. Several implementations of this idea have been investigated [118]. These algorithms derive their efficiency from the ability to remove or replace fragments already included in the ligand. If a growing ligand comes to a dead end in the active site, certain fragments could be removed and replaced while keeping the best positioned ones still as part of the structure.

When the *de novo* design algorithm Skelgen [119-120] was applied [121] to several crystal structures of the estrogen receptor, scientists at De Novo Pharmaceuticals were able to rapidly develop active compounds. The top-scoring structures from each design strategy were clustered and a retrosynthetic analysis algorithm was used to rank them. 35 compounds were chosen for synthesis, 17 were synthesized and assayed and five were found to be actives. Four of the five actives were novel structures, the fifth was a known compound designed *de novo* giving a 30% hit rate. The total time for design, synthesis and biological testing was six months. Such studies provide encouragement for further application of computational fragment-based design methods to targets of therapeutic interest to provide novel, patentable compounds.

Some FBDD design methods employ established molecular docking techniques to position the structures in the site. These methods separate the exploration of the different chemical structures from the search for the binding mode of the particular compound under investigation. These methods step through an individual compound, perform conformational and docking analysis and binding energy evaluation before moving to another compound to investigate. The advantages of this approach are that it is built on top of other established technologies and it is thermodynamically correct since it examines and evaluates each receptor-ligand system individually, but could be computationally intensive since each compound is examined in a separate docking run. Several examples of implementations of this strategy have been reported [21, 122, 123].

CombiSMoG [124] uses a fast and accurate knowledge-based scoring function to predict binding affinities of protein−ligand complexes, and a Monte Carlo combinatorial growth algorithm that generates low-free-energy ligands in the binding site of a protein. Human carbonic anhydrase II was one of the first examples used to demonstrate the potential of the fragment growing methods. Starting from a sulphonamide moiety taken from a known ligand, new structures were synthesized based on an interative FBDD *de novo* design method [124]. One of the suggestions turned out to be the most potent ligand ever found for that enzyme.

SYNOPSIS is a relatively recent addition to the repertoire of methods for *de novo* design with special emphasis on synthetic feasibility issues [21]. The application of this method has yielded a number of novel HIV reverse transcriptase ligands with activity in the μM range. These methods could be computationally intensive since each compound is examined in a separate docking run. While this is recommended for compounds likely to be active, the compounds examined at the start of the procedure could be completely random and resources spent on them could be wasteful. One way to alleviate this problem is to reduce the sampling of the docking method and accuracy of the scoring in the beginning of the search and then gradually bring it back to desired accuracy. If the molecules in the sequence are similar and have a large common sub-structure, the binding mode of one of them could be used to guide the position of the other and thus reduce the cost of searching.

Fragment Linking

In this approach, fragments are connected into lead compounds either directly or via link fragments. Several early examples of this approach have been demonstrated in the context of computational de novo design and are applicable as an aid to more recent FBDD experimental effort [125]. The experimentally verified favourable interactions of the fragments increase the chances that once constructed the ligand will also maintain those interactions.

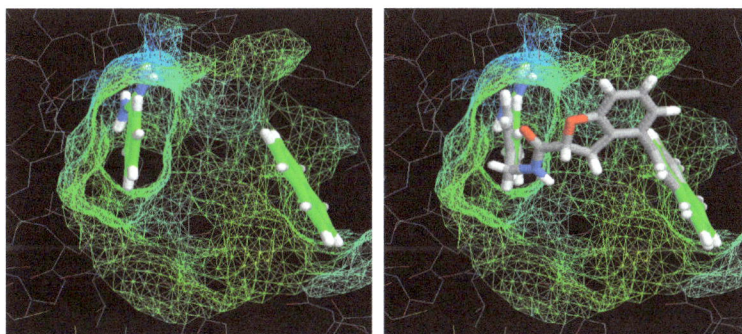

Figure 7: Fragment-linking example.

Several methods have been reported that store pre-computed conformations of the linkers. CAVEAT is a classic [126]; more recent examples include ReCore [127] and CONFIRM[128]. In FBDD, approaches using fragment linking could often lead to disturbing the positions of the individual fragments [129] or induced strain in the linker [130]. Displacements of the fragments from their ideal positions increase the number of binding modes and could allow alternative fragment arrangements where the interaction of the ligand is stronger. Larger fragments will have fewer binding modes, but at the cost of an increase in the number of fragments.

Since optimal ligand is unknown during the construction stage, it is unclear how close to its lowest energy conformation the proposed ligand would be. It may turn out, when considering ideal fragment placements, to be impossible to build a ligand in a low energy conformation. Rohrig *et al.* [131] used a set of linkers connect fragments known to bind to FK506 binding protein. Ester and thioester linkages resulted in high-affinity ligands, whereas an amide linkage decreased affinity remarkably and oxime and triazole were not tolerated. By investigating corresponding derivatized non-linked fragments and docking studies of linked fragments, they able evaluated the effect of the linker region on ligand binding affinity. If the fragments are far apart, long-chain linker fragments could be used to connect them [132-133] or alternatively the link could be grown starting from one of the fragments [134]. Sometimes the fragments could be too close to get connected within acceptably small geometrical deviations and alternative placements have to be considered. When more than two fragments have to be connected the simplest strategy is to work in steps [135].

Takano *et al.* illustrated the fragment-linking approach to efficiently find a novel inhibitor against activated coagulation Factor X by using FBDD design programs and the X-ray crystal structure of the enzyme. They used multiple copy simultaneous search (MCSS) and CAVEAT linker search techniques, which disclosed a novel Factor Xa inhibitor consisting of benzamidinyl and adamantyl groups and a carboxybenzylamine linker group. The inhibitory activity against Factor Xa was K_i of 48 nM. An X-ray crystal analysis of the complex with trypsin and docking studies of with trypsin and Factor Xa showed that the benzamidinyl group is a predominant binding moiety in the anionic pocket (S1 site) site, that the hydrophobic pocket (S4 site) is the binding site of the adamantyl group and that the carboxylate group of the linker contributes to the selectivity for Factor Xa against thrombin. Thus, the combination of the knowledge of the X-ray crystal structure of the target molecule with MCSS and CAVEAT linker search techniques proved to be an effective hit-finding method based on fragment linking that did not require the screening of huge compound libraries [136].

Fragment Merging

Several *in silico* fragment-merging methods have been reported. Fragment merging could be achieved by starting growth from different fragments that have been positioned in the site until the different components meet; then the components could be connected based on their overlapping fragments [137], the possibility of forming direct bonds or use linking fragments but with a reduced distance to span. Alternatively, acyclic chains of atoms of different lengths could be fitted between fragments and then the resulting ligand could be grown further by ring-bracing approach [138].

Another possibility is to dock multiple molecules in the site and recombine moieties from different ligands. An example of such method is BREED from Vertex [139]. It exchanges fragments from active compounds with complementary fragments from other active compounds. Because molecules used to derive the fragments have activity and fit in the binding site, it is likely that the fragments themselves are making reasonable interactions and could be used in different molecules with the same binding site. When the starting compounds are bioactive conformers from aligned crystal structures, the resulting shuffled compounds are already positioned in reasonable docked positions. When applied to HIV protease, the method generated 22 new molecules some of which with nM activity.

In a related class of methods fragments are allowed to move in the site and to connect in ways that favour good interactions with the receptor [140]. The connections between fragments are formed and broken dynamically to reflect the shape and the chemical nature of the receptor structure. These methods are attractive due to their simplicity, but the synthetic accessibility of the generated compounds tends to be more difficult to control.

Virtual fragment linking [141] uses fragment-based screening data to build statistical models based on connectivity fingerprints. A Bayesian model is derived from the screening data then used to rank compounds predicted to be active, as they contain desirable fragments or chemical features. The method is capable of identifying compounds of nM potency based on μM fragment binding; it succeeded in 57 out of 75 target classes. On seven target-screening programs virtual fragment linking captured between 28% and 67% of the hits in the top 5% of the ranked library for four of the targets. Proper coverage of chemical space by the fragment library appears crucial for the success of the methodology.

Douget *et al.* presented a program LEA3D developed to design organic molecules by combining 3D fragments [142] using genetic algorithms. Fragments are obtained from biological compounds and known drugs. The fitness function is build up by combining several independent property evaluations, including the score provided by a docking program. An example application uses of the structure of thymidine monophosphate kinase from *Mycobacterium tuberculosis* to generate analogues of one of its natural substrates. Among 22 tested compounds, 17 show inhibitory activity in the micromolar range.

Ligand-based Design

Development of fragments into leads in ligand-based design mode uses a pseudoreceptor or a superposition of known ligands and a defined pharmacophore. It could be performed using any of the three strategies, growing, linking or merging. Roche *et al.* [143] describe the design and synthesis of highly potent and selective non-imidazole inverse agonists for the histamine H3 receptor is described. The study validates a new pharmacophore model based on the merging of two previously described models and uses a fragment-growing approach [144].

GANDI (*G*enetic *A*lgorithm-based de *N*ovo *D*esign of *I*nhibitors) uses a genetic algorithm and a tabu search acting in concert to join pre-docked fragments with a user-supplied list of fragments [145]. GANDI performs simultaneous optimization of force field energy and 2D-similarity to known inhibitors or 3D-overlap to known binding modes. Top ranking molecules for cyclin-dependent kinase 2 exhibit a well-known interaction pattern observed in kinase inhibitors. A molecule with very favourable predicted binding affinity shares a 2-N-phenyl-1,3-thiazole-2,4-diamine moiety with a known nanomolar inhibitor. Eight of the 1809 designed molecules are found in the ZINC database of commercially available compounds; which also contains about 600 compounds have identical scaffolds as those in the top ranking molecules.

Several other reports are available on the use of FBDD *de novo* design methodology in ligand-based mode. Waszkowycz *et al.* describe PRO_LIGAND method for the design of novel molecules from molecular field analysis models and pharmacophores [146]. Fechner and Schneider introduce Flux as a virtual synthesis scheme for fragment-based *de novo* design [147]. The method fragments drug-like molecules by applying retrosynthesis rules and use a stochastic search algorithm in linked to a ligand-based similarity scoring. The algorithm succeeded in redesigning the templates of imantinib (Gleevec) and a Factor Xa inhibitor and suggested several alternative, chemically reasonable, molecular structures.

DISCUSSION

The application of computational fragment-based methods for the development of new lead compounds has become an accepted and useful technique in the pharmaceutical industry. *In silico* methods can be used both as an aid to experimental FBDD in the design, filtering and prioritization of fragment libraries as well as the computation equivalent of the FBDD process, developed as part of virtual screening and *de novo* design methods.

It has been shown that in certain cases, computational docking methods work for fragments as well as for lead-like compounds and could be successfully used for ranking before experimental screening with biophysical methods. This result should be validated in studies of additional targets in order to ascertain its validity. Such ranking will increase the number of available fragments that could be considered and potentially lower the cost of screening since more hits would be found for the same number of fragments screened.

Reliable affinity values from experimental FBDD puts lead generation on a firm basis. Although exploratory chemistry around the fragment hit is required, the effort is much smaller compared to a full-blown combinatorial exploration.

Emergent binding modes of fragments could seriously undermine rational efforts to design lead compounds by combining fragments with experimentally determined positions. Multiple binding modes could be generated and evaluated by computational methods and various combinations of fragment binding modes examined in an attempt to essentially reduce emergent binding to modular.

Another useful niche for computational methods is in the growth of from validated fragments and designing linkers. This aspect has been demonstrated to work successfully in a number of cases in pure *de novo* design but has not been considered extensively in conjunction with experimental FBDD.

Computational FBDD methods have to improve in several aspects. Scoring should incorporate multiple binding modes as demonstrated in recent developments based on rigorous statistical mechanics results. In addition, receptor flexibility and hydration should be considered both as part of the effort for improved scoring as well as taking into account the variability of the shape and chemical properties of the active site, now capable of accommodating a larger variety of ligands. Simultaneously, the possibility of additional false positive ligands should be considered.

The excitement of the drug discovery community regarding FBDD methods is contagious; these methods indeed seem to hold a lot of promise for solving some of the critical problems in the development of new drugs and making the process cost-effective and making a significant contribution towards improving healthcare.

ACKNOWLEDGEMENTS

We would like to thank David Bettinson for kindly reading the manuscript and providing comments and feedback. APH acknowledges generous funding from UCB and the BBSRC.

REFERENCES

[1]   Dimasi, J.; Hansen, R.; Grabowski, H. The price of innovation: new estimates of drug development costs. *Journal of Health Economics* **2003**, *22*(2), 151-185.

[2]   Warr, W. Fragment-based drug discovery. *Journal of Computer-Aided Molecular Design* **2009**, *23*(8), 453-458.

[3]   de Kloe, G.E.; Bailey, D.; Leurs, R.; de Esch, I.J.P. Transforming fragments into candidates: small becomes big in medicinal chemistry. *Drug Discovery Today* **2009**, *14*(13-14), 630-646.

[4]   Hann, M.M.; Leach, A.R.; Harper, G. Molecular Complexity and Its Impact on the Probability of Finding Leads for Drug Discovery. *Journal of chemical information and computer sciences* **2001**, *41*(3), 856-864.

[5]   Price Waterhouse, C. Pharma 2005 Silicon Rally: the Race to e-R&D. http://www.pwcglobal.com.

[6]   Fink, T.; Reymond, J.-L. Virtual Exploration of the Chemical Universe up to 11 Atoms of C, N, O, F: Assembly of 26.4 Million Structures (110.9 Million Stereoisomers) and Analysis for New Ring Systems, Stereochemistry, Physicochemical Properties, Compound Classes, and Drug Discovery. *Journal of chemical information and modeling* **2007**, *47*(2), 342-353.

[7]   Kuntz, I.D.; Chen, K.; Sharp, K.A.; Kollman, P.A. The maximal affinity of ligands. *Proceedings of the National Academy of Sciences of the United States of America* **1999**, *96*(18), 9997-10002.

[8]   Hajduk, P.J.; Huth, J.R.; Tse, C. Predicting protein druggability. *Drug Discov. Today* **2005**, *10*(23-24), 1675-1682.

[9]   Wyatt, P.G. Identification of N-(4-piperidinyl)-4-(2,6-dichlorobenzo- ylamino)-1H-pyrazole-3-carboxamide (AT7519), a novel cyclin dependent kinase in- hibitor using fragment-based X-ray crystallography and structure based drug design. *J. Med. Chem.* **2008**, *52*, 4986–4999.

[10] Hajduk, P.J.; Greer, J. A decade of fragment-based drug design: strategic advances and lessons learned. *Nat. Rev. Drug Discov.* **2007**, *6*(3), 211-219.

[11] Pitt, W.R.; Parry, D.M.; Perry, B.G.; Groom, C.R. Heteroaromatic Rings of the Future. *Journal of Medicinal Chemistry* **2009**, *52*(9), 2952-2963.

[12] Bohacek, R.; McMartin, C. Multiple Highly Diverse Structures Complementary to Enzyme Binding Sites: Results of Extensive Application of a de Novo Design Method Incorporating Combinatorial Growth. *Journal of the American Chemical Society* **1994**, *116*(13), 5560-5571.

[13] Hann, M.M.; Oprea, T.I. Pursuing the leadlikeness concept in pharmaceutical research. *Current Opinion in Chemical Biology* **2004**, *8*(3), 255-263.

[14] Fink, T.; Bruggesser, H.; Reymond, J.-L. Virtual Exploration of the Small-Molecule Chemical Universe below 160 Daltons13. *Angewandte Chemie International Edition* **2005**, *44*(10), 1504-1508.

[15] van der Horst, E.; Ijzerman, A.P. Computational Approaches to Fragment and Substructure Discovery and Evaluation. In *Fragment-Based Drug Discovery: A Practical Approach*, Zartler, E. R.; Shapiro, M. J., Eds. 2008; pp 199-222.

[16] Bemis, G.; Murcko, M. The Properties of Known Drugs. 1. Molecular Frameworks. *Journal of Medicinal Chemistry* **1996**, *39*(15), 2887-2893.

[17] Koch, M.; Schuffenhauer, A.; Scheck, M.; Wetzel, S.; Casaulta, M.; Odermatt, A.; Ertl, P.; Waldmann, H. Charting biologically relevant chemical space: a structural classification of natural products (SCONP). *Proceedings of the National Academy of Sciences of the United States of America* **2005**, *102*(48), 17272-17277.

[18] Fejzo, J.; Lepre, C.A.; Peng, J.W.; Bemis, G.W.; Ajay; Murcko, M.A.; Moore, J.M. The SHAPES strategy: an NMR-based approach for lead generation in drug discovery. *Chemistry & Biology* **1999**, *6*(10), 755-769.

[19] Aronov, A.; McClain, B.; Moody, C.; Murcko, M. Kinase-likeness and Kinase-Privileged Fragments: Toward Virtual Polypharmacology. *Journal of Medicinal Chemistry* **2008**, *51*(5), 1214-1222.

[20] Lewell, X.Q.; Judd, D.B.; Watson, S. P.; Hann, M.M. RECAPRetrosynthetic Combinatorial Analysis Procedure: A Powerful New Technique for Identifying Privileged Molecular Fragments with Useful Applications in Combinatorial Chemistry. *Journal of chemical information and computer sciences* **1998**, *38*(3), 511-522.

[21] Vinkers, H.M.; de Jonge, M.R.; Daeyaert, F.F. D.; Heeres, J.; Koymans, L.M. H.; van Lenthe, J.H.; Lewi, P.J.; Timmerman, H.; Van Aken, K.; Janssen, P.A.J. SYNOPSIS: SYNthesize and OPtimize System in Silico. *Journal of Medicinal Chemistry* **2003**, *46*(13), 2765-2773.

[22]   Kubinyi, H. Chemogenomics in Drug Discovery. In *Chemical Genomics*, 2006; pp 1-19.

[23]   Muller, G. Medicinal chemistry of target family-directed masterkeys. *Drug Discov. Today* **2003**, *8*(15), 681-91.

[24]   DeSimone, R.W.; Currie, K.S.; Mitchell, S.A.; Darrow, J.W.; Pippin, D.A. Privileged structures: applications in drug discovery. *Comb Chem High Throughput Screen* **2004**, *7*(5), 473-94.

[25]   Lipinski, C.; Hopkins, A. Navigating chemical space for biology and medicine. *Nature* **2004**, *432*(7019), 855-861.

[26]   Lipinski, C.; Lombardo, F.; Dominy, B.; Feeney, P. Experimental and computational approaches to estimate solubility and permeability in drug discovery and development settings. *In Vitro Models for Selection of Development Candidates* **1997**, *23*(1-3), 3-25.

[27]   Leeson, P.; Springthorpe, B. The influence of drug-like concepts on decision-making in medicinal chemistry. *Nature Reviews Drug Discovery* **2007**, *6*(11), 881-890.

[28]   Leach, A.R.; Hann, M.M.; Burrows, J.N.; Griffen, E.J. Fragment screening: an introduction. *Molecular bioSystems* **2006**, *2*(9), 429-446.

[29]   Teague, S.J.; Davis, A.M.; Leeson, P.D.; Oprea, T. The Design of Leadlike Combinatorial Libraries. *Angewandte Chemie International Edition* **1999**, *38*(24), 3743-3748.

[30]   Chen, I.; Hubbard, R. Lessons for fragment library design: analysis of output from multiple screening campaigns. *Journal of Computer-Aided Molecular Design* **2009**, *23*(8), 603-620.

[31]   Congreve, M.; Carr, R.; Murray, C.; Jhoti, H. A Rule of Three for fragment-based lead discovery? *Drug Discovery Today* **2003**, *8*(19), 876-877.

[32]   Blomberg, N.; Cosgrove, D.; Kenny, P.; Kolmodin, K. Design of compound libraries for fragment screening. *Journal of Computer-Aided Molecular Design* **2009**, *23*(8), 513-525.

[33]   Siegal, G.; Ab, E.; Schultz, J. Integration of fragment screening and library design. *Drug Discovery Today* **2007**, *12*(23-24), 1032-1039.

[34]   Brewer, M.; Ichihara, O.; Kirchhoff, C.; Schade, M.; Whittaker, M. Chapter 3: Assembling a Fragment Library. In *Fragment-BasedDrugDiscovery: APracticalApproach*, Zartler, E.R.; Shapiro, M.J., Eds. 2008JohnWiley&Sons,Ltd: 2008; pp 39-62.

[35]   Erlanson, D.; Lam, J.; Wiesmann, C.; Luong, T.; Simmons, R.; Delano, W.; Choong, I.; Burdett, M.; Flanagan, W.; Lee, D.; Gordon, E.; O'Brien, T. In situ assembly of enzyme inhibitors using extended tethering. *Nature Biotechnology* **2003**, *21*(3), 308-314.

[36]   Schuffenhauer, A.; Ruedisser, S.; Marzinzik, A.; Jahnke, W.; Blommers, M.; Selzer, P.; Jacoby, E. Library design for fragment based screening. *Current topics in medicinal chemistry* **2005**, *5*(8), 751-762.

[37]   Wetzel, S.; Klein, K.; Renner, S.; Rauh, D.; Oprea, T.I.; Mutzel, P.; Waldmann, H. Interactive exploration of chemical space with Scaffold Hunter. *Nat. Chem. Biol.* **2009**, *5*(8), 581-583.

[38]   Oprea, T. Chemical space navigation in lead discovery. *Current Opinion in Chemical Biology* **2002**, *6*(3), 384-389.

[39]   Schuffenhauer, A.; Brown, N.; Selzer, P.; Ertl, P.; Jacoby, E. Relationships between Molecular Complexity, Biological Activity, and Structural Diversity. *Journal of chemical information and modeling* **2005**, *46*(2), 525-535.

[40]   Shuker, S.B.; Hajduk, P.J.; Meadows, R.P.; Fesik, S.W. Discovering High-Affinity Ligands for Proteins: SAR by NMR. *Science* **1996**, *274*(5292), 1531-1534.

[41]   Pellecchia, M.; Bertini, I.; Cowburn, D.; Dalvit, C.; Giralt, E.; Jahnke, W.; James, T.; Homans, S.; Kessler, H.; Luchinat, C.; Meyer, B.; Oschkinat, H.; Peng, J.; Schwalbe, H.; Siegal, G. Perspectives on NMR in drug discovery: a technique comes of age. *Nature Reviews Drug Discovery* **2008**, *7*(9), 738-745.

[42]   Hartshorn, M.; Murray, C.; Cleasby, A.; Frederickson, M.; Tickle, I.; Jhoti, H. Fragment-Based Lead Discovery Using X-ray Crystallography. *Journal of Medicinal Chemistry* **2005**, *48*(2), 403-413.

[43]   Pröll, F.; Fechner, P.; Proll, G. Direct optical detection in fragment-based screening. *Analytical and Bioanalytical Chemistry* **2009**, *393*(6), 1557-1562.

[44]   Perspicace, S.; Banner, D.; Benz, J.; Muller, F.; Schlatter, D.; Huber, W. Fragment-Based Screening Using Surface Plasmon Resonance Technology. *J. Biomol. Screen.* **2009**, *14*(4), 337-349.

[45]   Lo, M.-C.; Aulabaugh, A.; Jin, G.; Cowling, R.; Bard, J.; Malamas, M.; Ellestad, G. Evaluation of fluorescence-based thermal shift assays for hit identification in drug discovery. *Analytical biochemistry* **2004**, *332*(1), 153-159.

[46]   Barker, J.; Courtney, S.; Hesterkamp, T.; Ullmann, D.; Whittaker, M. Fragment screening by biochemical assay. *Expert Opinion on Drug Discovery* **2006**, 225-236.

[47]   Erlanson, D.; Wells, J.; Braisted, A. TETHERING: Fragment-Based Drug Discovery. *Annual review of biophysics and biomolecular structure* **2004**, *33*(1), 199-223.

[48]   Chung, C.W.; Lowe, P.N.; Jhoti, H.; Leach, A.R. Biophysical Methods, Mechanism of Action Studies. In *Structure-Based Drug Discovery*, Springer: 2007; pp 155-199.

[49] Mattos, C.; Ringe, D. Locating and characterizing binding sites on proteins. *Nat Biotech* **1996,** *14*(5), 595-599.

[50] English, A.C.; Done, S.H.; Caves, L.S.; Groom, C.R.; Hubbard, R.E. Locating interaction sites on proteins: the crystal structure of thermolysin soaked in 2% to 100% isopropanol. *Proteins* **1999,** *37*(4), 628-640.

[51] Liepinsh, E.; Otting, G. Organic solvents identify specific ligand binding sites on protein surfaces. *Nat. Biotech.* **1997,** *15*(3), 264-268.

[52] Miranker, A.; Karplus, M. Functionality maps of binding sites: a multiple copy simultaneous search method. *Proteins: Structure, Function, and Genetics* **1991,** *11*, 29-34.

[53] Brooks, B.; Bruccoleri, R.; Olafson, B.; States, D.; Swaminathan, S.; Karplus, M. CHARMM: A program for macromolecular energy, minimization, and dynamics calculations. *Journal of Computational Chemistry* **1983,** *4*(2), 187-217.

[54] Schubert, C.R.; Stultz, C.M. The multi-copy simultaneous search methodology: a fundamental tool for structure-based drug design. *Journal of Computer-Aided Molecular Design* **2009,** 475–489.

[55] Goodford, P.J. A computational procedure for determining energetically favorable binding sites on biologically important macromolecules. *Journal of Medicinal Chemistry* **1985,** *28*(7), 849-857.

[56] Verdonk, M.; Cole, J.; Taylor, R. SuperStar: A Knowledge-based Approach for Identifying Interaction Sites in Proteins. *Journal of Molecular Biology* **1999,** *289*(4), 1093-1108.

[57] Allen, F.; Davies, J.; Galloy, J.; Johnson, O.; Kennard, O.; Macrae, C.; Mitchell, E.; Mitchell, G.; Smith, M.; Watson, D. The development of versions 3 and 4 of the Cambridge Structural Database System. *Journal of chemical information and computer sciences* **1991,** *31*(2), 187-204.

[58] Bernstein, F.C.; Koetzle, T.F.; Williams, G.J.; Meyer, E.F.; Brice, M.D.; Rodgers, J.R.; Kennard, O.; Shimanouchi, T.; Tasumi, M. The Protein Data Bank: a computer-based archival file for macromolecular structures. *Journal of Molecular Biology* **1977,** *112*(3), 535-542.

[59] English, A.C.; Groom, C.R.; Hubbard, R.E. Experimental and computational mapping of the binding surface of a crystalline protein. *Protein engineering* **2001,** *14*(1), 47-59.

[60] Clark, M.; Guarnieri, F.; Shkurko, I.; Wiseman, J. Grand canonical Monte Carlo simulation of ligand-protein binding. *J. Chem. Inf. Model.* **2006,** *46*(1), 231-42.

[61] Landon, M.; Lancia, D.; Yu, J.; Thiel, S.; Vajda, S. Identification of hot spots within druggable binding regions by computational solvent mapping of proteins. *Journal of Medicinal Chemistry* **2007,** *50*(6), 1231-1240.

[62] Guvench, O.; MacKerell, A. Computational fragment-based binding site identification by ligand competitive saturation. *PLoS Computational Biology* **2009,** *5*(7), e1000435.

[63] Ekonomiuk, D.; Su, X.-C.; Ozawa, K.; Bodenreider, C.; Lim, S.; Otting, G.; Huang, D.; Caflisch, A. Flaviviral Protease Inhibitors Identified by Fragment-Based Library Docking into a Structure Generated by Molecular Dynamics. *Journal of Medicinal Chemistry* **2009,** *52*(15), 4860-4868.

[64] Seco, J.; Luque, J.; Barril, X. Binding site detection and druggability index from first principles. *Journal of Medicinal Chemistry* **2009,** *52*(8), 2363-2371.

[65] Hajduk, P.J.; Huth, J.R.; Fesik, S.W. Druggability indices for protein targets derived from NMR-based screening data. *J. Med. Chem.* **2005,** *48*(7), 2518-2525.

[66] Taylor, J.; Gilbert, P.; Williams, M.; Pitt, W.; Ladbury, J. Identification of novel fragment compounds targeted against the pY pocket of v-Src SH2 by computational and NMR screening and thermodynamic evaluation. *Proteins: Structure, Function, and Bioinformatics* **2007,** *67*(4), 981-990.

[67] Teotico, D.G.; Babaoglu, K.; Rocklin, G.J.; Ferreira, R.S.; Giannetti, A.M.; Shoichet, B.K. Docking for fragment inhibitors of AmpC β-lactamase. *Proceedings of the National Academy of Sciences* **2009,** *106*(18), 7455-7460.

[68] Meng, E.; Shoichet, B.; Kuntz, I. Automated docking with grid-based energy evaluation. *Journal of Computational Chemistry* **1992,** *13*(4), 505-524.

[69] Irwin, J.; Shoichet, B. ZINC--a free database of commercially available compounds for virtual screening. *Journal of chemical information and modeling* **2005,** *45*(1), 177-182.

[70] Jones, G. Development and validation of a genetic algorithm for flexible docking. *Journal of Molecular Biology* **1997,** *267*(3), 727-748.

[71] Hall, R. Target Specific Binding Motifs for Improvement of Docking Performance. http://www.ccdc.cam.ac.uk/events/usergroupmeeting/d/presentations/RHall_approved.pdf (accessed sep 2009).

[72] Feng, B.; Shoichet, B. Synergy and Antagonism of Promiscuous Inhibition in Multiple-Compound Mixtures. *Journal of Medicinal Chemistry* **2006,** *49*(7), 2151-2154.

[73] Feng, B.Y.; Simeonov, A.; Jadhav, A.; Babaoglu, K.; Inglese, J.; Shoichet, B. K.; Austin, C.P. A High-Throughput Screen for Aggregation-Based Inhibition in a Large Compound Library. *Journal of Medicinal Chemistry* **2007,** *50*(10), 2385-2390.

[74]   Babaoglu, K.; Simeonov, A.; Irwin, J.J.; Nelson, M.E.; Feng, B.; Thomas, C. J.; Cancian, L.; Costi, M.P.; Maltby, D.A.; Jadhav, A.; Inglese, J.; Austin, C.P.; Shoichet, B.K. Comprehensive Mechanistic Analysis of Hits from High-Throughput and Docking Screens against β-Lactamase. *Journal of Medicinal Chemistry* **2008**, *51*(8), 2502-2511.

[75]   Roche, O.; Schneider, P.; Zuegge, J.; Guba, W.; Kansy, M.; Alanine, A.; Bleicher, K.; Danel, F.; Gutknecht, E.-M.; Rogers-Evans, M.; Neidhart, W.; Stalder, H.; Dillon, M.; Sjogren, E.; Fotouhi, N.; Gillespie, P.; Goodnow, R.; Harris, W.; Jones, P.; Taniguchi, M.; Tsujii, S.; von der Saal, W.; Zimmermann, G.; Schneider, G. Development of a Virtual Screening Method for Identification of "Frequent Hitters" in Compound Libraries. *Journal of Medicinal Chemistry* **2001**, *45*(1), 137-142.

[76]   Aronov, A.; Murcko, M. Toward a Pharmacophore for Kinase Frequent Hitters. *Journal of Medicinal Chemistry* **2004**, *47*(23), 5616-5619.

[77]   Hopkins, A.; Groom, C.; Alex, A. Ligand efficiency: a useful metric for lead selection. *Drug Discovery Today* **2004**, *9*(10), 430-431.

[78]   Abad-Zapatero, C.; Met, J.T. Ligand efficiency indices as guideposts for drug discovery. *Drug Discovery Today* **2005**, *10*, 464–469.

[79]   Bembenek, S.D.; Tounge, B.A.; Reynolds, C.H. Ligand efficiency and fragment-based drug discovery. *Drug Discovery Today* **2009**, *14*(5-6), 278-283.

[80]   Hajduk, P. Fragment-based drug design: how big is too big? *J. Med. Chem.* **2006**, *49*(24), 6972-6.

[81]   Eldridge, M.D.; Murray, C.W.; Auton, T.R.; Paolini, G.V.; Mee, R.P. Empirical scoring functions: I. The development of a fast empirical scoring function to estimate the binding affinity of ligands in receptor complexes. *Journal of Computer-Aided Molecular Design* **1997**, *11*, 425–45.

[82]   Stahl, M.; Rarey, M. Detailed Analysis of Scoring Functions for Virtual Screening. *Journal of Medicinal Chemistry* **2001**, *44*(7), 1035-1042.

[83]   Muegge, I.; Martin, Y. A General and Fast Scoring Function for Protein--Ligand Interactions: A Simplified Potential Approach. *Journal of Medicinal Chemistry* **1999**, *42*(5), 791-804.

[84]   Charifson, P.; Corkery, J.; Murcko, M.; Walters, P. Consensus Scoring: A Method for Obtaining Improved Hit Rates from Docking Databases of Three-Dimensional Structures into Proteins. *Journal of Medicinal Chemistry* **1999**, *42*(25), 5100-5109.

[85]   Wang, R.; Lu, Y.; Wang, S. Comparative evaluation of 11 scoring functions for molecular docking. *J. Med. Chem.* **2003**, *46*(12), 2287-303.

[86]   Srinivasan, J.; Cheatham, T.E.; Cieplak, P.; Kollman, P.A.; Case, D.A. Continuum Solvent Studies of the Stability of DNA, RNA, and Phosphoramidate−DNA Helices. *Journal of the American Chemical Society* **1998**, *120*(37), 9401-9409.

[87]   Aqvist, J.; Medina, C.; Samuelsson, J.-E. A new method for predicting binding affinity in computer-aided drug design. *Protein Eng.* **1994**, *7*(3), 385-391.

[88]   Rizzo, R.; Toba, S.; Kuntz, I. A molecular basis for the selectivity of thiadiazole urea inhibitors with stromelysin-1 and gelatinase-A from generalized born molecular dynamics simulations. *Journal of Medicinal Chemistry* **2004**, *47*(12), 3065-74.

[89]   Kuhn, B.; Gerber, P.; Schulz-Gasch, T.; Stahl, M. Validation and Use of the MM-PBSA Approach for Drug Discovery. *Journal of Medicinal Chemistry* **2005**, *48*(12), 4040-4048.

[90]   Ali, M.; Bhogal, N.; Findlay, J.; Fishwick, C. The first de novo-designed antagonists of the human NK(2) receptor. *Journal of Medicinal Chemistry* **2005**, *48*(18), 5655-8.

[91]   Kallblad, P.; Mancera, R.L.; Todorov, N.P. Assessment of multiple binding modes in ligand-protein docking. *Journal of Medicinal Chemistry* **2004**, *47*(13), 3334-7.

[92]   Ruvinsky, A. Calculations of protein-ligand binding entropy of relative and overall molecular motions. *Journal of Computer-Aided Molecular Design* **2007**, *21*(7), 361-370.

[93]   Clark, M.; Meshkat, S.; Talbot, G.; Carnevali, P.; Wiseman, J. Fragment-Based Computation of Binding Free Energies by Systematic Sampling. *J. Chem. Inf. Model.* **2009**, *49*(8), 1901-13.

[94]   Verkhivker, G.M.; Bouzida, D.; Gehlhaar, D.K.; Rejto, P.A.; Freer, S.T.; Rose, P.W. Complexity and simplicity of ligand-macromolecule interactions: the energy landscape perspective. *Current Opinion in Structural Biology* **2002**, *12*, 197–203.

[95]   Wang, J.; Verkhivker, G.M. Energy Landscape Theory, Funnels, Specificity, and Optimal Criterion of Biomolecular Binding. *Physical Review Letters* **2003**, *90*, 1–4.

[96]   Lloyd, D.G.; Buenemann, C.L.; Todorov, N.P.; Manallack, D.T.; Dean, P.M. Scaffold hopping in de novo design. Ligand generation in the absence of receptor information. *Journal of Medicinal Chemistry* **2004**, *47*(3), 493-6.

[97]   Todorov, N.P.; Alberts, I.L.; Esch, I.J.P.D.; Dean, P.M. QUASI: a novel method for simultaneous superposition of multiple flexible ligands and virtual screening using partial similarity. *Journal of chemical information and modeling* **2007**, *47*,

1007–20.

[98] Hahn, M. Receptor Surface Models. 1. Definition and Construction. *Journal of Medicinal Chemistry* **1995,** *38*(12), 2080-2090.

[99] Snyder, J.P.; Rao, S.N.; Koehler, K.F.; Vedani, A. Minireceptors and Pseudoreceptors. In *3D QSAR in Drug Design: Theory, Methods and Applications*, Escom: Leiden, 1993; pp 336-354.

[100] Zbinden, P.; Dobler, M.; Folkers, G.; Vedani, A. PrGen: Pseudoreceptor Modeling Using Receptor-mediated Ligand Alignment and Pharmacophore Equilibration. *Quantitative Structure-Activity Relationships* **1998,** *17*(02), 122-130.

[101] Tanrikulu, Y.; Schneider, G. Pseudoreceptor models in drug design: bridging ligand- and receptor-based virtual screening. *Nature Reviews Drug Discovery* **2008,** *7*(8), 667-677.

[102] Dill, K.A. Additivity principles in biochemistry. *Journal of Biological Chemistry* **1997,** *272*, 701–704.

[103] Jencks, W. On the attribution and additivity of binding energies. *Proceedings of the National Academy of Sciences of the United States of America* **1981,** *78*(7), 4046-4050.

[104] Murray, C.W.; Verdonk, M.L. The consequences of translational and rotational entropy lost by small molecules on binding to proteins. *Journal of Computer-Aided Molecular Design* **2002,** *16*, 741–53.

[105] Andrews, P.R.; Craik, D.J.; Martin, J.L. Functional group contributions to drug-receptor interactions. *J. Med. Chem.* **1984,** *27*(12), 1648-1657.

[106] Ciulli, A.; Williams, G.; Smith, A.G.; Blundell, T.L.; Abell, C. Probing hot spots at protein-ligand binding sites: a fragment-based approach using biophysical methods. *Journal of Medicinal Chemistry* **2006,** *49*, 4992–5000.

[107] Lewis, R.; Leach, A. Current methods for site-directed structure generation. *Journal of Computer-Aided Molecular Design* **1994,** *8*(4), 467-475.

[108] Schneider, G.; Clément-Chomienne, O.; Hilfiger, L.; Schneider, P.; Kirsch, S.; Böhm, H.-J.; Neidhart, W. Virtual Screening for Bioactive Molecules by Evolutionary De Novo Design. *Angewandte Chemie* **2000,** *39*(22), 4130-4133.

[109] Schneider, G.; Fechner, U. Computer-based de novo design of drug-like molecules. *Nature Reviews Drug Discovery* **2005,** *4*(8), 649-663.

[110] Dean, P.M.; Lloyd, D.G.; Todorov, N.P. De novo drug design: integration of structure-based and ligand-based methods. *Current opinion in drug discovery & development* **2004,** *7*(3), 347-53.

[111] Congreve, M.; Murray, C.W.; Blundell, T.L. Structural biology and drug discovery. *Drug Discovery Today* **2005,** *10*(13), 895-907.

[112] Mauser, H.; Guba, W. Recent developments in de novo design and scaffold hopping. *Current opinion in drug discovery & development* **2008,** *11*(3), 365-374.

[113] Hajduk, P.J. Puzzling through fragment-based drug design. *Nature chemical biology* **2006,** *2*, 658–659.

[114] Babaoglu, K.; Shoichet, B. Deconstructing fragment-based inhibitor discovery. *Nature chemical biology* **2006,** *2*(12), 720-723.

[115] Besong, G.E.; Bostock, J.M.; Stubbings, W.; Chopra, I.; Roper, D.I.; Lloyd, A.J.; Fishwick, C.W.G.; Johnson, A.P. A De Novo Designed Inhibitor of D-Ala-D-Ala Ligase from E. coli. *Angewandte Chemie International Edition* **2005,** *44*(39), 6403-6406.

[116] Nishibata, Y.; Itai, A. Confirmation of usefulness of a structure construction program based on three-dimensional receptor structure for rational lead generation. *Journal of Medicinal Chemistry* **1993,** *36*(20), 2921-2928.

[117] Honma, T.; Hayashi, K.; Aoyama, T.; Hashimoto, N.; Machida, T.; Fukasawa, K.; Iwama, T.; Ikeura, C.; Ikuta, M.; Suzuki-Takahashi, I.; Iwasawa, Y.; Hayama, T.; Nishimura, S.; Morishima, H. Structure-based generation of a new class of potent Cdk4 inhibitors: new de novo design strategy and library design. *J. Med. Chem.* **2001,** *44*(26), 4615-27.

[118] Glen, R.C.; Payne, A.W.R. A genetic algorithm for the automated generation of molecules within constraints. *Journal of Computer-Aided Molecular Design* **1995,** *9*(2), 181-202.

[119] Dean, P.M.; Firth-Clark, S.; Harris, W.; Kirton, S.B.; Todorov, N.P. SkelGen: a general tool for structure-based de novo ligand design. *Expert Opin. Drug Discov.* **2006,** *1*, 179–189.

[120] Stahl, M.; Todorov, N.P.; James, T.; Mauser, H.; Boehm, H.J.; Dean, P.M. A validation study on the practical use of automated de novo design. *Journal of Computer-Aided Molecular Design* **2002,** *16*(7), 459-78.

[121] Firth-Clark, S.; Willems, H.T.; Williams, A.; Harris, W. Generation and selection of novel estrogen receptor ligands using the de novo structure-based design tool, SkelGen. *J. Chem. Inf. Model.* **2006,** *46*(2), 642-7.

[122] Makino, S.; Ewing, T.J.A.; Kuntz, I.D. DREAM++: Flexible docking program for virtual combinatorial libraries. *Journal of Computer-Aided Molecular Design* **1999,** *13*(5), 513-532.

[123] Durrant, J.; Amaro, R.; McCammon, J. AutoGrow: A Novel Algorithm for Protein Inhibitor Design. *Chemical Biology & Drug Design* **2009,** *73*(2), 168-178.

[124] Grzybowski, B.; Ishchenko, A.; Kim, C.-Y.; Topalov, G.; Chapman, R.; Christianson, D.; Whitesides, G.; Shakhnovich, E.

Combinatorial computational method gives new picomolar ligands for a known enzyme. *Proceedings of the National Academy of Sciences of the United States of America* **2002**, *99*(3), 1270-1273.

[125] Böhm, H.J. LUDI: rule-based automatic design of new substituents for enzyme inhibitor leads. *Journal of Computer-Aided Molecular Design* **1992**, *6*(6), 593-606.

[126] Lauri, G.; Bartlett, P.A. CAVEAT: a program to facilitate the design of organic molecules. *Journal of Computer-Aided Molecular Design* **1994**, *8*(1), 51-66.

[127] Maass, P.; Schulz-Gasch, T.; Stahl, M.; Rarey, M. Recore: A Fast and Versatile Method for Scaffold Hopping Based on Small Molecule Crystal Structure Conformations. *J. Chem. Inf. Model.* **2007**, *47*(2), 390-399.

[128] Thompson, D.; Denny, A.; Nilakantan, R.; Humblet, C.; Joseph-Mccarthy, D.; Feyfant, E. CONFIRM: connecting fragments found in receptor molecules. *J. Comput. Aided Mol. Des.* **2008**, *22*(10), 761-72.

[129] Howard, N.; Abell, C.; Blakemore, W.; Chessari, G.; Congreve, M.; Howard, S.; Jhoti, H.; Murray, C.; Seavers, L.; Montfort, R.V. Application of Fragment Screening and Fragment Linking to the Discovery of Novel Thrombin Inhibitors. *Journal of Medicinal Chemistry* **2006**, *49*(4), 1346–1355.

[130] Huth, J.R.; Park, C.; Petros, A.M.; Kunzer, A.R.; Wendt, M.D.; Wang, X.; Lynch, C.L.; Mack, J.C.; Swift, K.M.; Judge, R.A.; Chen, J.; Richardson, P.L.; Jin, S.; Tahir, S.K.; Matayoshi, E.D.; Dorwin, S.A.; Ladror, U.S.; Severin, J.M.; Walter, K.A.; Bartley, D.M.; Fesik, S.W.; Elmore, S.W.; Hajduk, P.J. Discovery and Design of Novel HSP90 Inhibitors Using Multiple Fragment-based Design Strategies. *Chemical Biology & Drug Design* **2007**, *70*(1), 1-12.

[131] Röhrig, C.H.; Loch, C.; Guan, J.-Y.; Siegal, G.; Overhand, M. Fragment-Based Synthesis and SAR of Modified FKBP Ligands: Influence of Different Linking on Binding Affinity. *ChemMedChem* **2009**, *2*(7), 1054 - 1070.

[132] Coutsias, E.A.; Seok, C.; Jacobson, M.P.; Dill, K.A. A kinematic view of loop closure. *Journal of Computational Chemistry* **2004**, *25*(4), 510-528.

[133] Shenkin, P.S.; Yarmush, D.L.; Fine, R.M.; Wang, H.; Levinthal, C. Predicting antibody hypervariable loop conformation. I. Ensembles of random conformations for ringlike structures. *Biopolymers* **1987**, *26*(12), 2053-2085.

[134] Lewis, R.A. Automated site-directed drug design: a method for the generation of general three-dimensional molecular graphs. *Journal of Molecular Graphics* **1992**, *10*(3), 131-143.

[135] Leach, A.R.; Kilvington, S.R. Automated molecular design: a new fragment-joining algorithm. *Journal of Computer-Aided Molecular Design* **1994**, *8*(3), 283-298.

[136] Takano, Y.; Koizumi, M.; Takarada, R.; Kamimura, M.T.; Czerminski, R.; Koike, T. Computer-aided design of a factor Xa inhibitor by using MCSS functionality maps and a CAVEAT linker search. *Journal of Molecular Graphics and Modelling* **2003**, *22*(2), 105-114.

[137] Law, J.M.S.; Fung, D.Y.K.; Zsoldos, Z.; Simon, A.; Szabo, Z.; Csizmadia, I. G.; Johnson, A.P. THEOCHEM : Validation of the SPROUT de novo design program. *Journal of Molecular Structure* **2003**, 651-657.

[138] Leach, A.R.; Lewis, R.A. A ring-bracing approach to computer-assisted ligand design. *Journal of Computational Chemistry* **1994**, *15*(2), 233-240.

[139] Pierce, A.C.; Rao, G.; Bemis, G.W. BREED: Generating novel inhibitors through hybridization of known ligands. Application to CDK2, p38, and HIV protease. *Journal of Medicinal Chemistry* **2004**, *47*, 2768–75.

[140] Pearlman, D.A.; Murcko, M.A. CONCEPTS: New dynamic algorithm for de novo drug suggestion. *Journal of Computational Chemistry* **1993**, *14*(10), 1184-1193.

[141] Crisman, T.J.; Bender, A.; Milik, M.; Jenkins, J.L.; Scheiber, J.; Sukuru, S. C. K.; Fejzo, J.; Hommel, U.; Davies, J.W.; Glick, M. "Virtual Fragment Linking": An Approach To Identify Potent Binders from Low Affinity Fragment Hits. *Journal of Medicinal Chemistry* **2008**, *51*(8), 2481-2491.

[142] Douguet, D.; Munier-Lehmann, H.; Labesse, G.; Pochet, S. LEA3D: a computer-aided ligand design for structure-based drug design. *J. Med. Chem* **2005**, *48*(7), 2457-2468.

[143] Roche, O.; Rodriguez Sarmiento, R. A new class of histamine H3 receptor antagonists derived from ligand based design. *Bioorganic & Medicinal Chemistry Letters* **2007**, *17*(13), 3670-5.

[144] Todorov, N.P.; Dean, P.M. Evaluation of a method for controlling molecular scaffold diversity in de novo ligand design. *Journal of Computer-Aided Molecular Design* **1997**, *11*(2), 175-92.

[145] Dey, F.; Caflisch, A. Fragment-Based de Novo Ligand Design by Multiobjective Evolutionary Optimization. *J. Chem. Inf. Model.* **2008**, *48*(3), 679-90.

[146] Waszkowycz, B.; Clark, D.; Frenkel, D.; Li, J.; Murray, C.; Robson, B.; Westhead, D. PRO_LIGAND: An Approach to de Novo Molecular Design. 2. Design of Novel Molecules from Molecular Field Analysis (MFA) Models and Pharmacophores. *Journal of Medicinal Chemistry* **1994**, *37*(23), 3994-4002.

[147] Fechner, U.; Schneider, G. Flux (1): A Virtual Synthesis Scheme for Fragment-Based de Novo Design. *Journal of chemical information and modeling* **2006**, *46*(2), 699-707.

CHAPTER 5

Analyzing and Predicting Protein Binding Pockets

Rooplekha C. Mitra and Emil Alexov*

Computational Biophysics and Bioinformatics, Department of Physics & Astronomy, Clemson University, Clemson, SC 29642, USA, tel: (864) 656-5307, Fax: (864) 656-0805, ealexov@clemson.edu

Abstract. This chapter outlines the progress made in analyzing and predicting binding pockets. Typical structure-based drug discovery project begins with 3D structure of the target protein, identifies putative binding pocket(s), analyses its properties and performs virtual screening to find suitable candidate drug molecules. The success crucially depends on both the correct binding pocket prediction and revealing essential biophysical characteristics of the binding site. These two tasks are intertwined, since many of the binding pocket prediction methods relay of previous studies on binding site physico-chemical properties. In this chapter the most popular methods and approaches for both analysis and prediction of binding sites are reviewed and the corresponding URLs are provided. The emerging picture is the most successful methods of binding site prediction are based on many components analysis and thus reflect the complex nature of the receptor-ligand interactions.

INTRODUCTION

Molecular biology aims to decipher the function of biological macromolecules in the cell and this goal ultimately includes understanding receptor-ligand interactions. On one hand, revealing the details of a particular biochemical reaction involving macromolecular binding is a crucial step toward being able to control and alter it in a desired fashion. On another hand, i.e. on proteomic scale, better understanding of the forces that drive receptor-ligand interactions will elucidate how molecular recognition processes govern complex biological systems. In both cases, however, combined experimental and modeling approaches are needed to predict the binding site and its biophysical properties. Even more, frequently the goal is to alter either the function of a given macromolecule or particular metabolic network by drugs in order to reduce the effect of a disease. This will require targeted binding of a drug molecule to a pre-selected macromolecule, a task involving binding pocket analysis and prediction [1-4].

The phenomenon of molecular recognition, which is ubiquitously important for almost all biological processes, is dynamic, complex and subtle. Interaction between the receptor and the ligand involves mutual structural rearrangements guided by a highly convoluted energy landscape. Several different models have attempted to explain protein binding mechanisms [5-10]. The lock and key concept [11, 12] assumes that one protein has a cavity or indentation that the ligand perfectly fits into. Thus the receptor and the ligand interact with almost no change in conformations. However, this model cannot account for proteins that can bind various substrates that have different shape in unbound form. The induced fit model [13-15] accounts for this by assuming that there is a certain amount of plasticity in the active site to accommodate its ligand, analogous to a hand and a glove. The ligand would induce a conformational change at the binding site, shifting it toward an active state. Even more, for proteins that exhibit allosteric behavior, the binding of a ligand in one area of a protein can affect the conformation of the protein at a distant region away from the binding site. A more recent popular model for cooperative regulation, the dynamic population shift model [16], assumes that proteins exist in a population of conformations. Upon ligand binding, the probability distribution of the ensemble of native states will be redistributed, changing the stability of certain residues throughout the protein molecule and propagating a conformational change at specific residues.

The interactions of a protein with ligands or other proteins occur physically on the surface of the protein molecule and frequently the interaction requires a specific shape (either pre-existing or induced) of the local binding region. The ligand- or substrate-binding regions of macromolecules are often solvent-accessible manifolds, and in most cases, they are in the form of a cavity (i.e., a pocket). Identification of "druggable" (a term coined to denote protein surface pockets) [17-22] protein-binding pockets and allosteric sites [23-26] is the most important starting point when working on a potential target. However, while large number of protein structures have already been solved, many of them have unknown binding sites and functions [27-30]. Successful prediction of putative binding pockets/interfaces requires that their properties are well understood, which in turn requires further analysis of the biophysical characteristics of binding epitopes.

Of specific interest are cases for which the 3D structure of either the receptor or the receptor-ligand complex is experimentally available. Analyzing the 3D structure of the receptor is a crucial step in any drug design. Typically the first step is to predict the binding site(s), which may be different for different ligands. Further, the 3D structure can be used to predict plausible structural changes induced by the binding by the means of normal mode analysis or another technique. At the end, the researcher is armed with 3D structure of the target protein, the location of the binding site(s) and possible changes of the geometry of the binding pocket, and could begin *in silico* screening for appropriate drugs. If the structure of the receptor-ligand complexes is available, it can be used to analyze the receptor-ligand interactions and to provide insights of the driving force causing the binding. New ligands can be designed by either mimicking the physico-chemical properties and geometry of the native ligand or by growing/deleting atoms from the original ligand.

This chapter focuses on methods and algorithms to analyze and predict binding sites. To achieve broader representation of the field of receptor-ligand interaction, we will not distinguish between binding pockets, binding sites and binding interfaces. The analysis of the binding sites is the necessary step toward understanding the principles of binding and delivering either statistical measurements or semi-empirical rules to guide structure-based drug design. The knowledge accumulated in the binding sites analysis together with first principle-based approaches is then used to develop methods for predicting binding pockets. This chapter follows this logical path and first outlines the major achievements in analyzing binding pockets and then focuses on the approaches for predicting the corresponding sites.

ANALYSIS OF BINDING POCKETS

The amount of experimentally determined 3D structures of receptor-ligand complexes is constantly growing and thus provides a unique opportunity for assessing the common features driving molecular recognition. Frequently the 3D structure of a particular protein is experimentally solved with different ligands [31, 32]. Studying and comparing such cases reveals which properties are important for the binding, and which are not. However, always the question is how representative the test set is? The last question points out to the necessity of either manually curated or purged by other means databases that can serve as clean set of cases that can be used to deliver common principles for molecular recognition. Below we outline existing databases and search engines that can be used either directly to study binding pockets properties or to extract representative set of cases.

Several databases contain information about binding sites and allow comparison of binding sites properties. The PROSITE [33, 34] is a particular example and it consists of documentation entries describing protein domains, families and functional sites, as well as associated patterns and profiles to identify them. The CavBase [35] provides useful classification of the amino acids in the binding pocket depending on their physico-chemical properties. Patterns the In Non-homologous Tertiary Structures (PINTS) Site Engine [36] allows for search of sequence signatures and provides a measure of statistical significance for any similarity uncovered. In addition, it predicts what residues may the functionally important by comparing them with pre-defined patterns in databases of complete structures. The eF-site [37-39] (electrostatic surface of Functional site) is a newly constructed molecular surface database that includes pre-computed potential at the molecular surface. It is done by numerically solving Poisson-Boltzmann equation. The ProFunc [40] is a web server that is developed to help functional annotation of proteins with unknown function. However, it can be also used to predict functionally important groups and thus in many practical cases the catalytic amino acids. The ProFunc [40] is an web server that allows for predicting functional motifs by comparing the input sequence/structure to functionally characterized proteins. The SitesBase [41] can be used for molecular recognition of small molecules to the target protein. It holds information about structural similarities between known ligand binding sites found in the Protein Data Bank. These and other, some of them in-house built databases, were used to study characteristics of the binding pockets. Below we conveniently group these investigations into several topics.

Amino Acid Frequencies within Binding Pockets

It is a straightforward question to address if the amino acid frequencies (preferences) within binding interfaces are similar to the amino acids distribution over the entire protein structure or only for the accessible surface of the proteins. Depending on the dataset used in particular investigation slightly controversial results were reported. Thus, on a set of 37 protein-protein complexes was shown [42] that the occurrence frequencies of amino acids on the complex interfaces do not differ considerably from the compositions of both entire complexes and of all known proteins in Protein Data Bank with several hydrophobic (alanine, valine and leucine) and polar (threonine) groups found to be underrepresented while other polar (tyrosine, asparagine and tryptophan) groups and glycine were overrepresented. Similar conclusion was reached in another study [43] on much larger set of dimers (768 entries) that included also homodimers and interfaces created from the multi-chain entries in the PDB databank. The absolute numbers and trends for interfacial residues being under- or overrepresented (except for tryptophane and glycine) are very similar to those observed in the previous study [42]. Similar values for the occurrence frequencies of interfacial residues were also reported in the work [44] on the non-redundant set of 621 protein-protein interfaces. However different conclusion was reached in the study [45] where six types pf protein-protein interfaces were analyzed for the set of 1812 PDB structures. For this set of proteins, the residue composition on all types of interface (including interfaces of hetero- and homo-complexes) was found to be considerably different form the "background" frequencies for majority of the amino acids, which most likely is related to the different methodology of defining interfacial residues.

Geometric Properties of the Binding Pocket

Many of the current methods for comparing binding pockets focus on geometric properties [46, 47]. The shape and properties of binding site are determining factors for binding of a ligand to a protein pocket [48-50]. Geometric descriptions of the 'depth' or 'size' of binding pockets are method-dependent and vague. Indeed, several commonly used pocket detection methodologies, give different and potentially conflicting descriptions of a pocket's size and location [51].

The geometrical properties of the binding site can also be evaluated from the point of view of the ligand. Thus, a promising approach is to use cubic 'quantization' of small molecules where each conformation of a compound is represented as a set of contiguous 4.24 Å cubes that encompasses the van der Waals (vdW) volume of the molecule's atoms [52, 53]. Each of the cubes of the quantized molecule is assigned functionalities based on the attributes of the atoms it contained. Once in a cubically quantized format, the small-molecules are then mapped to a basis set of theoretical target surfaces: a complete enumeration under a given set of rules of all possible shapes containing 6–14 4.24 Å cubes. This process coined as Quantized Surface Complementarity Diversity (QCSD) [54, 55] has been shown to be a valid method of classifying small molecules; when small molecules were grouped according to the similarity of their QSCD mappings, these groupings correlated well with the actual biological activity of the molecules. An extension of the QSCD approach has been implemented to map actual protein pockets to the same basis set of theoretical target surfaces, thus bringing both proteins and small molecules into the same frame of reference by developing an algorithm that dissected a potentially large protein pocket into a set of constituent 'binding volumes'. The set of binding volumes that results from dissection of a given protein pocket must both be able to characterize the protein pocket in comparison to other protein pockets, and must also be complementary to the small molecules that bind to the protein pocket.

Physical Characteristics of the Binding Pocket

Physical properties of the binding sites are one of the most important characteristics of molecular recognition. The dynamic nature of ligand binding as well as the thermodynamic contribution of solvation effects have contributed to the difficulty of predicting a priori specific residues or binding pockets that are hot spots for ligand binding affinity. In addition to the above geometrical parameters, other physical properties include density of contacts, hydrophobicity and hydrophilicity, net charge and many others. A review of the empirical structural rules for specificity of protein-protein recognition was recently published [56].

A typical approach is to compare physico-chemical characteristics across several protein families [57, 58]. A complete Multiple Solvent Crystal Structure (MSCS) analysis of a protein surface takes into account patterns of organic solvent molecules indicating the location of hot spots, binding pockets and the areas of plasticity observed by superimposing the protein models and the distribution of explicit water molecules [59-61]. The results are consistent with binding site properties deciphered form database analysis of large number of proteins as the properties being probed are the component features of binding sites. A novel approach which takes into account specific information on geometric, physicochemical and evolutionary characteristics for local surface assignments, is the SPLITPOCKET [62-65] algorithm which uses the alpha shape theory, and develops analytical approach to identify protein functional surfaces by the geometric concept of a split pocket, where a set of all pockets in the protein is computed with its ligand(s) and a set of all pockets with the ligand(s) removed. The pocket is termed a split pocket if the comparison between these set shows altered properties of the pocked.

BINDING POCKETS PREDICTION

Below we review the existing methods for predicting binding pockets, beginning with experimental techniques and ending with computational approaches. Although they are discussed separately, frequently they are used in tandem in practical investigations.

Experimental Techniques

Straightforward method of identifying the binding pocket is simply to experimentally determine the 3D structure of the receptor-ligand complex by either X-ray or NMR methods. This could be tedious and difficult process and in addition, will result in localizing the binding site for that particular ligand. In cases of a receptor with single, well defined binding pocket, as many enzymes, this will provide ultimate detection of the binding site since almost for sure all potential ligands will bind to the same pocket. However, in case of receptor with multiple binding sites, which in addition can not be detected from geometrical considerations only, such an approach will not result to complete picture and will identify one out of many binding pockets. Instead, the Multiple Solvent Crystal Structures (MSCS) method is based on solving the X-ray crystal structure of a protein in aqueous solutions of various compounds; primarily organic solvents [60]. The method is based on superimposing different receptor structures that have been solved in different solvents. Typically the organic molecules cluster at the binding sites forming 'consensus' sites of the binding pockets. This method allows also studying the selectivity of alternative binding sites, singe sites with no preference will bind all types of organic molecules present in the solvent, while more selective sites will bind only a subset of the molecules.

Mapping of proteins by NMR is a well established method which is particularly well suited for identifying and characterizing localized protein hot spots which are strongly affected by ligand binding [66-69]. Measuring the NMR spectra of the protein in isolation (prior binding) and following the spectra changes as the protein is being titrated with a particular ligand will result in identifying the results at the binding epitope. The major strength of NMR-based screening is the high sensitivity of the method to chemical shifts, and fragments that bind to the protein with even millimolar affinity can be detected using heteronuclear single-quantum correlation (HSQC) NMR spectroscopy [70, 71]. As the chemical shift is highly sensitive to the local environment each amino acid has a unique chemical shift if the NMR frequency is sufficiently high. Upon addition of organic fragments to the protein, changes in the HSQC parameters will occur in the amino acids that form the binding site. Chemical shift data are so sensitive to perturbations in the local environment that variations of the HSQC parameters as a function of concentration of added ligand can be used to generate dose response curves and estimates of binding constants [72].

Biochemical methods as alanine mutagenesis can be also used to identify amino acids contributing to the binding, which in vast majority of the cases will be the groups situated in the binding epitope. Thus, each alanine-substituted protein must be separately constructed, expressed and sometimes refolded, and the change of binding affinity is then assessed [73-77]. Alanine substitution of an amino acid situated at the binding interface almost always should cause a change of the binding affinity, since the substitution will either leave a hole at the interface (replacement of a bulky side chain), remove a charge (replacement of a charged residue), or delete a hydrogen bond (replacement of a polar group). The change of the affinity could be in both directions: it can make it more favorable by replacing residue opposing the binding or to lead to decrease of the affinity by mutating an amino acid contributing favorably to the binding. Specific cases are mutations of charged residues which contribute to the binding through long range electrostatic interactions and therefore do not necessary have to be at the binding interface to contribute to the binding affinity. Alanine mutagenesis will show a change of the binding affinity, but the charged residue may not be located in the binding pocket [78].

Table 1: Software and databases for binding pocket analysis.

PROGRAMS & DATABASES	URL	DESCRIPTION
APROPOS	http://www.csb.yale.edu/	Generates a list of atoms arranged in clusters which represent pockets at the molecule envelope.
AUTOLIGAND	http://mgltools.scripps.edu/	Searches the space surrounding the protein and finds the contiguous envelope with the specified volume of atoms, which has the largest possible interaction energy with the protein.
CAST	http://cast.engr.uic.edu.	Locating and measuring protein pockets and cavities, based on precise computational geometry methods, including alpha shape, Voronoi diagrams, Delaunay triangulation and discrete flow theory.
CAVBASE	http://www.ccdc.cam.ac.uk/	Describing and comparing protein binding pockets on the basis of the geometrical and physicochemical properties of the active sites.
DOCK	http://dock.compbio.ucsf.edu/	Semigrid Body Method takes into account possible chemical interactions, that analyses shape & chemical complimentarity of ligand & binding site.
eFSITE	http://ef-site.hgc.jp/	Database for molecular surfaces of proteins' functional sites, displaying the electrostatic potentials and hydrophobic properties together on the Connolly surfaces of the active sites, for analyses of the molecular recognition mechanisms.
FINDSITE	http://cssb.biology.gatech.edu/skolnick/files/FINDSITE/	Threading based method for ligand-binding site prediction and functional annotation based on binding-site similarity across groups of weakly homologous template structures
FPOCKET	http://fpocket.sourceforge.net/run_online.html	Fpocket is an open source pocket detection package based on Voronoi tessellation and alpha spheres built on top of the publicly available package Qhull. The modular source code is organised around a central library of functions, i) Fpocket, (ii) Tpocket and (iii) Dpocket
GRID	http://www.bestgrid.org/	Binding Site Embedded in a regular grid and energy interactions between protein and probe (small molecule) placed at lattice interactions calculated for favourable positions
ICM	http://www.bestgrid.org/	Performs grid-based rigid-receptor flexible-ligand docking through a modified Monte-Carlo searching procedure and a rigorous empirical scoring function
LIGSITE	http://projects.biotec.tu-dresden.de/	Automatic and time-efficient detection of pockets, using a series of simple operations on a cubic grid on the surface of proteins that may act as binding sites for small molecule ligands
PASS	http://www.ccl.net/	Uses geometry to characterize regions of buried volume in proteins and to identify positions likely to represent binding sites based upon the size, shape, and burial extent of these volumes
PINTS	http://www.russell.emblheidelberg.de/pints/	Finds similarities between protein structures containing structural patterns and provides a statistical measure of local structural similarities
POCKET	http://pocket.ekmd.huji.ac.il/	Operates on a rectangular grid, which is constructed around the protein and denotes grid points as either solvent-accessible or inaccessible to the solvent and searches for cavities by scanning along the x-, y and z-axes to locate groups of solvent accessible gridpoints that are enclosed by grid points not accessible to solvent
POCKETPICKER	http://gecco.org.chemie.uni-frankfurt.de/pocketpicker/	A Pymol plugin for cavity detection and binding site prediction
PROSITE	http://www.expasy.ch/prosite	Database of protein families and domains which consists of entries describing the domains, families and functional sites as well as amino acid patterns, signatures, and profiles in them.
QSITE FINDER	http://bmbpcu36.leeds.ac.uk/qsitefinder/	Energy-based method for predicting protein-ligand binding sites by evaluating the interaction energy between the protein and a simple van der Waals probe to locate energetically favourable binding sites
SITE ENGINE	http://bioinfo3d.cs.tau.ac.il/SiteEngine/	Receives as an input two protein structures and searches the complete surface of one protein for regions similar to the binding site of the other.
SPLITPOCKET	http://pocket.uchicago.edu/	Web server to identify functional surfaces of protein from structure coordinates.
SURFNET	www.biochem.ucl.ac.uk/~roman/surfnet/	Geometric algorithms, which detects the gap regions in proteins by fitting spheres into the spaces between protein atoms. The fitting process results in a number of separate groups of interpenetrating spheres, which correspond to the cavities and clefts of the protein.
VISGRID	http://dragon.bio.purdue.edu/	Uses visibility of surface points to find pockets, aimed to identify large protrusions, hollows, and flat regions, which can characterize geometric features of a protein structure

Computational Methods

Common methods consider pocket size, surface roughness or polar/apolar surface area as key descriptors for druggability [51, 79]. Previous results indicate that the endogenous binding site is usually the largest, most hydrophobic and geometrically most complex pocket of a protein, though none of these parameters alone is sufficient to predict binding site druggability [51, 80, 81]. A chart with list of the programs and databases, which are mainly used for binding pocket analysis and detection is shown below.

Below we outline the existing approaches grouping them into several distinctive categories.

Geometric Approaches

Typically the algorithms search protein surface for pockets assuming that the binding site is usually in the largest pocket. From an algorithm point of view, for identifying and measuring binding pockets, the discrete-flow method can be employed [82]. For the two-dimensional model, discrete flow is defined only for empty triangles, that is, those Delaunay triangles that are not part of the dual complex [83, 84]. An obtuse empty triangle "flows" to its neighboring triangle, whereas an acute empty triangle is a sink that collects flow from neighboring empty triangles. All flows are stored, and empty triangles are later merged when they share dotted edges (dual, non-complex edges). Ultimately, the pocket is delineated as a collection of empty triangles. The actual size of the molecular pocket is computed by subtracting the fractions of atom disks contained within each empty triangle. The two-dimensional mouth is an edge on the boundary of the pocket. All features of the two-dimensional description have more complex three-dimensional counterparts. The convex hull in three dimensions is a convex polytope instead of a polygon, and its Delaunay triangulation is a tessellation of the polytope with three dimensional tetrahedra. When atoms have different radii, the weighted Delaunay triangulation is required, and the corresponding weighted Voronoi cells are also different. An explicit realization of the algorithm is the program Delaney, a technique in which the protein is placed within a three-dimensional grid. Where grid points intersect the protein, they are set to 'true', otherwise they are set to 'false'. The protein surface (and cavity boundaries) is defined as to be grid points set to 'true' that are adjacent to grid points set to 'false'. A monolayer of particles is then added to the protein surface (a surface expansion) and the true/false representation is recalculated to redefine the surface. A surface contraction then takes place, where a monolayer of particles is removed. Another method, Automatic Protein Pocket Search (APROPOS) [82, 85, 86] looks for characteristic patterns of small "caves" into which molecular groups can fit into, and has a high reported success rate. The specific algorithm for this technique is based upon creating alpha-shape representation of the protein. The algorithm is used to generate the alpha-shape which creates a Delaunay representation of the protein. The nature of the alpha-shape is dependent on a parameter. The probe can erase the sides and edges of the triangles, except the vertices. When the parameter approaches infinity, the convex hull is formed and the pockets are identified by comparing the structures of the alpha-shape and convex hull. Protein pockets are revealed where the structures of the two representations differ significantly. Another popular predictor is FINDSITE[87], a recently developed algorithm for ligand binding site prediction, ligand screening and molecular function prediction, which is based on binding site conservation across evolutionary distant proteins identified by threading. One key feature of the FINDSITE is that it gives comparable results when high-resolution experimental structures as well as predicted protein models are used. Some approaches operate specifically on a rectangular grid (POCKET) or hinges upon geometric criteria of initial gap sphere (SURFNET) [59, 88, 89]. Fig. **1** illustrates the SURFNET algorithm.

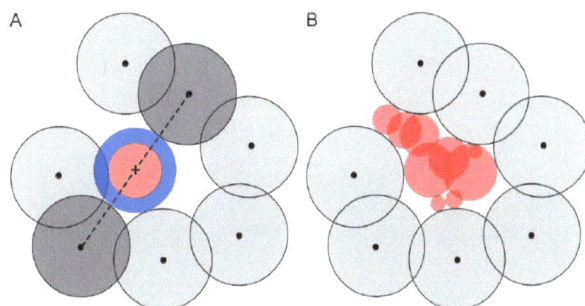

Figure 1: Two-dimensional depiction of the pocket detection process of SURFNET. A: An initial gap sphere (blue disc) is placed midway between the van der Waals surfaces of a pair of atoms. The radius of this gap sphere is then reduced until it is not penetrated by any of the neighboring atoms. The resulting final gap sphere is shown in red. B: The arrangement of final gap spheres is used to describe the shapes and sizes of protein cavities in SURFNET.

VOLUME BASED APPROACHES

The most common approach to ligand binding site localization is volumetric search for large cavities; more recent methods also use electrostatic potential and conservation. The SURFNET performs a gap search by fitting spheres inside protein convex hull. The PocketPicker and LIGSITE [85] methods consist in creating a grid and scanning it for protein-void protein events in many directions, whereas the VisGrid uses visibility of surface points to find pockets. From an algorithmic perspective, pockets are identified by using ligand-binding sites in proteins. Briefly, the algorithm is based on a van der Walls grid potential map using a carbon probe [4, 90]. Bracketing and successive smoothing of that map with two parameter sets produces two distinct maps. One indicates volume considered inside the molecule and the other indicates cavities of empty space. Applying a threshold to the product of those two maps creates solid geometrical objects, which represent the identified pockets. Pockets below a minimum volume are discarded. The algorithm and all pocket analysis routines made use of the ICM software [91, 92] package.

Another approach has been used [93] to detect similarities between the query binding site and recomputed cavity's descriptors from Cavbase database. PASS [94-96], is an alternative approach, a purely geometrical method in which the protein is covered with spherical probes. These probes are then filtered and are given weights proportionally to the number of their neighboring spheres and the extent to which they are buried. Among the probes with the highest weights, the active-site points are determined, which represent the center of potential binding sites. The algorithm looks at all possible combinations of three protein atoms. If the three atoms are close enough together, the algorithm calculates two possible positions for a probe sphere. The LIGSITE is an extension and implementation of the LIGSITE algorithm, adding information on the degree of conservation of the involved surface residues and thus becoming a combination of pure geometrical approach with homology-based prediction. In this method the identification of protein pockets is less dependent on the orientation of the protein in the three-dimensional grid. The LIGSITE has a variable known as the MINPSP (minimum protein-site-protein) threshold. A single grid point has seven probe lines passing through it (x, y, z and the four cubic diagonals) (Fig. **2**). The grid point can be defined to be a pocket (PSP event) up to seven times. The MINPSP threshold defines how many PSP events must occur for a grid point to be defined as being part of a pocket. By setting the threshold higher, shallow pockets are excluded. The accuracy, speed and simplicity of this type of algorithm have made it ideal for use in several subsequent developments, including CavBase and SuperStar. Pockets were ranked by a scoring function [97-99] with six terms, with lower scores indicating a better pocket. The general equation of the scoring function is

$$S = -C_A - C_R + \frac{1}{1.0001 - C_H} + (V - V_0)^2 + (A - A_0)^2 + R \tag{1}$$

where C_A is the absolute residue conservation and C_R is the relative residue conservation. C_H is the absolute conservation within *Homo sapiens*. V and A are the pocket volume and surface area, respectively. Both V_0 and A_0 were set to a specific value and R is the resolution of the crystallographic structure.

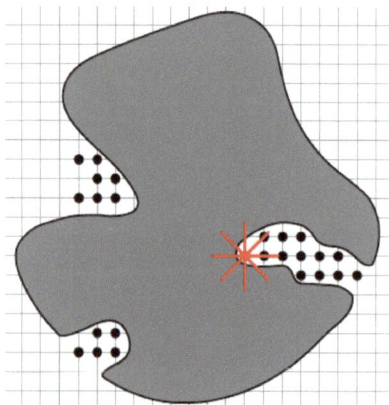

Figure 2: Pocket detection method used in POCKET, LIGSITE and its derivatives. Grid probes are installed at the edges of an artificial grid generated around the protein (shaded area). A scanning process is applied to detect protein-solvent-protein events (POCKET and LIGSITE) or surface-solvent-surface events (LIGSITE).

Another combination of geometric descriptors and homology (conservation) is implemented in PocketPicker which uses a buriedness index at grid points, and then clusters those with values lying within a cutoff range. SURFNET-ConSurf use the SURFNET program to identify pockets on the protein surface, and retains regions that are more highly conserved using the ConSurf-HSSP database [59]. The SURFNET algorithm identifies the clefts on the surface of a protein by placing a sphere between all pairs of atoms such that the sphere just touches each atom and is between some predefined minimum and maximum radius. Each sphere is progressively reduced in size if any other atoms intersect it until it either intersects with no further atoms, in which case it is retained, or its radius drops below the minimum size, in which case it is discarded. Once the clefts on the surface have been filled by spheres it is possible to cluster the spheres into separate regions and calculate a volume for each cleft. The second part of SURFNET-ConSurf involves discarding the spheres that are distant from highly conserved residues, resulting in a trimming of the clusters representing each cleft. Residue conservation scores are obtained from the ConSurf-HSSP database which provides estimates for the rate of evolution of each amino acid in a PDB structure. The scores are calculated using HSSP's multiple sequence alignments (MSA) as the input for the Rate4Site algorithm. Another technique is the POCKET algorithm, where a probe sphere of radius 3Å is passed across the protein along each line of a Cartesian three-dimensional grid and an interaction between the protein and probe sphere occurs if the centre of a protein atom is found to be within the probe sphere. A pocket is identified if an interaction occurs followed by a period of no interaction, followed by another interaction.

Separate class of approaches are (a) the CS-Map [100] which is a method that performs computational solvent mapping in an attempt to recreate the multiple solvent crystal structures (MSCS) results and (b) the CAST [82] is a method for measuring protein pockets and cavities, based on precise computational geometry methods, including alpha shape and discrete flow theory and it identifies and measures pockets and pocket mouth openings, as well as cavities. The program specifies the atoms lining pockets, pocket openings, and buried cavities; the volume and area of pockets and cavities; and the area and circumference of mouth openings.

From geometrical point of view, the binding site is typically considered to be made of core and peripheral residues (Fig. 3). The core residues are completely buried upon binding of the ligand (shown in green in Fig. 3), white peripheral areas are partially exposed to the water (shown in blue in Fig. 3). Frequently the entire complex surface is taken out of the receptor and analyzed independently (middle panel in Fig. 3), or projected on a flat surface as a kind of contour map (right panel in Fig. 3).

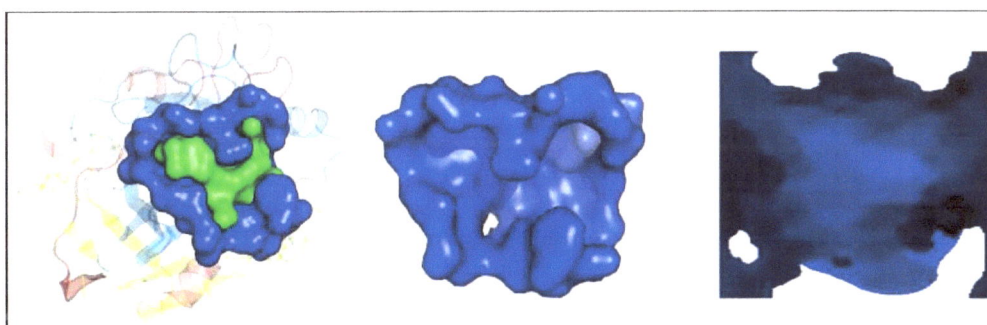

Figure 3: Overview of the binding pocket representation process: the ligand binding site of a protein (on the left, PDB:1dwd protein) is represented by the whole cavity surface (middle), which is sphere-mapped from its center of gravity and projected (right).

BINDING ENERGY APPROACHES

The binding energy-based algorithms analyze the binding energies of probes placed on a grid around the protein. Representatives of such a class of methods are the Q-fit which creates a series of three-dimensional energy grids around the binding pocket for a variety of probe's types [101] and the AutoLigand which is based on finding energetically favorable binding sites. Another energy-based method, Q-SiteFinder, describes a volume envelope where an atom can interact favorably at the protein surface and the pockets are ranked according to interaction energy, and it is assumed that these relate to locations where a putative ligand could bind and optimize its van der Waals interaction energy [102, 103]. Alternative technique, called the pocketome algorithm, is similar to that of Q-

SiteFinder and creates a three-dimensional grid around the protein and calculates van der Waals potentials at each point. The potential map is then smoothed, envelopes of favorable binding energy are identified and only envelopes with volume exceeding $100Å^3$ are retained. These approaches could be modified to take advantage of probe clustering and energy contour analysis to predict ligand binding sites [104, 105].

HOMOLOGY BASED APPROACHES

The homology approach is based on the presumption that sequence and structural features that are important for the function of proteins should be conserved [21, 59, 106-113]. Thus, if two proteins perform their function by binding the same ligand, the binding site properties in terms of sequence and structure should be similar. This statement needs clarification: not all amino acids within the binding interface have to be conserved, but only that have specific role in the molecular recognition.

Recently a two-stage template-based ligand binding site prediction method was applied to CASAP8 targets and reported results were very encouraging [114]. Homology approach was utilized in developing another method (FINDSITE) for ligand-binding site prediction based on binding site similarity across groups of weakly homologous template structures [87].

Another unique binding pocket detection method is local similarity search on protein surface. It consists in using a database of known binding sites and scanning the surface of a protein to find surface patch matches in the database [115, 116]. Current implementations include patch analysis combined with a Bayesian network [117], support vector machine [118, 119], neural network [120], analysis of biophysical surface properties [121, 122], and regression approaches as implemented in the Multilayer Perceptron algorithm [123].

EVOLUTIONARY METHODS

Evolutionary methods are based on understanding that the binding sites properties may be evolutionary conserved. In general this means that the network of important interactions and the shape of the binding pocket should be similar across protein families binding similar ligands.

Preserving important interactions is the core of the correlated mutation analysis. Thus, if two macromolecules interact, and a spontaneous mutation occurs at the binding site of one of them, the other molecule will evolutionary accept a mutation that compensates the effect. Thus, by monitoring mutations that are correlated within multiple sequence alignment of the receptor and the ligand, the amino acids important for the binding can be predicted [50, 89, 124-128]. Such an approach, however, can be applied only to protein interaction with relatively large ligands, but not to the case of protein binding small molecules.

Evolutionary trace (ET) method does not have such a limitation since is it based on the evolutionary conserved properties of the target protein only [129]. The ET method can be used to identify residues of functional importance [130, 131], to design proteins with broader substrate specificity [132] and to predict new drug binding sites [133].

Amino acid propensity, or in general, the propensity at different levels can be used to predict binding sites [128, 134]. These levels could include curvature of the binding pocket, amino acid identity and electrostatic potential [135], just the contact preferences [136] or the propensities of surface residues (InterProSurf) [137].

CONCLUSION

The object of drug design is to find a drug molecule that tightly binds to the target protein, moderates function and/or competes with natural substrates of the protein. Such a drug can be best found on the basis of knowledge of the structure of the target protein. Even more, if the binding site is known or can be reliably predicted, then docking methods can be applied to select suitable lead compounds that have the potential of being refined to drugs. Because of that, developing methods and software for binding site prediction and analysis is a very important task.

This chapter outlines the most popular algorithms and software for prediction of binding sites without making an attempt to rank their performance, but rather providing a short description of their algorithms and letting the users to decide the best approach for their own applications. Obviously any ranking will be benchmarking set dependent and therefore the success will vary from application to application. Perhaps the best strategy is to apply several methods and to elect consensus prediction.

ACKNOWLEDGEMENTS

The work was supported by Award Number R03LM009748 from the National Library of Medicine. The content is solely the responsibility of the authors and does not necessary represents the official views of the National Library of Medicine or the National Institutes of Health.

REFERENCES

[1]   Salemme, F.R.; Spurlino, J.; Bone, R. Serendipity meets precision: the integration of structure-based drug design and combinatorial chemistry for efficient drug discovery. *Structure*, **1997**, *5*(3),319-24.

[2]   Aung, Z.; Tong, J.C. BSAlign: a rapid graph-based algorithm for detecting ligand-binding sites in protein structures. *Genome Inform.*, **2008**, *21*,65-76.

[3]   Gong, Y.; Pace, T.C.; Castillo, C.; Bohne, C.; O'Neill, M.A.; Plettner, E. Ligand-interaction kinetics of the pheromone-binding protein from the gypsy moth, L. dispar: insights into the mechanism of binding and release. *Chem. Biol.*, **2009**, *16*(2),162-72.

[4]   Oda, A.; Yamaotsu, N.; Hirono, S. Evaluation of the searching abilities of HBOP and HBSITE for binding pocket detection. *J. Comput. Chem.*, **2009**, *30*, 2728-37.

[5]   Breckenridge, R.J. Molecular recognition: models for drug design. *Experientia*, **1991**, *47*(11-12), 1148-61.

[6]   Brooijmans, N.; Kuntz, I.D. Molecular recognition and docking algorithms. *Annu. Rev. Biophys. Biomol. Struct.*, **2003**, *32*, 335-73.

[7]   Jorgensen, J.L.; Reay, P.A.; Ehrich, E.W.; Davis, M.M. Molecular components of T-cell recognition. *Annu. Rev. Immunol.*, **1992**, *10*, 835-73.

[8]   Ehrhardt, M.R.; Erijman, L.; Weber, G.; Wand, A.J. Molecular recognition by calmodulin: pressure-induced reorganization of a novel calmodulin-peptide complex. *Biochemistry*, **1996**, *35*(5), 1599-605.

[9]   Neuberger, M.S. Novartis Medal Lecture. Antibodies: a paradigm for the evolution of molecular recognition. *Biochem. Soc. Trans.*, **2002**, *30*(4), 341-50.

[10]   Caceres, R.A.; Pauli, I.; Timmers, L.F.; de Azevedo, W.F., Jr. Molecular recognition models: a challenge to overcome. *Curr. Drug Targets*, **2008**, *9*(12), 1077-83.

[11]   Jorgensen, W.L. Rusting of the lock and key model for protein-ligand binding. *Science*, **1991**, *254*(5034), 954-5.

[12]   Chen, B.; Piletsky, S.; Turner, A.P. High molecular recognition: design of "Keys". *Comb. Chem. High Throughput Screen.*, **2002**, *5*(6), 409-27.

[13]   Bosshard, H.R. Molecular recognition by induced fit: how fit is the concept? *News Physiol Sci*, **2001**, *16*, 171-3.

[14]   Bostock-Smith, C.E.; Harris, S.A.; Laughton, C.A.; Searle, M.A. Induced fit DNA recognition by a minor groove binding analogue of Hoechst 33258: fluctuations in DNA A tract structure investigated by NMR and molecular dynamics simulations. *Nucleic Acids Res.*, **2001**, *29*(3), 693-702.

[15]   Hazy, E.; Tompa, P. Limitations of induced folding in molecular recognition by intrinsically disordered proteins. *Chemphyschem.*, **2009**, *10*(9-10), 1415-9.

[16]   Okazaki, K.; Takada, S. Dynamic energy landscape view of coupled binding and protein conformational change: induced-fit versus population-shift mechanisms. *Proc. Natl. Acad. Sci. U S A*, **2008**, *105*(32), 11182-7.

[17]   Hopkins, A.L.; Groom, C.R. The druggable genome. *Nat. Rev. Drug Discov.*, **2002**, *1*(9), 727-30.

[18]   Russ, A.P.; Lampel, S. The druggable genome: an update. *Drug Discov. Today*, **2005**, *10*(23-24), 1607-10.

[19]   Sioud, M.; Leirdal, M. Druggable signaling proteins. *Methods Mol. Biol*, **2007**, *361*, 1-24.

[20]   Hambly, K.; Danzer, J.; Muskal, S.; Debe, D.A. Interrogating the druggable genome with structural informatics. *Mol. Divers*, **2006**, *10*(3), 273-81.

[21]   Fuller, J.C.; Burgoyne, N.J.; Jackson, R.M. Predicting druggable binding sites at the protein-protein interface. *Drug Discov. Today*, **2009**, *14*(3-4), 155-61.

[22]   Guntas, G.; Mansell, T.J.; Kim, J.R.; Ostermeier, M. Directed evolution of protein switches and their application to the creation of ligand-binding proteins. *Proc. Natl. Acad. Sci. U S A*, **2005**, *102*(32), 11224-9.

[23]   Malherbe, P.; Kratochwil, N.; Muhlemann, A.; Zenner, M.T.; Fischer, C.; Stahl, M.; Gerber, P.R.; Jaeschke, G.; Porter, R.H. Comparison of the binding pockets of two chemically unrelated allosteric antagonists of the mGlu5 receptor and identification of crucial residues involved in the inverse agonism of MPEP. *J. Neurochem.*, **2006**, *98*(2), 601-15.

[24]   Oldham, W.M.; Van Eps, N.; Preininger, A.M.; Hubbell, W.L.; Hamm, H.E. Mapping allosteric connections from the receptor to the nucleotide-binding pocket of heterotrimeric G proteins. *Proc. Natl. Acad. Sci. U S A*, **2007**, *104*(19), 7927-32.

[25]   Ragona, L.; Catalano, M.; Luppi, M.; Cicero, D.; Eliseo, T.; Foote, J.; Fogolari, F.; Zetta, L.; Molinari, H. NMR dynamic studies suggest that allosteric activation regulates ligand binding in chicken liver bile acid-binding protein. *J. Biol. Chem.*, **2006**, *281*(14), 9697-709.

[26]   Tu, J.L.; Chin, K.H.; Wang, A.H.; Chou, S.H. Unique GTP-binding pocket and allostery of uridylate kinase from a gram-negative phytopathogenic bacterium. *J Mol Biol*, **2009**, *385*(4),1113-26.

[27]   Laurie, A.T.; Jackson, R.M. Methods for the prediction of protein-ligand binding sites for structure-based drug design and virtual ligand screening. *Curr. Protein Pept. Sci.*, **2006**, *7*(5), 395-406.

[28]   Kufareva, I.; Budagyan, L.; Raush, E.; Totrov, M.; Abagyan, R. PIER: protein interface recognition for structural proteomics. *Proteins*, **2007**, *67*(2), 400-17.

[29]   Tewari, A.K.; Dubey, R. Emerging trends in molecular recognition: utility of weak aromatic interactions. *Bioorg. Med. Chem.*, **2008**, *16*(1), 126-43.

[30]   Ghersi, D.; Sanchez, R. Improving accuracy and efficiency of blind protein-ligand docking by focusing on predicted binding sites. *Proteins*, **2009**, *74*(2), 417-24.

[31]   Everse, S.J.; Spraggon, G.; Veerapandian, L.; Riley, M.; Doolittle, R.F. Crystal structure of fragment double-D from human fibrin with two different bound ligands. *Biochemistry*, **1998**, *37*(24), 8637-42.

[32]   Hogner, A.; Greenwood, J.R.; Liljefors, T.; Lunn, M.L.; Egebjerg, J.; Larsen, I.K.; Gouaux, E.; Kastrup, J.S. Competitive antagonism of AMPA receptors by ligands of different classes: crystal structure of ATPO bound to the GluR2 ligand-binding core, in comparison with DNQX. *J. Med. Chem.*, **2003**, *46*(2), 214-21.

[33]   Yoon, S.; Ebert, J.C.; Chung, E.Y.; De Micheli, G.; Altman, R.B. Clustering protein environments for function prediction: finding PROSITE motifs in 3D. *BMC Bioinformatics*, **2007**, *8 Suppl 4*,S10.

[34]   Hulo, N.; Bairoch, A.; Bulliard, V.; Cerutti, L.; Cuche, B.A.; de Castro, E.; Lachaize, C.; Langendijk-Genevaux, P.S.; Sigrist, C.J. The 20 years of PROSITE. *Nucleic Acids Res.*, **2008**, *36*(Database issue), D245-9.

[35]   Kuhn, D.; Weskamp, N.; Schmitt, S.; Hullermeier, E.; Klebe, G. From the similarity analysis of protein cavities to the functional classification of protein families using cavbase. *J. Mol. Biol.*, **2006**, *359*(4), 1023-44.

[36]   Stark, A.; Russell, R.B. Annotation in three dimensions. PINTS: Patterns in Non-homologous Tertiary Structures. *Nucleic Acids Res.*, **2003**, *31*(13), 3341-4.

[37]   Kinoshita, K.; Nakamura, H. eF-site and PDBjViewer: database and viewer for protein functional sites. *Bioinformatics*, **2004**, *20*(8), 1329-30.

[38]   Kinoshita, K.; Nakamura, H. Identification of protein biochemical functions by similarity search using the molecular surface database eF-site. *Protein Sci.*, **2003**, *12*(8), 1589-95.

[39]   Kinoshita, K.; Furui, J.; Nakamura, H. Identification of protein functions from a molecular surface database, eF-site. *J. Struct. Funct. Genomics*, **2002**, *2*(1), 9-22.

[40]   Laskowski, R.A.; Watson, J.D.; Thornton, J.M. ProFunc: a server for predicting protein function from 3D structure. *Nucleic Acids Res.*, **2005**, *33*(Web Server issue), W89-93.

[41]   Gold, N.D.; Jackson, R.M. SitesBase: a database for structure-based protein-ligand binding site comparisons. *Nucleic Acids Res.*, **2006**, *34*(Database issue), D231-4.

[42]   Kundrotas, P.J.; Alexov, E. Electrostatic properties of protein-protein complexes. *Biophys. J.*, **2006**, *91*(5), 1724-36.

[43]   Lu, H.; Lu, L.; Skolnick, J. Development of unified statistical potentials describing protein-protein interactions. *Biophysical Journal*, **2003**, *84*(3), 1895-1901.

[44]   Glaser, F.; Steinberg, D.M.; Vakser, I.A.; Ben-Tal, N. Residue frequencies and pairing preferences at protein-protein interfaces. *Proteins-Structure Function and Genetics*, **2001**, *43*(2), 89-102.

[45]   Ofran, Y.; Punta, M.; Schneider, R.; Rost, B. Beyond annotation transfer by homology: novel protein-function prediction methods to assist drug discovery. *Drug Discov. Today*, **2005**, *10*(21), 1475-82.

[46]   Xie, L.; Bourne, P.E. A robust and efficient algorithm for the shape description of protein structures and its application in predicting ligand binding sites. *BMC Bioinformatics*, **2007**, *8 Suppl 4*,S9.

[47]   Vacic, V.; Oldfield, C.J.; Mohan, A.; Radivojac, P.; Cortese, M.S.; Uversky, V.N.; Dunker, A.K. Characterization of molecular recognition features, MoRFs, and their binding partners. *J. Proteome Res.*, **2007**, *6*(6), 2351-66.

[48]   Erickson, J.A.; Jalaie, M.; Robertson, D.H.; Lewis, R.A.; Vieth, M. Lessons in molecular recognition: the effects of ligand and protein flexibility on molecular docking accuracy. *J. Med. Chem.*, **2004**, *47*(1), 45-55.

[49]   Nobeli, I.; Spriggs, R.V.; George, R.A.; Thornton, J.M. A ligand-centric analysis of the diversity and evolution of protein-ligand relationships in E.coli. *J. Mol. Biol.*, **2005**, *347*(2), 415-36.

[50] Kauffman, C.; Rangwala, H.; Karypis, G. Improving homology models for protein-ligand binding sites. *Comput. Syst. Bioinformatics Conf.*, **2008**, *7*, 211-22.

[51] Weisel, M.; Proschak, E.; Kriegl, J.M.; Schneider, G. Form follows function: shape analysis of protein cavities for receptor-based drug design. *Proteomics*, **2009**, *9*(2), 451-9.

[52] Jain, T.; Jayaram, B. An all atom energy based computational protocol for predicting binding affinities of protein-ligand complexes. *FEBS Lett.*, **2005**, *579*(29), 6659-66.

[53] Barratt, E.; Bingham, R.J.; Warner, D.J.; Laughton, C.A.; Phillips, S.E.; Homans, S.W. Van der Waals interactions dominate ligand-protein association in a protein binding site occluded from solvent water. *J. Am. Chem. Soc.*, **2005**, *127*(33), 11827-34.

[54] Wintner, E.A.; Moallemi, C.C. Quantized surface complementarity diversity (QSCD): a model based on small molecule-target complementarity. *J. Med. Chem.*, **2000**, *43*(10), 1993-2006.

[55] Mason, K.; Patel, N.M.; Ledel, A.; Moallemi, C.C.; Wintner, E.A. Mapping protein pockets through their potential small-molecule binding volumes: QSCD applied to biological protein structures. *J. Comput. Aided Mol. Des.*, **2004**, *18*(1), 55-70.

[56] Bahadur, R.P.; Zacharias, M. The interface of protein-protein complexes: analysis of contacts and prediction of interactions. *Cell Mol. Life Sci.*, **2008**, *65*(7-8), 1059-72.

[57] Briand, L.; Nespoulous, C.; Huet, J.C.; Takahashi, M.; Pernollet, J.C. Ligand binding and physico-chemical properties of ASP2, a recombinant odorant-binding protein from honeybee (Apis mellifera L.). *Eur. J. Biochem.*, **2001**, *268*(3), 752-60.

[58] Stewart, J.M.; Blakely, J.A.; Karpowicz, P.A.; Kalanxhi, E.; Thatcher, B.J.; Martin, B.M. Unusually weak oxygen binding, physical properties, partial sequence, autoxidation rate and a potential phosphorylation site of beluga whale (Delphinapterus leucas) myoglobin. *Comp. Biochem. Physiol. B Biochem. Mol. Biol.*, **2004**, *137*(3), 401-12.

[59] Glaser, F.; Morris, R.J.; Najmanovich, R.J.; Laskowski, R.A.; Thornton, J.M. A method for localizing ligand binding pockets in protein structures. *Proteins*, **2006**, *62*(2), 479-88.

[60] Mattos, C.; Bellamacina, C.R.; Peisach, E.; Pereira, A.; Vitkup, D.; Petsko, G.A.; Ringe, D. Multiple solvent crystal structures: probing binding sites, plasticity and hydration. *J. Mol. Biol.*, **2006**, *357*(5), 1471-82.

[61] Bialas, A.; Grembecka, J.; Krowarsch, D.; Otlewski, J.; Potempa, J.; Mucha, A. Exploring the Sn binding pockets in gingipains by newly developed inhibitors: structure-based design, chemistry, and activity. *J. Med. Chem.*, **2006**, *49*(5), 1744-53.

[62] Tseng, Y.Y.; Dupree, C.; Chen, Z.J.; Li, W.H. SplitPocket: identification of protein functional surfaces and characterization of their spatial patterns. *Nucleic Acids Res.*, **2009**, *37*(Web Server issue), W384-9.

[63] Kawabata, T.; Go, N. Detection of pockets on protein surfaces using small and large probe spheres to find putative ligand binding sites. *Proteins*, **2007**, *68*(2), 516-29.

[64] Montero-Cabrera, L.A.; Rohrig, U.; Padron-Garcia, J.A.; Crespo-Otero, R.; Montero-Alejo, A.L.; Garcia de la Vega, J.M.; Chergui, M.; Rothlisberger, U. CNDOL: A fast and reliable method for the calculation of electronic properties of very large systems. Applications to retinal binding pocket in rhodopsin and gas phase porphine. *J. Chem. Phys.*, **2007**, *127*(14), 145102.

[65] Tseng, Y.Y.; Li, W.H. Identification of protein functional surfaces by the concept of a split pocket. *Proteins*, **2009**, *76*(4), 959-76.

[66] Constantine, K.L.; Davis, M.E.; Metzler, W.J.; Mueller, L.; Claus, B.L. Protein-ligand NOE matching: a high-throughput method for binding pose evaluation that does not require protein NMR resonance assignments. *J. Am. Chem. Soc.*, **2006**, *128*(22), 7252-63.

[67] Polshakov, V.I.; Birdsall, B.; Feeney, J. Effects of co-operative ligand binding on protein amide NH hydrogen exchange. *J. Mol. Biol.*, **2006**, *356*(4), 886-903.

[68] Shimotakahara, S.; Furihata, K.; Tashiro, M. Application of NMR screening techniques for observing ligand binding with a protein receptor. *Magn. Reson. Chem.*, **2005**, *43*(1), 69-72.

[69] Patching, S.G.; Herbert, R.B.; O'Reilly, J.; Brough, A.R.; Henderson, P.J. Low 13C-background for NMR-based studies of ligand binding using 13C-depleted glucose as carbon source for microbial growth: 13C-labeled glucose and 13C-forskolin binding to the galactose-H+ symport protein GalP in Escherichia coli. *J. Am. Chem. Soc.*, **2004**, *126*(1), 86-7.

[70] Fesik, S.W. NMR studies of molecular complexes as a tool in drug design. *J. Med. Chem.*, **1991**, *34*(10), 2937-45.

[71] Shimada, I. [An NMR method for the determination of the interface of large protein-protein complexes]. *Tanpakushitsu Kakusan Koso*, **2002**, *47*(10), 1285-91.

[72] West, G.M.; Tang, L.; Fitzgerald, M.C. Thermodynamic analysis of protein stability and ligand binding using a chemical modification- and mass spectrometry-based strategy. *Anal. Chem.*, **2008**, *80*(11), 4175-85.

[73] Gkountelias, K.; Tselios, T.; Venihaki, M.; Deraos, G.; Lazaridis, I.; Rassouli, O.; Gravanis, A.; Liapakis, G. Alanine scanning mutagenesis of the second extracellular loop of type 1 corticotropin-releasing factor receptor revealed residues critical for peptide binding. *Mol. Pharmacol.*, **2009**, *75*(4), 793-800.

[74]   Nakayama, T., *et al.* Identification of amino acid residues responsible for von Willebrand factor binding to sulfatide by charged-to-alanine-scanning mutagenesis. *Int. J. Hematol.*, **2008**, *87*(4), 363-70.

[75]   Chen, Y.; Gollnick, P. Alanine scanning mutagenesis of anti-TRAP (AT) reveals residues involved in binding to TRAP. *J. Mol. Biol.*, **2008**, *377*(5), 1529-43.

[76]   Sorensen, H.; Whittaker, L.; Hinrichsen, J.; Groth, A.; Whittaker, J. Mapping of the insulin-like growth factor II binding site of the Type I insulin-like growth factor receptor by alanine scanning mutagenesis. *FEBS Lett.*, **2004**, *565*(1-3), 19-22.

[77]   Heredia, V.V.; Cooper, W.C.; Kruger, R.G.; Jin, Y.; Penning, T.M. Alanine scanning mutagenesis of the testosterone binding site of rat 3 alpha-hydroxysteroid dehydrogenase demonstrates contact residues influence the rate-determining step. *Biochemistry*, **2004**, *43*(19), 5832-41.

[78]   Clemenza, L.; Isenman, D.E. Structure-guided identification of C3d residues essential for its binding to complement receptor 2 (CD21). *J. Immunol.*, **2000**, *165*(7), 3839-48.

[79]   Hopkins, A.L.; Groom, C.R. Target analysis: a priori assessment of druggability. *Ernst Schering Res. Found. Workshop*, **2003**, (42), 11-7.

[80]   An, J.; Totrov, M.; Abagyan, R. Comprehensive identification of "druggable" protein ligand binding sites. *Genome Inform.*, **2004**, *15*(2), 31-41.

[81]   Neidle, S.; Parkinson, G.N. Quadruplex DNA crystal structures and drug design. *Biochimie*, **2008**, *90*(8), 1184-96.

[82]   Liang, J.; Edelsbrunner, H.; Woodward, C. Anatomy of protein pockets and cavities: measurement of binding site geometry and implications for ligand design. *Protein Sci.*, **1998**, *7*(9), 1884-97.

[83]   Singh, R.K.; Tropsha, A.; Vaisman, II. Delaunay tessellation of proteins: four body nearest-neighbor propensities of amino acid residues. *J. Comput. Biol.*, **1996**, *3*(2), 213-21.

[84]   Wen, Z.; Li, M.; Li, Y.; Guo, Y.; Wang, K. Delaunay triangulation with partial least squares projection to latent structures: a model for G-protein coupled receptors classification and fast structure recognition. *Amino Acids*, **2007**, *32*(2), 277-83.

[85]   Hendlich, M.; Rippmann, F.; Barnickel, G. LIGSITE: automatic and efficient detection of potential small molecule-binding sites in proteins. *J. Mol. Graph. Model.*, **1997**, *15*(6), 359-63, 389.

[86]   Brady, G.P., Jr.; Stouten, P.F. Fast prediction and visualization of protein binding pockets with PASS. *J. Comput. Aided Mol. Des.*, **2000**, *14*(4), 383-401.

[87]   Brylinski, M.; Skolnick, J. A threading-based method (FINDSITE) for ligand-binding site prediction and functional annotation. *Proc. Natl. Acad. Sci. U S A*, **2008**, *105*(1), 129-34.

[88]   Laskowski, R.A. SURFNET: a program for visualizing molecular surfaces, cavities, and intermolecular interactions. *J. Mol. Graph.*, **1995**, *13*(5), 323-30, 307-8.

[89]   Huang, B. MetaPocket: a meta approach to improve protein ligand binding site prediction. *Omics*, **2009**, *13*(4), 325-30.

[90]   Barbany, M.; Gutierrez-de-Teran, H.; Sanz, F.; Villa-Freixa, J. Towards a MIP-based alignment and docking in computer-aided drug design. *Proteins*, **2004**, *56*(3), 585-94.

[91]   Smielewski, P.; Czosnyka, M.; Steiner, L.; Belestri, M.; Piechnik, S.; Pickard, J.D. ICM+: software for on-line analysis of bedside monitoring data after severe head trauma. *Acta Neurochir. Suppl.*, **2005**, *95*, 43-9.

[92]   Guendling, K.; Smielewski, P.; Czosnyka, M.; Lewis, P.; Nortje, J.; Timofeev, I.; Hutchinson, P.J.; Pickard, J.D. Use of ICM+ software for on-line analysis of intracranial and arterial pressures in head-injured patients. *Acta Neurochir. Suppl.*, **2006**, *96*, 108-13.

[93]   Schmitt, S.; Kuhn, D.; Klebe, G. A new method to detect related function among proteins independent of sequence and fold homology. *J. Mol. Biol.*, **2002**, *323*(2), 387-406.

[94]   Jain, A.N.; Dietterich, T.G.; Lathrop, R.H.; Chapman, D.; Critchlow, R.E., Jr.; Bauer, B.E.; Webster, T.A.; Lozano-Perez, T. A shape-based machine learning tool for drug design. *J. Comput. Aided Mol. Des.*, **1994**, *8*(6), 635-52.

[95]   Tzeng, Y.L.; Hoch, J.A. Molecular recognition in signal transduction: the interaction surfaces of the Spo0F response regulator with its cognate phosphorelay proteins revealed by alanine scanning mutagenesis. *J. Mol. Biol.*, **1997**, *272*(2), 200-12.

[96]   Matter, H.; Baringhaus, K.H.; Naumann, T.; Klabunde, T.; Pirard, B. Computational approaches towards the rational design of drug-like compound libraries. *Comb. Chem. High Throughput Screen.*, **2001**, *4*(6), 453-75.

[97]   Arnold, J.R.; Burdick, K.W.; Pegg, S.C.; Toba, S.; Lamb, M.L.; Kuntz, I.D. SitePrint: three-dimensional pharmacophore descriptors derived from protein binding sites for family based active site analysis, classification, and drug design. *J. Chem. Inf. Comput. Sci.*, **2004**, *44*(6), 2190-8.

[98]   Deng, W.; Verlinde, C.L. Evaluation of different virtual screening programs for docking in a charged binding pocket. *J. Chem. Inf. Model.*, **2008**, *48*(10), 2010-20.

[99]   Qiu, Z.; Wang, X.; Abagyan, R.; Kufareva, I. Identification of ligand-binding pockets in proteins using residue preference methods. The flexible pocketome engine for structural chemogenomics. *Protein Pept. Lett.*, **2009**, *16*(8), 984-90.

[100] Landon, M.R.; Lancia, D.R., Jr.; Yu, J.; Thiel, S.C.; Vajda, S. Identification of hot spots within druggable binding regions by computational solvent mapping of proteins. *J. Med. Chem.*, **2007**, *50*(6), 1231-40.

[101] Jackson, R.M. Q-fit: a probabilistic method for docking molecular fragments by sampling low energy conformational space. *J. Comput. Aided Mol. Des.*, **2002**, *16*(1), 43-57.

[102] Griffith, R.; Luu, T.T.; Garner, J.; Keller, P.A. Combining structure-based drug design and pharmacophores. *J. Mol. Graph. Model.*, **2005**, *23*(5), 439-46.

[103] Wass, M.N.; Sternberg, M.J. Prediction of ligand binding sites using homologous structures and conservation at CASP8. *Proteins*, **2009**, *77*, 147-51.

[104] Kroemer, R.T. Structure-based drug design: docking and scoring. *Curr. Protein Pept. Sci.*, **2007**, *8*(4), 312-28.

[105] De Benedetti, P.G.; Fanelli, F. Ligand-receptor communication and drug design. *Curr. Protein Pept. Sci.*, **2009**, *10*(2), 186-93.

[106] Stewart, P.L. Integrating cryo-electron microscopy into drug-design strategies. *Trends. Biotechnol.*, **1994**, *12*(11), 429-31.

[107] Dalgarno, D.C.; Botfield, M.C.; Rickles, R.J. SH3 domains and drug design: ligands, structure, and biological function. *Biopolymers*, **1997**, *43*(5), 383-400.

[108] Neve, K.A.; Cumbay, M.G.; Thompson, K.R.; Yang, R.; Buck, D.C.; Watts, V.J.; DuRand, C.J.; Teeter, M.M. Modeling and mutational analysis of a putative sodium-binding pocket on the dopamine D2 receptor. *Mol. Pharmacol.*, **2001**, *60*(2), 373-81.

[109] Taylor, J.D.; Ababou, A.; Fawaz, R.R.; Hobbs, C.J.; Williams, M.A.; Ladbury, J.E. Structure, dynamics, and binding thermodynamics of the v-Src SH2 domain: implications for drug design. *Proteins*, **2008**, *73*(4), 929-40.

[110] Weskamp, N.; Hullermeier, E.; Klebe, G. Merging chemical and biological space: Structural mapping of enzyme binding pocket space. *Proteins*, **2009**, *76*(2), 317-30.

[111] Munoz-Torrero, D. Exploiting multivalency in drug design. *Curr. Pharm. Des.*, **2009**, *15*(6), 585-6.

[112] Cooke, R.M.; Dalvit, C.; Narula, S.S.; Wright, P.E. NMR studies of the heme pocket conformations of monomeric hemoglobins from Glycera dibranchiata. Implications for ligand binding. *Eur. J. Biochem.*, **1987**, *166*(2), 399-408.

[113] Gooptu, B., *et al.* Crystallographic and cellular characterisation of two mechanisms stabilising the native fold of alpha1-antitrypsin: implications for disease and drug design. *J. Mol. Biol.*, **2009**, *387*(4), 857-68.

[114] Oh, M.; Joo, K.; Lee, J. Protein-binding site prediction based on three-dimensional protein modeling. *Proteins*, **2009**, *77* Suppl 9:152-6.

[115] Keil, M.; Exner, T.E.; Brickmann, J. Pattern recognition strategies for molecular surfaces: III. Binding site prediction with a neural network. *J. Comput. Chem.*, **2004**, *25*(6), 779-89.

[116] Exner, T.E.; Keil, M.; Brickmann, J. Pattern recognition strategies for molecular surfaces. II. Surface complementarity. *J. Comput. Chem.*, **2002**, *23*(12), 1188-97.

[117] Bradford, J.R.; Needham, C.J.; Bulpitt, A.J.; Westhead, D.R. Insights into protein-protein interfaces using a Bayesian network prediction method. *J. Mol. Biol.*, **2006**, *362*(2), 365-86.

[118] Bradford, J.R.; Westhead, D.R. Improved prediction of protein-protein binding sites using a support vector machines approach. *Bioinformatics*, **2005**, *21*(8), 1487-94.

[119] Liu, R.; Zhou, Y. Using Support Vector Machine Combined with Post-processing Procedure to Improve Prediction of Interface Residues in Transient Complexes. *Protein J*, **2009**, *28*, 369-374.

[120] Chen, H.; Zhou, H.X. Prediction of interface residues in protein-protein complexes by a consensus neural network method: test against NMR data. *Proteins*, **2005**, *61*(1), 21-35.

[121] Neuvirth, H.; Raz, R.; Schreiber, G. ProMate: a structure based prediction program to identify the location of protein-protein binding sites. *J. Mol. Biol.*, **2004**, *338*(1), 181-99.

[122] Taroni, C.; Jones, S.; Thornton, J.M. Analysis and prediction of carbohydrate binding sites. *Protein Eng.*, **2000**, *13*(2), 89-98.

[123] Giard, J.; Ambroise, J.; Gala, J.L.; Macq, B. Regression applied to protein binding site prediction and comparison with classification. *BMC Bioinformatics*, **2009**, *10*, 276.

[124] Cummins, P.L.; Gready, J.E. Computer-aided drug design: a free energy perturbation study on the binding of methyl-substituted pterins and N5-deazapterins to dihydrofolate reductase. *J. Comput. Aided Mol. Des.*, **1993**, *7*(5), 535-55.

[125] Yang, A.Y.; Kallblad, P.; Mancera, R.L. Molecular modelling prediction of ligand binding site flexibility. *J. Comput. Aided Mol. Des*, **2004**, *18*(4), 235-50.

[126] Huang, N.; Kalyanaraman, C.; Bernacki, K.; Jacobson, M.P. Molecular mechanics methods for predicting protein-ligand binding. *Phys. Chem. Chem. Phys.*, **2006**, *8*(44), 5166-77.

[127] Glazer, D.S.; Radmer, R.J.; Altman, R.B. Combining molecular dynamics and machine learning to improve protein function recognition. *Pac. Symp. Biocomput.*, **2008**, 332-43.

[128] Liu, B.; Wang, X.; Lin, L.; Dong, Q.; Wang, X. Exploiting three kinds of interface propensities to identify protein binding sites. *Comput. Biol. Chem.*, **2009**, *33*(4), 303-11.

[129] Ward, R.M.; Venner, E.; Daines, B.; Murray, S.; Erdin, S.; Kristensen, D.M.; Lichtarge, O. Evolutionary Trace Annotation Server: automated enzyme function prediction in protein structures using 3D templates. *Bioinformatics*, **2009**, *25*(11),1426-7.

[130] Sheng, C., *et al.* Evolutionary trace analysis of CYP51 family: implication for site-directed mutagenesis and novel antifungal drug design. *J. Mol. Model.* **2010** *16,* 279-84.

[131] Nagarajan, S.; Marimuthu, P. Binding site prediction of galanin peptide using evolutionary trace method. *Bioinformation*, **2006**, *1*(5), 180-3.

[132] Mohamad, S.B.; Ong, A.L.; Ripen, A.M. Evolutionary trace analysis at the ligand binding site of laccase. *Bioinformation*, **2008**, *2*(9), 369-72.

[133] Song, Y., *et al.* Evolutionary trace analysis of eukaryotic DNA topoisomerase I superfamily: identification of novel antitumor drug binding site. *Sci. China C. Life. Sci.*, **2005**, *48*(4), 375-84.

[134] Dong, Q.; Wang, X.; Lin, L.; Guan, Y. Exploiting residue-level and profile-level interface propensities for usage in binding sites prediction of proteins. *BMC Bioinformatics*, **2007**, *8*, 147.

[135] Joughin, B.A.; Tidor, B.; Yaffe, M.B. A computational method for the analysis and prediction of protein:phosphopeptide-binding sites. *Protein Sci.*, **2005**, *14*(1), 131-9.

[136] Nissink, J.W.; Taylor, R. Combined use of physicochemical data and small-molecule crystallographic contact propensities to predict interactions in protein binding sites. *Org. Biomol. Chem.*, **2004**, *2*(22), 3238-49.

[137] Negi, S.S.; Schein, C.H.; Oezguen, N.; Power, T.D.; Braun, W. InterProSurf: a web server for predicting interacting sites on protein surfaces. *Bioinformatics*, **2007**, *23*(24), 3397-9.

CHAPTER 6

Receptor Flexibility in Ligand Docking and Virtual Screening

Maria A. Miteva[1], Charles H. Robert[2], Jean Didier Maréchal[3] and David Perahia[4,*]

[1]*MTI, Inserm UMR-S 973, University Paris Diderot, 35 Rue Helene Brion 75013 Paris, France;* [2]*CNRS Laboratoire de Biochimie Théorique, IBPC, 13 rue Pierre et Marie Curie, 75005 Paris, France;* [3]*Departament de Química, Universitat Autònoma de Barcelona, 08193Bellaterra, Spain and* [4] *Institut de Biochimie et Biophysique Moléculaire et Cellulaire, Université Paris-Sud, Bât 430, 91405 Orsay, France; Fax: +33-1-69 85 37 15 ; david.perahia@u-psud.fr*

Abstract: It is now well recognized that receptor flexibility plays an important role in protein-ligand binding. This flexibility can concern not only the mobility and reorganization of side chains in the binding site but also conformational changes of the whole molecule, which makes modeling of ligand docking more challenging. We give here an overview of existing approaches to treating receptor flexibility in protein-ligand docking and virtual screening that range from approximate to more accurate methods, including the use of normal modes in accounting for global conformational changes of the receptor or the use of more precise force fields. In addition, we describe recent successful applications of such approaches that have led to the design or discovery of new lead compounds with therapeutic relevance. These new developments in protein-ligand docking and screening are increasingly applied for better prediction of binding affinities. In view of the ever-increasing power of machine computation, a better accounting of the flexibility of the receptor is justified in order to improve the prediction of ligand binding in the search for new drug candidates by virtual screening.

INTRODUCTION

Protein flexibility is crucial for all cell processes and has been extensively explored during the last decades. A protein's movements are required for most of its functions: folding, assembly, enzyme catalyzed reactions (substrate binding or product desorbing), changes between active and non-active conformations triggered by agonist, antagonist or allosteric ligand binding, transport across membranes (channels opening and closing for various ligands), etc. [1, 2]. Motions in macromolecules can occur over picoseconds to hours and with amplitudes as large as 1.5 nm. The relation between these motions and their functions has been investigated using various structural (NMR, X-ray etc.), biophysical and biochemical methods [3].

Studying protein flexibility has a considerable impact on drug design projects in order to understand how drugs act in the body: interactions with the target, their binding mode and site, their metabolism, pharmacokinetics and pharmacodynamics [4]. The growing amount of structural information at the atomic level has demonstrated that a large number of pharmaceutically relevant targets undergo conformational changes upon small ligand binding [1, 5-8]. Hence, handling protein flexibility would allow structure-based drug design to expand the conformational and chemical space in which potentially active molecules could be found. Although an enormous number of degrees of freedom must be considered [2], dealing with protein flexibility is an essential issue in current structure-based drug design and virtual screening enterprises [8-11]. During the last decade, tremendous effort has been devoted to the design of small drug-like compounds aimed at inhibiting flexible proteins in the treatment of various diseases. In this context numerous examples can be given. Among these one can find anti-cancer targets: thymidylate synthase [12-16], p53 [17]; anti-pathogenic infection targets: dihydrofolate reductase [18]; anti-AIDS targets [19-21], anti-viral Influenza targets [22, 23], targets for anti-tropical diseases: RNA-editing ligase [24], etc.

PROTEIN-LIGAND BINDING

Through the years different mechanisms have been suggested to describe binding of a small ligand to a protein receptor. In the earliest view, Fisher's model [25], the ligand was supposed to fit into the receptor like a key into a lock. A more accurate view [26], the so called "induced-fit" idea, suggested that the structures of the ligand and receptor adapt to each other upon binding. The modern understanding of proteins as dynamic macromolecules existing in a statistical ensemble at thermal equilibrium [2, 27, 28] offers the possibility for alternative approaches to deal with protein receptor flexibility, many introduced below, for molecular docking and virtual screening. Changes in protein structure due to ligand binding can be provoked by side chain, loop or domain movements [1, 29]. In Fig. **1** can be seen examples for three different cases of important therapeutic targets undergoing such conformational changes upon small ligand binding. The changes occurring in the active site of the cyclin dependent kinase CDK2 (an anti-cancer target) due to binding of various inhibitors is exemplified by two X-ray crystal structures of ligand-protein complexes (Fig. **1A**). In this case, domain movement occurs close to the binding pocket triggered by interaction with the inhibitor. Fig. **1B** shows an example of structural differences at the interface of the protein-protein complex MDM2-p53 (involved in cancer diseases) when a small ligand binds at the protein-protein interaction (PPI) region. The changes in the MDM2 binding site when bound to a small inhibitor or a protein partner can be seen in Fig. **1B**. PPIs represent a highly populated class of essentially untouched potential targets for drug discovery [30, 31]. Small molecules blocking protein-protein complexes have enormous potential for future drug discovery in order to target regions outside of the catalytic sites (review [32-34]. In the case of PPI, the receptor flexibility is even more challenging because of the difficulty in defining accurately druggable pockets.

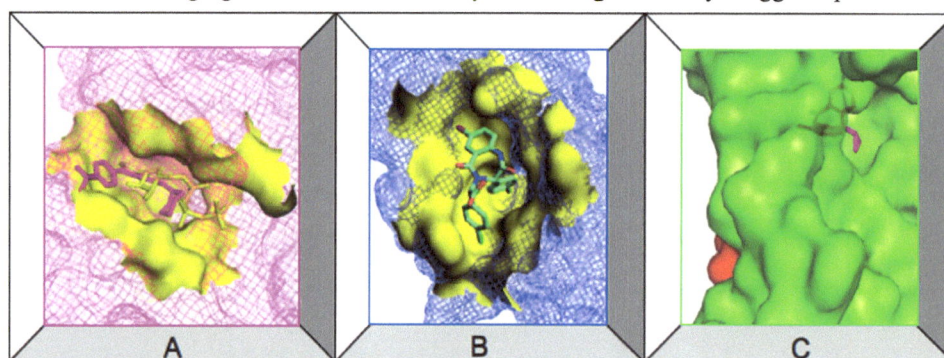

Figure 1: Conformational changes for therapeutic protein targets upon ligand binding. A.Two CDK2 X-ray structures with co-crystallized inhibitors: codes pdb 1fin (yellow) and 2c6t (magenta); B.Two X-ray structures of MDM2: in complex with a benzodiazepine inhibitor (code pdb 1t4e, solid surface in yellow, the inhibitor in sticks) and in complex with a p53 peptide (pdb code 1t4f, mesh surface in blue). C. Alosteric inhibitors: nevirapine (in sticks) bound into HIV-RT (surface in green; the active site in red; pdb code 1vrt).

In the context of receptor plasticity, other interesting examples are allosteric inhibitors, where binding can be remote from the active site (or PPI) [35-37]. For example, the binding of nevirapine (a member of an important class of non-nucleosidic inhibitors) to HIV-1 reverse transcriptase (HIV-RT) occurs at a site 10 Å distant from the active site [38] (Fig. **1C**).

The examples provided here, together with the results of numerous other studies, have demonstrated the impact of target flexibility on the docking of small ligands as well on the identification of new binders by virtual screening protocols [5, 6, 29, 39-45]. Small movements in the receptor structure can lead to considerable errors in the binding energy estimation [46]. However, diversity in ligand binding mechanisms and receptor conformational changes makes it difficult to handle receptor conformational changes during the docking simulation, especially when screening large collections of chemical compounds. In addition, it has been shown that incorrect consideration of receptor flexibility can worsen the virtual screening performance by increasing the number of false positives [47-49].

APPROACHES FOR PROTEIN FLEXIBILITY AND LIGAND BINDING PREDICTION

Receptor flexibility should be taken into account in order to generate energetically acceptable poses of a ligand when searching or building new drug candidates towards a protein target. Either local or global conformational protein changes can affect the binding site, the former consisting mostly of side-chain motions and the latter global motions consisting of large amplitude loop or domain motions. These two kinds of motions influence the shape and volume of the binding pocket as well as the location of atomic groups that may potentially interact with the ligand. When the binding site is located at the interface between two domains, the relative motions of these domains can critically affect the binding site (an example can be seen in Fig. **1A**). Depending on which kinds of changes are to be considered, different methods can be employed: local changes are better studied by applying Monte Carlo (MC) or Molecular Dynamics simulations (MD), global changes by carrying out normal mode analysis (NMA) or large scale Molecular Dynamics (though this still requires a considerable computational power). Hybrid methods using MD or NMA with molecular docking can constitute a powerful approach for generating relevant ligand poses.

Molecular Dynamics

Molecular dynamics consists in resolving the Newton Equation of motion, $F_i = M_i \gamma_i$, for all the atoms; here F_i is the force applied to atom i, M_i its mass, and γ_i the acceleration which depends on the potential energy and therefore on the 3D structure considered. Many simulation programs exist to perform MD simulations, the commonly used ones are CHARMM [50], AMBER [51], GROMACS [52] and NAMD [53]. Simulations in vacuum may be sufficient to explore movements and thus to generate an ensemble of different conformations. Implicit solvent models such as the Generalized Born (GB) method to account for the solvation of polar groups and the accessible surface area (SA) approximation for non-polar solvation, provide a more correct solvent approximation for estimating protein flexibility than a simple coulomb potential, although they are computationally more demanding. The simulations are usually carried out on an experimental structure (X-ray or NMR) of a receptor which is not complexed to a ligand in order to avoid biases arising from a specific bound ligand and to generate a large variety of open binding site conformers. These conformers can be thus used in subsequent docking simulations. It has been shown that using such multiple receptor conformations in ligand docking and virtual screening increases the probability of finding good ligand poses in the binding pocket [54, 55].

Monte Carlo Methods

Monte Carlo (MC) methods can be used to generate protein conformations by sampling a statistically consistent ensemble at a given temperature. Random perturbations of the atomic positions or other variables are applied in order to explore the conformational space of the molecular system [56]. The energy is evaluated for each newly generated conformation, which is accepted either if its energy is lower than the one from the previous step or, if higher, is within an energy range defined by the so-called Boltzmann factor and Metropolis criteria [57]. The classical MC Metropolis sampling has been improved using the simulated annealing concept (Monte Carlo simulated annealing, MCSA or MCA [58, 59] among others (J-walking, multicanonical MC, configurational biased MC method, or grand canonical MC simulations) in different programs like MMC (http://fulcrum.physbio. mssm.edu/<mezei/mmc/), ISIM (http://mccammon.ucsd.edu/isim/), or BIGMAC (http://molsim.chem.uva.nl/bigmac/).

To treat receptor flexibility for ligand binding the program ICM (internal coordinate mechanics) (www.molsoft.com) [5] employs the ICM-flexible receptor docking algorithm (IFREDA) via a Monte Carlo simulation based on energy function optimization of a flexible ligand in a receptor with flexible side-chains and/or flexible main-chains. The algorithm consists of a random change of two types, torsional or positional, followed by a local minimization. A biased probability methodology is applied for torsion moves of the amino-acid side chains where the biased probabilities are derived from known protein structures (*e.g.* amino acid dihedral angles).

MC methods have been applied in a number of studies to deal with ligand binding in a flexible receptor. Caflisch *et al.* [60] have predicted the binding mode of a tetrapeptide to the FK506 binding protein (FKBP) using a standard MC method incorporating random displacements of the substrate with minimization of the complex and employing a rapid method for calculating polar and nonpolar solvation effects by finite difference Poisson-Boltzmann method as well as soft Lennard-Jones and coulombic potentials. In another study, the impact of receptor flexibility has been assessed by using MC simulation and the Generalized Born/surface area (GB/SA) continuum solvent model on various protein-ligand complexes with and without side-chain movements [61]. The results suggested that including protein flexibility may lead to a strong energy landscape and less distinguishable multiple minima.

Recently, the new approach ROSETTALIGAND has been described [62] for accounting the protein flexibility. ROSETTALIGAND explicitly models full side-chain, backbone, and ligand flexibility with ligand and receptor degrees of freedom explored simultaneously. The method employs a Monte Carlo sampling and the Rosetta full atom energy function.

Genetic Algorithms

Genetic Algorithm (GA) methods mimic the evolution process by considering a set of individuals having a chromosome in which each gene corresponds to a given conformational state (*i.e.* a gene describing a given rotamer for a side chain). Successive generations are produced by mutating the genes randomly and performing cross-overs between the chromosomes of different individuals. The chromosomes for which the corresponding structures are energetically the most favored can also be duplicated. The next generation of individuals that are generated by applying the various GA operators are composed of those presenting the most preferred energies, but also those that are moderately less preferred in view of the consideration that they may evolve to preferred conformations in next generations. A large number of generations are needed to converge to reliable structures. Such a technique allows a more powerful exploration of ligand poses that a conventional MD simulation [63]. GAs are employed in many docking simulation methods such as AutoDock [64] and GOLD [65] to account for flexibility of the ligand as well as side-chains.

There are several programs that use genetic algorithms for protein-ligand docking. AutoDock is open source and one of the most popular packages. Since its 4.0 release, AutoDock incorporates an approach for dealing with local receptor flexibility [66]. For selected residues, all possible rotations can be considered in the conformational search. The side-chain conformational search is similar to the conformational space search of the ligand via a Lamarckian genetic algorithm. AutoDock pre-calculates interaction points for the protein-ligand complementarity before placing the ligand into the binding pocket. The flexible residues are separated from the rest of the macromolecule and the interaction grid is generated on the remaining structure. GOLD also offers consideration of receptor flexibility. Initially, only a very limited degree of receptor flexibility was provided that corresponded to optimization of the hydrogen bond network when performing the docking. In the latest releases of the program, the conformational space associated with the flexibility of a given number of residues is based on a library of known rotamers. All scoring functions implemented in GOLD are now compatible with this scheme. The recent program FITTED [67] employs a genetic algorithm to sample different combinations of side-chain rotamers and backbone conformations.

Normal Mode Analysis

The best approaches for accounting for global changes in proteins are Normal Mode and Principal Component analysis. Normal Mode analysis (NMA) allows the exploration of large-scale changes of the target protein [43]. It is based on calculation of the intrinsic vibrations of a molecule. It has been shown that the lowest frequency vibrations correspond to movements that involve large parts of the molecule, *e.g.* domain motions in multi-domains proteins. In general, the higher the frequency the more localized the motion is. For example, different kinds of loop motions may be studied by analyzing relatively higher frequency vibrations. As such, NMA allows a large variety of motions to be characterized and ultimately retained for subsequent docking simulations. The modes are obtained by diagonalizing the Hessian matrix, which is the matrix of the second derivatives of the potential function. The eigenvalues are related to frequencies, and the eigenvectors to the associated direction of motions of the atoms. As the dimension of the Hessian matrix can be quite large (3Nx3N, N being the total number of atoms), new approaches have been developed using simplified potentials in which Calpha atoms are connected by harmonic springs, usually defined for pairs of atoms within a given cutoff distance of each other. This approach is called the ENM (Elastic Normal Modes) method [68]. It has been shown that such a simplified representation provides global motions that are similar to those computed with an all atoms using a complete physical force field. However for cases where the structure is relatively compact ENM can fail to deliver closed to open conformational changes, making it necessary to consider all-atom NMA calculations using physical force fields. The DIMB method implemented in CHARMM allows the computation of all-atom NMA for large macromolecules without a requirement of a large computer memory [69].

The consideration of global motions has shown an important impact in finding good ligand poses via docking [42]. Modes have been used to generate multiple receptor conformers in various docking simulations [45, 70-74] showing the usefulness of this technique. Using all-atom NMA we obtained a very good exploration of conformational changes occurring in the active site of CDK2 compared to the experimentally observed ones in numerous X-ray CDK2 structures bound or not to different inhibitors (Fig. **2**). Recently, it has been suggested that the binding site displacements for best positioning of the ligand may in some cases be independent of the global collective motions of proteins. Based on this observation it was proposed to take into consideration only elastic network normal modes in a zone of ~10-15 Å around the binding pocket [44].

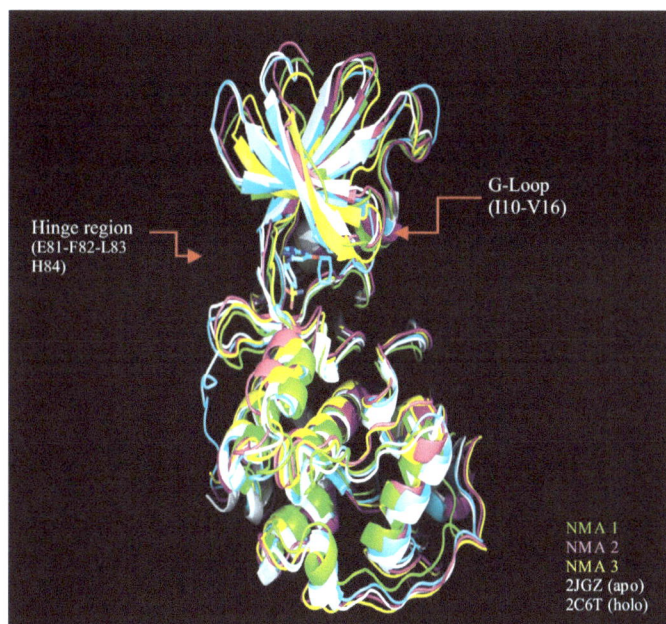

Figure 2: Superimposition of the three best conformations of CDK2 found via NMA showing a good exploration of the conformational changes of the binding site compared to the changes observed in several CDK2 X-ray structures; Two X-ray structures of an apo- (code PDB 2JGZ, in white) and a holo- (code PDB 2C6T in cyan) [45] structure of CDK2 are also shown.

Principal Component Analysis

Another approach which has been used for generating multiple conformers of a protein is to calculate the preferential large-scale directions of motions obtained directly from the frames of an MD trajectory [75]. Such an approach uses Principal Component Analysis (PCA) and considers the mean correlation matrix between the movements of atoms. The diagonalization of such a matrix furnishes the preferential correlated motion of atoms, of which those corresponding to the largest eigenvalues involve mostly the motion of large parts of the protein. The PCA approach has been used in several docking studies to account for large-scale changes in the target protein [76-78].

Hybrid Approaches

Several hybrid methods have also been reported to be appropriate for studying protein-ligand complex conformations, which mainly involve MD along with other techniques. Multiple Copy Simultaneous Search (MCSS) approach [79] combines MD with Locally Enhanced Sampling [80] and has been applied for *de novo* design of novel molecules constructed in a binding site. MCSS samples low-energy conformations of the target protein and functional groups of a ligand that can be subsequently linked to build a novel molecule.

CDOCKER [81] is a hybrid method using MD and grid-based docking. In general, CDOCKER generates ligand "seeds" to populate the binding pocket. Each seed is then subjected to high temperature MD using a modified version of CHARMM. This approach has been used for assessing flexible protein-ligand docking in a number of studies [63, 82] where it has been shown that it can significantly improve predicted binding modes compared to those achieved with rigid receptor docking methods.

Lately SDOCKER [83] has been developed to combine classical force field energies by MD with a 3D similarity search, by means of the correspondence between the ligand pose and a predefined template derived from the protein-ligand complex crystal structures. This method has been validated on a number of known structures of ligands bound to protein targets (thrombin, CDK2 and HIV-1 protease), which has shown that it provides improved docking accuracy in comparison to the methods based only on a classical energy treatment.

Other approaches that combine MD with various other methods are emerging, such as applying MD, docking, and QM/MM methods [84] or molecular dynamics with Quantum-Refined Force-Field [85].

STRATEGIES TO HANDLE RECEPTOR FLEXIBILITY FOR VIRTUAL SCREENING

In the last decade different strategies have been suggested in order to consider receptor flexibility for structure-based virtual screening in identifying new potential binders (drug candidates or chemical probes for exploration of protein functions) among a large number of small molecules. A commonly used protocol consists of first using rapid but less precise methods for finding compounds susceptible to bind into a single druggable pocket, and in a second step carrying out more precise calculations using detailed physical force fields and free energy calculation for a better estimation of binding affinities [86, 87], as well as induced-fit predictions [88, 89]. Further, many studies have shown that a successful strategy can be to generate multiple conformers of the protein target displaying different configurations for the binding site [90]. Fig. **3** shows commonly used techniques to deal with receptor flexibility [29] for predicting ligand binding modes via docking and for carrying out virtual screening experiments.

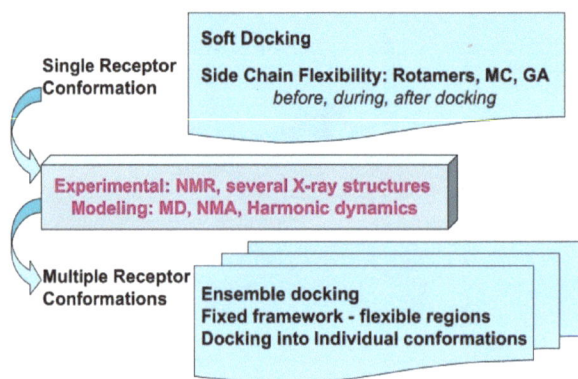

Figure 3: Approaches dealing with receptor flexibility for docking and virtual screening.

One simple approach to accommodating receptor flexibility problem when many ligands are under consideration is the development of "softer" energy potentials [55, 91] or the inclusion of rotamer libraries [92, 93]. For example, a soft potential incorporating alternative receptor side-chain conformations has been shown to improve the scoring of generated poses and to provide a remarkably good correlation with experimentally-determined free energy changes for ligand binding [94].

Several docking programs applied to virtual screening can take into consideration side-chain receptor flexibility. A partially flexible protein can be treated with the docking program SLIDE (http://www.bch.msu.edu/labs/kuhn/web/index.html) [95, 96]. This program is based on an iterative matching of interaction centers within the receptor and interaction points within the ligand. The receptor site is analyzed in terms of interaction points: hydrogen-bond donor, hydrogen-bond acceptor and hydrophobic. Steric hindrance between the protein and the ligand anchor fragment is resolved by using rigid body translations, and then the ligand is optimized by rotating all single bonds. This includes some side-chain flexibility, as well as water molecule displacements. The generated poses are evaluated with an empirical scoring accounting for hydrogen bonds and hydrophobic contacts. Side-chain flexibility can also be treated with the programs AutoDock and GOLD mentioned above.

However, while side-chain conformational imprecision may be taken into account in these methods, backbone flexibility cannot be so straightforwardly described [11, 97]. The simplest goal of more "realistic" modeling is to generate conformational diversity in the receptor, whether free or in the presence of the ligand. Backbone and side chain degrees of freedom, collective coordinates such as normal modes mentioned above, positions and orientations of solvating water molecules, etc., can all be explored by Monte Carlo approaches or by molecular dynamics/harmonic dynamics [71] simulations. One advantage of MD is that continuous changes can be made in atom coordinates, allowing small but critical adjustments during the simulation. Other methods of generating diverse ensembles also exist. For example, graph-theoretical approaches such as FIRST [98] can also be used [99] to generate diverse structural sampling comparable to that obtained by MD.

The conformers sampled using these techniques can each be used for individual docking trials, with a corresponding multiplication of the computational time required for the docking studies. Docking on individual multiple receptor conformations is based on the assumption that proteins exist in an equilibrated ensemble of structures, and ligand binding involves an equilibrium shift toward the structure of the protein-ligand complex [36, 100]. Clustering procedures can be applied to decrease the number of receptor conformations, and thus the computational cost. Such a strategy is employed for example in the Relaxed Complex Scheme (RCS) method [101, 102] (shown in Fig. **4**) demonstrated to be very efficient when applied to identify novel hits for important therapeutic targets [22, 24]. Docking on multiple receptor conformations has been accepted today as an attractive alternative [49] to the induced-fit approximations and has been successfully applied in virtual screening experiments [23, 103, 104]. It has been demonstrated that the use of multiple structures of the target protein, either experimental (multiple crystal structures [47, 105, 106], NMR ensembles [41, 107, 108], etc.) or modeled, increases the chances for finding a larger set of drug like compounds having an appreciable affinity for the protein considered [54].

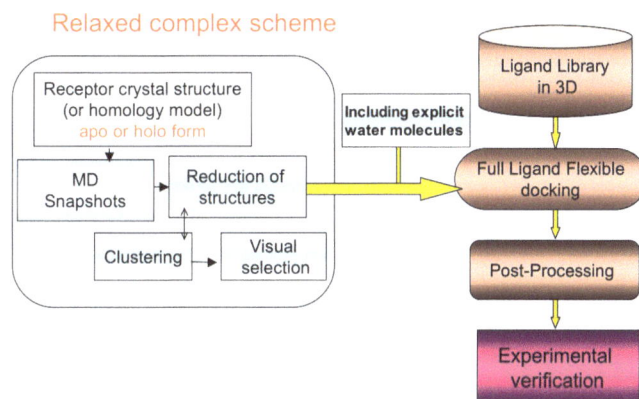

Figure 4: Multiple receptor conformations by molecular dynamics simulations: Relaxed complex scheme [102].

Docking to conformers generated for the free receptor rely on the appearance of bound or bound-like conformations in the free receptor ensembles even in the absence of ligand: conformations that only occur with reasonable probability when ligand is present may thus remain unsampled. The two views of conformational change on binding, via conformer selection or via induced fit by the ligand, have been amply addressed elsewhere in the literature since their inception over 40 years ago (*e.g.*, see references in [11]). The use of both free (apo) and "bound" (holo) free-receptor conformations as targets, the latter consisting of a receptor conformation obtained from a complex with a different ligand, have also been explored and compared [6]. While certain biases can be reduced using the bound conformation, the value of this approach is diminished as more dissimilar ligands are considered. On the other hand, docking to such forms, bound or unbound, has been shown to be improved by exploration of collective coordinates in the free receptor target [42, 44].

Alternatively, averaging can be used in creating a single target for docking trials [109], *i.e.* by grid averaging [54, 110, 111]. Multiple modeled or experimental structures may be used for this purpose [54, 107]. Huang and Zou employed ensemble docking on NMR [107] or multiple crystal protein structures [106] by simultaneously docking a ligand into the receptor structure ensemble and selecting an optimal protein structure that best fits the ligand. Such approaches must be used with care in order to avoid artefacts due to unrealistic average geometries; other properties besides geometry can also be averaged [11, 54, 107]. In this context, in a recent study it has been suggested to select receptor structures among MD or NMR ensembles showing a preservation of the geometry of the binding site [112].

In the same context, FlexE [113], an extension of the widely used FlexX program (http://www.biosolveit.de/) (http://www.tripos.com/), treats the receptor flexibility through discrete alternative conformations of selected parts of the protein taken from structures that have a very similar backbone trace, which are merged in a combinatorial way and considered directly during the rigid ligand docking. The method superimposes multiple receptor structures to represent a flexible binding site. The similar parts are merged whereas the rests are treated as separate alternatives. Assessment of this method showed comparable success to that of the sequential docking experiments.

Recently, an original approach employing a four-dimensional (4D) docking method has been reported which incorporates a receptor conformational ensemble in a single docking simulation [114]. The receptor flexibility is represented as the fourth discrete dimension of the small molecule conformational space, with multiple 3D grids recomputed from optimally superimposed conformers. The separate receptor conformer entities are expressed as conformational variants of the same object. In this way, the ligand would move from one 4D "plane" to another, thus, switching between the different receptor conformations via a Monte Carlo procedure.

RECEPTOR FLEXIBILITY AND BINDING FREE ENERGY PREDICTION

Most current high-throughput docking methods neglect detailed aspects of the ligand-receptor interaction. Even when receptor flexibility is taken into account in initial docking trials to some degree, inadequate receptor structures may result in poor correlation of docking scores with experimentally determined binding affinities. Inadequacies may include errors in the coordinate data due to bias from existing ligands or from crystal disorder. Modelling of conformational changes creates inaccuracies when only discrete regions of the conformational space are sampled. The use of rotamer libraries for modelling flexibility in a protein binding site, for example, leads to low precision in side-chain atom placements and entails a corresponding risk of false negatives in docking studies [115]. Such is the price to pay for the speed of ranking a large number of compounds and poses. However, once the search for acceptable compounds has narrowed, more detailed modelling of ligand-receptor interactions can be justified. Monte Carlo and Molecular Dynamics simulations are often introduced at this point in order to refine promising candidates.

The binding of ligand to a receptor is characterized by its affinity, which expresses the free energy of association of the two components. Experimental values of the affinity are most commonly determined using techniques such as isothermal titration calorimetry (ITC), whose role in docking studies has been reviewed recently [116]. Databases such as the BindingDB [117] compile known receptor-ligand binding affinities from the literature (currently more than 30,000 molecules in BindingDB) along with links to related structural information, and can be used in assessing binding free energies predicted using the methods described in this section. MD calculations in particular permit realistic simulations of the receptor-ligand-water system under given thermodynamic conditions, which can then be used for accurate prediction of the free energy of association between the protein and the ligand. Rigorous procedures for calculating binding affinities using MD simulations are well established, along with approximations that have been introduced in order to reduce computational effort, and have been reviewed elsewhere [118-120]. We describe these approaches here briefly in order to describe recent progress.

Pathway Approaches

Current free energy calculation methodologies can be divided into pathway and endpoint approaches. Pathway approaches are used to calculate the change of the free energy along a given path, which can correspond to a physical or a non-physical process. In the potential of mean force method [121, 122], the path is a physical one, in which the ligand is taken from the unbound to the bound state, or more typically *vice versa*. In the alchemical approaches a non-physical path is used, which corresponds to the decoupling of a ligand from its surroundings (the receptor or the bulk solvent) via a coupling parameter that "dials in" the new potential and dials out the old. In both approaches, the calculated free energy change, if sufficiently converged, will be independent of the path used to calculate it.

PMF APPROACH

The standard expression for the free-energy change for ligand binding to a receptor in statistical mechanics involves integration of the Boltzmann factor, written in terms of the potential energy, over the entire configurational space. Alternatively, this integral can be expressed in terms of a potential of mean force (PMF) [121-123] acting on the ligand as a function of distance along the binding coordinate. The PMF can be estimated by the use of umbrella sampling during a series of MD simulations, which involves the use of artificial restraining potentials to maintain the ligand-receptor distance at intermediate values in order to enable proper configurational sampling. The simulations corresponding to different intermediate distances are then typically combined using the WHAM approach to calculate the free energy change (*e.g.*, [124]). More detailed descriptions of the method can be found in the references [121, 122, 125]. The principal advantage of the PMF approach is that an absolute binding affinity is calculated, although the structure of the bound complex must generally be known in advance and the binding reaction coordinate chosen carefully.

Applications of this method include the calculation of the affinity of the phosphotyrosine peptide pYEEI for a human SH2 domain in explicit solvent [121] in good agreement with the experimental value. A related approach using an implicit solvent model allowed description of a consensus path for thymol binding to a homology model of an odorant binding protein [126] together with an estimation of the binding free energy change, again providing good agreement with the value obtained by titration microcalorimetry. Another recent explicit-solvent study used a series of one-dimensional PMFs to provide good estimates of the affinity of benzamidine for trypsin as well as information regarding the binding path [122]. The PMF approach has also been modified to accommodate repeated series of non-equilibrium unbinding simulations [125, 127]. Such approaches may be of somewhat lower accuracy but have the advantage of being compatible with highly parallel operation for computational efficiency [125].

ALCHEMICAL APPROACHES

A second approach to estimating the free energy change involves the use of a non-physical coordinate, which generally corresponds to the decoupling of a ligand from its surroundings (the receptor or the bulk solvent) via a coupling parameter (λ). This approach is often used to calculate the difference in free-energy changes, $\Delta\Delta G$, for related ligands: the coupling parameter linearly scales the potential from that of one ligand to the other. It is also possible to use the scaling parameter to "annihilate" the ligand in the two environments to obtain absolute free energy change. In either case, thermodynamic integration or free energy perturbation permit calculation of the overall free energy change for the transformation [128]. The procedure must be carried out twice (two sets of simulations): once for the ligand(s) in the binding site of the protein and once free in solvent. The relevant thermodynamic cycle is shown in the following schema describing a relative free-energy calculation.

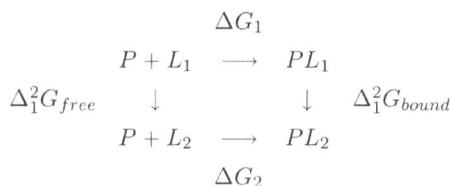

$$
\begin{array}{ccc}
 & \Delta G_1 & \\
P + L_1 & \longrightarrow & PL_1 \\
\Delta_1^2 G_{free} \quad \downarrow & & \downarrow \quad \Delta_1^2 G_{bound} \\
P + L_2 & \longrightarrow & PL_2 \\
 & \Delta G_2 &
\end{array}
$$

Figure 5: Thermodynamic cycle employed in alchemical free energy calculation methods.

The vertical pathway represented on the left side of this schema describes the alchemical transformation of ligand 1 to 2 in solution. The resulting free energy change is subtracted from that obtained for the corresponding transformation in the receptor binding site, in order to obtain $\Delta\Delta G_{1\rightarrow2}$ for the two ligands. The approach has the disadvantage of requiring long MD simulation, much of which is devoted to simulating non-physical configurations of the system (*e.g.*, mixtures of half-formed ligand 1 and 2). Further, the differences between the two ligands must be small for the calculations to converge properly [129].

Nevertheless, thermodynamic integration currently provides some of the highest accuracy theoretical estimates of (relative) binding free energies. It has been used to obtain $\Delta\Delta G$ values for pairs non-nucleosidic inhibitors of HIV protease [130] within about 1 kcal.mol^{-1} of the experimental values. The approach has also been used to predict binding properties of new members of the same class of molecules based on alchemical transformation from lead compounds [131]. A study of 16 structurally diverse putative inhibitors of the estrogen receptor α binding domain employed thermodynamic integration coupled with replica exchange methods [132]. This study also explored the use of a continuum solvent method based on the Generalized Born formalism (GBSA). The continuum solvent approach sped up the calculations with respect to explicit solvent but resulted in lower overall quality of the predicted affinities. Continuing efforts are also devoted to improving numerical aspects of the integration technique itself in order to achieve higher accuracy [133, 134].

A promising recent technique from the Dill group allows the calculation of absolute binding affinities by alchemical decoupling of a ligand from its binding site along with a corresponding procedure in solvent. The methodology was first applied in identifying new binders to a hydrophobic binding site in T4 lysozyme as model system, with good success in ranking binding affinities and in predicting the geometry of binding [135]. The method has also been applied in blind tests of prospective binders to the same protein, with a more complex binding site that subsumes the first and also includes polar groups from a Met $\rightarrow$ Gln mutation [136]. In comparing to experimental binding affinity data obtained for the different compounds by isothermal titration calorimetry and reported in the same study, these researchers found that their calculation resulted in 11 correct binding predictions out of 13 compounds and an RMS error in the free energy of association of less than 2 kcal.mol^{-1}. Crystal structures determined for the bound complexes permitted the assessment of RMSD positional errors that were less than 2 A in all cases.

Endpoint Approaches

The statistical thermodynamic definition of the binding free-energy change depends on properly taking into account the Boltzmann factors for all relevant configurational states of the receptor-ligand-water system. For tightly binding ligands, the average free energy difference can be expressed as the difference in average enthalpy and entropy contributions calculated over the bound and the free states of the receptor and ligand (see, *e.g.*, [120]). Such averages can be calculated from MD or MC simulations of these states without consideration of any particular binding pathway, thus giving rise to the so-called endpoint approaches.

LINEAR INTERACTION ENERGY APPROXIMATION

The Linear Interaction Energy (LIE) approach (*e.g.*, [137]) uses a truncated expansion of the solvation free energy change for the ligand in aqueous solution and in its binding site in the receptor. The method assumes that the free energy scales linearly with the electrostatic and van der Waals interactions between the ligand and its environment. These interactions are calculated from MD simulations of the free ligand in solvent and the (solvated) bound complex. The scaling parameters used for weighting the different contributions are not independent of the ligand, however, and must be adapted using data for known systems [128], although some studies suggests that more universal parameters can be used [138]. One recent application employed a continuum solvent model in a study of binding affinities of tubulin polymerization inhibitors [139]. In this work, known inhibitor binding free energies were reproduced to within 0.6 kcal.mol^{-1}, comparable to errors in thermodynamic integration or FEP protocols. Other applications include the rationalization of the experimentally-measured affinity changes arising from modification of alpha-substituted norstatine inhibitors of malarial aspartate protease [140].

MM-PB(GB)/SA APPROACH

The Molecular Mechanics - Poisson-Boltzmann Surface Area (MM-PBSA) approach also employs MD simulations of the two endpoint states, free and bound, to estimate enthalpy and entropy changes for the overall binding process [141]. A series of structures are extracted from MD simulations of the two states, and for each the enthalpy is estimated using molecular mechanics terms and the solvation energy using continuum solvent models such as the Poisson-Boltzmann or Generalized Born formulations. The averages are used in approximating the free energy change. Solute entropies can be included by using the harmonic approximation from Normal Mode calculations [141] or from quasiharmonic analysis of the MD simulations, similar to the PCA approaches described above [141, 142]. In many applications, further simplifications have been made with reasonable accuracy. The first is to neglect the explicit calculation of the entropy when $\Delta\Delta G$s are desired, as they may be assumed to cancel for similar systems. This assumption has been tested and shown to be valid in comparative mutational studies of the T-cell receptor $V\beta$ domain binding to staphyococcal enterotoxin (SEC) [143]. It may not be appropriate in all cases, however, in particular for small ligands [144]. A second simplifying assumption often used in MM-PBSA calculations is to perform simulations of the bound complex only, in order to approximately sample both free and bound conformational degrees of freedom. This approach has, too, been investigated and partially justified, at least when extensive simulation time is not possible, in light of the fact that statistical sampling error is significantly reduced with respect to separate simulations [145]. Nevertheless, caution concerning the use of single-simulation technique has been expressed in other applications when there are significant flexibility changes on binding [146].

Future Directions

Standard methodologies in MD simulation already furnish a degree of accuracy that, due to steady growth in computational power, will be increasingly exploited for lead compound identification. Yet active research is continuing into still more realistic treatments of the physical principles governing ligand binding to biological receptors that will no doubt provide even higher levels of accuracy to the prediction of binding affinities in the future. The role of water in ligand binding, for example, is currently not well understood. The effects of solvent are complex, and their properties can be very slowly convergent in explicit solvent simulations due to poor accessibility of the binding site [147, 148]. However, their neglect can lead to erroneous predictions of ligand orientation and binding affinity [147]. Other effects are also not yet well described by current forcefields. Ligand binding may be accompanied by a shift of the pKa of either protein or a ligand groups [149], as has been implicated, for instance, in inhibitor binding to the tRNA-guanine glycosylase target for bacillary dysentary (Shigellosis) [150]. The correct treatment of pKa shifts is a challenge to MD simulations. Aside from complications arising from charge redistribution, significant modification in the binding of hydrogen ion at constant pH formally requires simulation in the grand canonical ensemble, as has been described for the related problem of incorporating water equilibration into simulations [148]. Polarization changes in the ligand and in the receptor can strongly influence interaction energetics. One recent study incorporated polarization effects in the calculation of binding free energies using a decoupling approach for benzamidine and diazamidine to trypsin, with the resulting free energies predicted to within 0.5 kcal.mol^{-1} [151]. Khoruzhii *et al.* [152] also used a polarizable potential and thermodynamic integration to study the binding of five ligands in complex with three serine proteases. This study also explored the transferability of the potential to different systems, which will be essential to larger scale application. Both these studies demonstrated the necessity for correct treatment of polarization effects in order to achieve the highest accuracy affinity predictions in many important ligand-binding problems.

SUCCESS STORIES OF VIRTUAL SCREENING DEALING WITH RECEPTOR FLEXIBILITY

The availability of important computational power at relative low expense as well as recent successes in virtual screening employing molecular dynamics at different stages has changed the earlier practice of ignoring receptor flexibility in virtual screening experiments. In the following section, we give recent examples of successful applications of protein-ligand docking algorithms accounting for receptor conformational changes, with a focus on selecting striking evidence of the relevance of such approaches for identification of novel molecules. The studies cited here are only representative, and the number of successful examples in the literature is rapidly increasing.

We first detail a noticeable case of successful application of virtual screening employing molecular dynamics performed by Schames and coworkers [153]. In this work, the authors apply the "Relaxed Complex Scheme" [154] (see Fig. **4**) for identification of new inhibitors of the HIV-integrase (ref 2). This study served as a base for the development of a commercially available HIV-integrase inhibitor approved by the United States Food and Drug Administration in 2007. In this study, a 2 ns molecular dynamics simulation of HIV-integrase bound to the inhibitor 5CITEP was initially performed, followed by docking with AutoDock3 of the same ligand to 40 snapshots of the trajectory. Interestingly, these docking experiments revealed the presence of a second binding site in a close vicinity of the initial location of the ligand, whose "lifetime" was regulated by the motion of a 4 residue loop. This loop is unstructured in the X-ray structure. While the X-ray structure corresponds to a single closed conformation, the molecular dynamics results showed the possibility of open-close transitions. The impact of this extended cavity in drug design prospects has been studied. Compounds including a diphenil moiety as a core and two 5CITEP as substituents ("butterfly" compounds) have been tested that fit both cavities at once. The most active butterfly compounds identified *in silico* displayed noticeably higher predicted affinity to the target in the open binding site conformation and were therefore expected to have physiological relevance. These results led Merck Laboratories to the design of raltegravir (commercial name Isentress©), a compound that represents a major step forward in the fight against AIDS.

Recently, Amaro *et al.* [24, 102] discovered new and potent inhibitors of RNA-editing ligase in *Trypanosoma brucei*, the parasite responsible for African sleeping sickness. Using an improved version of the RCS protocol [102], screening of a dataset of ~ 1800 compounds from the U.S. National Cancer Institut (NCI) databank was performed on about 400 individual snaphots derived from 20 ns of molecular dynamics of the target. The docking stage was performed with AutoDock 4.0. The dynamics were performed on a crystal structure of a holo form of the protein. After verification by *in vitro* assays, 3 of the 7 best predicted binders displayed a low-micromolar affinity toward the RNA-editing ligase. The identification of these novel scaffolds as potential lead compounds opens possibilities for biomedical applications. A similar strategy has been applied to find novel antiviral compounds for Avian Influenza Neuraminidase [23].

Another interesting example is the study performed by Tang and coworkers [155]. The general framework is the design of small molecules able to inhibit an endopeptidase of the anaerobic bacterium *Clostridium botulinum*. Such molecules would have important therapeutic applications in the fight against botulism. The drug target is the Zinc endopeptidase of the botulinum Neurotoxin Serotype A. In a previous study, the authors identified an initial weak hit compound able to inhibit the endopeptidase with a K_i of about 15 µM. To carry out lead optimization, the authors performed multiple molecular dynamics simulations on the complex between the peptidase and the initial inhibitor. These simulations showed a vacancy close to the position *meta* of one of the phenyl substituents of the ligand. Because of the positive nature of the receptor residues surrounding this cavity, an additional hydroxyl group was added to *meta* position of the corresponding phenyl. Additional molecular dynamics simulation followed by synthesis and *in vitro* assay proved the correctness of the hypothesis.

An original approach combining a machine learning method, homology modeling, and multiple-conformation high-throughput docking has been applied for the discovery of *Yersinia* protein kinase A (YpkA) inhibitors [156]. YpkA is an essential virulence determinant in *Yersinia spp.*, which includes the causative agent of plague. The authors constructed homology models of YpkA based on the MAPK templates and further performed MD simulations to sample different protein conformations to account for protein flexibility. Finally, with an ensemble of protein structures and the kinase inhibitor-enriched library, multiple-conformation high-throughput docking was performed and a number of potent and selective inhibitors of YpkA was successfully identified. These first reported inhibitors of YpkA may facilitate studies of the pathogenic mechanism of YpkA and serve as a starting point for development of anti-plague drugs.

Novel and potent inhibitors of the enzyme plasmepsin II from *Plasmodium falciparum* have been recently identified [157]. *Plasmodium falciparum* is one of the parasites that causes malaria in human; plasmepsins are promising drug targets since they are key for the metabolism of these species. Taking full advantage of grid computation technologies, this work included an initial screening performed using a static x-ray structure. Post-docking refinement was then undertaken via the MM-PBSA approach. The automated protocol was called BEAR (binding estimation after refinement) and was based on the AMBER package allowing MM-PBSA and MM-GBSA calculations in the rescoring process. The initial 200 best hits were subsequently analyzed and found to be sufficiently represented by a subset of 30 compounds. This latter set of compounds was selected for *in vitro* assay and about half of them displayed nanomolar inhibition constants. This represents a promising step towards discovering new anti-malarial drugs.

One of cornerstones in the development of clinically efficient drugs is their ability to remain active at physiological concentrations. This activity is a function of a number of physiological events that are conventionally designed as Administration, Disposition, Metabolism, Excretion and Toxicity (ADME/Tox). Numerous experimental and theoretical approaches have been designed to predict as accurately and fast as possible the ADME/Tox profile of drug candidates. Drug metabolizing cytochromes P450 represent one of the most important family of enzymes involved in ADME/Tox. Human metabolizing P450 are known flexible proteins with low substrate specificity, and the understanding of the effect of induced fit in their molecular mechanism is crucial for efficient drug development. In this context, the study by Hritz *et al.* [158] represents a major step forward in both decoding the specificity of cytochromes P450 but also in allowing faster *in silico* drug metabolism prediction by virtual screening. Centered on Cytochrome P450 2D6 (CYP2D6), this study relied on the fact that drugs can be assembled in families as a function of their chemical similarities. Therefore, the investigators docked 5 drug family representatives to the unbound structure of CYP2D6 and performed short molecular dynamics on the resulting complexes. From the different dynamics trajectories, 2500 protein snapshots were selected and re-docking was performed for a full set of 65 ligands. The authors were able to reduce the conformational space of protein structures so that the best representative conformations for the docking of drug families were defined. These representatives could therefore be used for fast virtual screening approaches. This study represents an exploration of CYP2D6 specificity and how to connect such fundamental question to an applicative biomedical focus.

In addition to docking on protein ensembles, many examples exist demonstrating the power of undertaking dynamic pharmacophore models. In a recent study [103], considerable progress has been made from the simple ligand-based to a dynamic, receptor-based pharmacophore model of MDM2 (see Fig. **1B**). The complementary interactions with the binding site were studied using probe molecules that identified conserved regions of the protein. This is equivalent to identifying binding hotspots that are conserved throughout the protein dynamics. The models have been used to search the University of Michigan's Center for Chemical Genomics database of 35,000 commercially available compounds and to identify compounds which have features that fulfill the conserved binding pattern. Snapshots were taken every 100 ps from a 2 ns molecular dynamics simulation of human MDM2 bound to p53. This resulted in 21 structures for MDM2 for which a pharmacophore model was created by overlapping the snapshots. Probes from each structure were combined and clustered to give "consensus clusters" represented as a spherical pharmacophore element. The MDM2 pharmacophore models were screened against the database using MOE (Chemical Computing Group). This resulted in several attractive hits which went on for experimental testing. In the end, five nonpeptide, small-molecule inhibitors with new scaffolds disrupting the protein-protein human MDM2-p53 interactions were found. The most potent compound exhibited a K_i of 110 nM. Recently, a similar approach has been used to identify new class of inhibitors of HIV-1 protease [21].

In other work, rather than using MD simulations, the conformational states of the protein receptor have been taken from X-ray crystallographic structures [17] or an NMR ensemble [104].

CONCLUSIONS

The presented overview of methods handling protein receptor flexibility in a drug discovery context demonstrates the progress in current virtual screening techniques. Most often, ensembles of multiple receptor conformations, either modeled or experimental, are used in this respect. Still, receptor flexibility in drug discovery is a very challenging issue. While docking on multiple conformations may improve binding mode prediction, increase in the numbers of false positives in virtual screening computations is also possible. Thus, the choice of the most suitable receptor conformations is key to the success of virtual screening projects. Despite these difficulties, recent successful examples are presented in this chapter in which methods dealing with receptor flexibility have been applied to important therapeutic targets, resulting in promising hit compounds. Increasingly precise methods are applied as an option for predicting relative binding free energies in drug discovery with an acceptable level of accuracy, such as MM-PBSA/GBSA. More precise methods exist, like TI/FEP, however, these approaches are still computationally expensive and are not realistically available to virtual screening experiments today. One can only be optimistic about the future prospects for such methodologies.

ACKNOWLEDGEMENTS

Support from the French national research institutes Inserm and the CNRS are greatly appreciated, as well as that of the University Paris Diderot and University Paris-sud 11. J.-D.M. thanks the Spanish "Ministerio de Ciencia e Innovación" for funding through the project CTQ2008-06866-C02-01 and Consolider-Ingenio 2010 project CSD2007-00006.

REFERENCES

[1]   Teague, S.J. Implications of protein flexibility for drug discovery. *Nat. Rev. Drug Discov.* **2003**, *2*, 527-541.

[2]   Cavasotto, C.N.; Orry, A.J.; Abagyan, R. The challenge of considering receptor flexibility in ligand docking and virtual screening. *Curr. Comput-Aided Drug Design* **2005**, *1*, 423-440.

[3]   Falke, J.J. ENZYMOLOGY: A moving story. *Science* **2002**, *295*, 1480–1481.

[4]   Testa, B.; Vistoli, G.; Pedretti, A. Musings on ADME predictions and structure-activity relations. *Chem. Biodivers* **2005**, *2*, 1411-1427.

[5]   Cavasotto, C.N.; Abagyan, R.A. Protein flexibility in ligand docking and virtual screening to protein kinases. *J. Mol. Biol.* **2004**, *337*, 209-225.

[6]   Erickson, J.A.; Jalaie, M.; Robertson, D.H.; Lewis, R.A.; Vieth, M. Lessons in molecular recognition: the effects of ligand and protein flexibility on molecular docking accuracy. *J. Med. Chem.* **2004**, *47*, 45-55.

[7]   Alberts, I.L.; Todorov, N.P.; Dean, P.M. Receptor flexibility in de novo ligand design and docking. *J. Med. Chem.* **2005**, *48*, 6585-6596.

[8]   Cozzini, P.; Kellogg, G.E.; Spyrakis, F.; Abraham, D.J.; Costantino, G.; Emerson, A.; Fanelli, F.; Gohlke, H.; Kuhn, L.A.; Morris, G.M.; Orozco, M.; Pertinhez, T.A.; Rizzi, M.; Sotriffer, C.A. Target flexibility: an emerging consideration in drug discovery and design. *J. Med. Chem.* **2008**, *51*, 6237-6255.

[9]   McInnes, C. Virtual screening strategies in drug discovery. *Curr. Opin. Chem. Biol.* **2007**, *11*, 494-502.

[10] Cavasotto, C.N.; Singh, N. Docking and High Throughput Docking: Successes and the Challenge of Protein Flexibility. *Current Computer-Aided Drug Design* **2008**, *4*, 221-234.

[11] B-Rao, C.; Subramanian, J.; Sharma, S.D. Managing protein flexibility in docking and its applications. *Drug Discov. Today* **2009**, *14*, 394-400.

[12] Fritz, T.A.; Tondi, D.; Finer-Moore, J.S.; Costi, M.P.; Stroud, R.M. Predicting and harnessing protein flexibility in the design of species-specific inhibitors of thymidylate synthase. *Chem. Biol.* **2001**, *8*, 981-995.

[13] Sielecki, T.M.; Boylan, J.F.; Benfield, P.A.; Trainor, G.L. Cyclin-dependent kinase inhibitors: useful targets in cell cycle regulation. *J. Med. Chem.* **2000**, *43*, 1-18.

[14] Davies, T.G.; Bentley, J.; Arris, C.E.; Boyle, F.T.; Curtin, N.J.; Endicott, J.A.; Gibson, A.E.; Golding, B.T.; Griffin, R.J.; Hardcastle, I.R.; Jewsbury, P.; Johnson, L.N.; Mesguiche, V.; Newell, D.R.; Noble, M.E.; Tucker, J.A.; Wang, L.; Whitfield, H.J. Structure-based design of a potent purine-based cyclin-dependent kinase inhibitor. *Nat. Struct. Biol.* **2002**, *9*, 745-749.

[15] Hardcastle, I.R.; Arris, C.E.; Bentley, J.; Boyle, F.T.; Chen, Y.; Curtin, N.J.; Endicott, J.A.; Gibson, A.E.; Golding, B.T.; Griffin, R.J.; Jewsbury, P.; Menyerol, J.; Mesguiche, V.; Newell, D.R.; Noble, M.E.; Pratt, D.J.; Wang, L.Z.; Whitfield, H.J. N2-substituted O6-cyclohexylmethylguanine derivatives: potent inhibitors of cyclin-dependent kinases 1 and 2. *J. Med. Chem.* **2004**, *47*, 3710-3722.

[16] Frembgen-Kesner, T.; Elcock, A.H. Computational sampling of a cryptic drug binding site in a protein receptor: explicit solvent molecular dynamics and inhibitor docking to p38 MAP kinase. *J. Mol. Biol.* **2006**, *359*, 202-214.

[17] Bowman, A.L.; Lerner, M.G.; Carlson, H.A. Protein flexibility and species specificity in structure-based drug discovery: dihydrofolate reductase as a test system. *J. Am. Chem. Soc.* **2007**, *129*, 3634-3640.

[18] Birdsall, B.; Feeney, J.; Tendler, S.J.; Hammond, S.J.; Roberts, G.C. Dihydrofolate reductase: multiple conformations and alternative modes of substrate binding. **1989**, *7*, 2297-2305.

[19] Di Santo, R.; Costi, R.; Roux, A.; Artico, M.; Lavecchia, A.; Marinelli, L.; Novellino, E.; Palmisano, L.; Andreotti, M.; Amici, R.; Galluzzo, C.M.; Nencioni, L.; Palamara, A.T.; Pommier, Y.; Marchand, C. Novel bifunctional quinolonyl diketo acid derivatives as HIV-1 integrase inhibitors: design, synthesis, biological activities, and mechanism of action. *J. Med. Chem.* **2006**, *49*, 1939-1945.

[20] Hornak, V.; Simmerling, C. Targeting structural flexibility in HIV-1 protease inhibitor binding. *Drug Discov. Today* **2007**, *12*, 132-138.

[21] Damm, K.L.; Ung, P.M.; Quintero, J.J.; Gestwicki, J.E.; Carlson, H.A. A poke in the eye: inhibiting HIV-1 protease through its flap-recognition pocket. *Biopolymers* **2008**, *89*, 643-652.

[22] Amaro, R.E.; Minh, D.D.; Cheng, L.S.; Lindstrom, W.M., Jr.; Olson, A.J.; Lin, J.H.; Li, W.W.; McCammon, J.A. Remarkable loop flexibility in avian influenza N1 and its implications for antiviral drug design. *J. Am. Chem. Soc.* **2007**, *129*, 7764-7765.

[23] Cheng, L.S.; Amaro, R.E.; Xu, D.; Li, W.W.; Arzberger, P.W.; McCammon, J.A. Ensemble-based virtual screening reveals potential novel antiviral compounds for avian influenza neuraminidase. *J. Med. Chem.* **2008**, *51*, 3878-3894.

[24] Amaro, R.E.; Schnaufer, A.; Interthal, H.; Hol, W.; Stuart, K.D.; McCammon, J.A. Discovery of drug-like inhibitors of an essential RNA-editing ligase in Trypanosoma brucei. *Proc. Natl. Acad. Sci. U S A* **2008**, *105*, 17278-17283.

[25] Fischer, E. Einfluss der configuration auf die wirkung derenzyme. *Ber. Dtsch. Chem. Ges.* **1894**, *27*, 2985-2993.

[26] Koshland, D.E. Application of a theory of enzyme specificity to protein synthesis. *Proc. Natl. Acad. Sci. USA* **1958**, *44*, 98-104.

[27] Linderstrøm-Lang, K.U.; Schellman, J.A. *Protein structure and enzyme activity*. Academic Press: New York: **1959**; Vol. *1*, pp. 443-510.

[28] Straub, F.B. Formation of the secondary and tertiary structure of enzymes. *Adv. Enzymol. Relat. Areas Mol. Biol.* **1964**, *26*, 89-114.

[29] Alonso, H.; Bliznyuk, A.A.; Gready, J.E. Combining docking and molecular dynamic simulations in drug design. *Med. Res. Rev.* **2006**, *26*, 531-568.

[30] Stumpf, M.P.; Thorne, T.; de Silva, E.; Stewart, R.; An, H. J.; Lappe, M.; Wiuf, C. Estimating the size of the human interactome. *Proc. Natl. Acad. Sci. U S A* **2008**, *105*, 6959-6964.

[31] Kelly, W.; Stumpf, M. Protein-protein interactions: from global to local analyses. *Curr. Opin. Biotechnol.* **2008**, *19*, 396-403.

[32] Sperandio, O.; Miteva, M.A.; Segers, K.; Nicolaes, G.; Villoutreix, B.O. Screening outside the catalytic site: Inhibition of macromolecular interactions through structure-based virtual ligand screening experiments. *The Open Biochemistry Journal* **2008**, *2*, 29-37.

[33] Ruffner, H.; Bauer, A.; Bouwmeester, T. Human protein-protein interaction networks and the value for drug discovery.

Drug Discov. Today **2007**, *12*, 709-716

[34] Villoutreix, B.O.; Bastard, K.; Sperandio, O.; Fahraeus, R.; Poyet, J.L.; Calvo, F.; Deprez, B.; Miteva, M.A. In silico-in vitro screening of protein-protein interactions: towards the next generation of therapeutics. *Curr. Pharm. Biotechnol.* **2008**, *9*, 103-122.

[35] Rahuel, J.; Priestle, J.P.; Grutter, M.G. The crystal structures of recombinant glycosylated human renin alone and in complex with a transition state analogue inhibitor. *J. Struct. Biol.* **1991**, *107*, 227–236

[36] Ma, B.; Shatsky, M.; Wolfson, H.J.; Nussinov, R. Multiple diverse ligands binding at a single protein site: a matter of pre-existing populations. *Protein Sci.* **2002**, *11*, 184–197

[37] Mallya, M.; Phillips, R.L.; Saldanha, S.A.; Gooptu, B.; Brown, S.C.; Termine, D.J.; Shirvani, A.M.; Wu, Y.; Sifers, R.N.; Abagyan, R.; Lomas, D.A. Small molecules block the polymerization of Z alpha1-antitrypsin and increase the clearance of intracellular aggregates. *J. Med. Chem.* **2007**, *50*, 5357-5363.

[38] Esnouf, R.; Ren, J.; Ross, C.; Jones, Y.; Stammers, D.; Stuart, D. Mechanism of inhibition of HIV-1 reverse transcriptase by non-nucleoside inhibitors. *Nat. Struct. Biol.* **1995**, *2*, 303-308.

[39] Murray, C.W.; Baxter, C.A.; Frenkel, A.D. The sensitivity of the results of molecular docking to induced fit effects: Application to thrombin, thermolysin and neuraminidase. *J. Comput. Aided Mol. Des.* **1999**, *13*, 547-562.

[40] McGovern, S.L.; Shoichet, B.K. Information decay in molecular docking screens against holo, apo, and modeled conformations of enzymes. *J. Med. Chem.* **2003**, *46*, 2895 – 2907.

[41] Kallblad, P.; Todorov, N.P.; Willems, H.M.; Alberts, I.L. Receptor flexibility in the in silico screening of reagents in the S1' pocket of human collagenase. *J. Med. Chem.* **2004**, *47*, 2761-2767.

[42] Floquet, N.; Marechal, J.D.; Badet-Denisot, M.A.; Robert, C.H.; Dauchez, M.; Perahia, D. Normal mode analysis as a prerequisite for drug design: application to matrix metalloproteinases inhibitors. *FEBS Lett.* **2006**, *580*, 5130-5136.

[43] Floquet, N.; Durand, P.; Maigret, B.; Badet, B.; Badet-Denisot, M.A.; Perahia, D. Collective motions in glucosamine-6-phosphate synthase: influence of ligand binding and role in ammonia channelling and opening of the fructose-6-phosphate binding site. *J. Mol. Biol.* **2009**, *385*, 653-664.

[44] Rueda, M.; Bottegoni, G.; Abagyan, R. Consistent improvement of cross-docking results using binding site ensembles generated with elastic network normal modes. *J. Chem. Inf. Model.* **2009**, *49*, 716-725.

[45] Sperandio, O.; Mouawad, L.; Pinto, E.; Villoutreix, B.O.; Perahia, D.; Miteva, M.A. How to choose relevant multiple receptor conformations for virtual screening: a test case of Cdk2 and normal mode analysis. **2010**, DOI 10.1007/s00249-010-0592-0.

[46] Bouzida, D.; Rejto, P.A.; Arthurs, S.; Colson, A.B.; Freer, S.T.; Gehlhaar, D.K.; Larson, V.; Luty, B.A.; Rose, P.W.; Verkhivker, G.M. Computer simulations of ligand-protein binding with ensembles of protein conformations: a Monte Carlo study of HIV-1 protease binding energy landscapes. *Int. J. Quantum. Chem.* **1999**, *72*, 73–84.

[47] Barril, X.; Morley, S.D. Unveiling the full potential of flexible receptor docking using multiple crystallographic structures. *J. Med. Chem.* **2005**, *48*, 4432-4443.

[48] Polgar, T.; Keseru, G.M. Ensemble docking into flexible active sites. Critical evaluation of FlexE against JNK-3 and beta-secretase. *J. Chem. Inf. Model.* **2006**, *46*, 1795-1805.

[49] Totrov, M.; Abagyan, R. Flexible ligand docking to multiple receptor conformations: a practical alternative. *Curr. Opin. Struct. Biol.* **2008**, *18*, 178-184.

[50] Brooks, B.R.; Bruccoleri, R.E.; Olafson, B.D.; States, D.J.; Swaminathan, S.; Karplus, M. CHARMM: A program for macromolecular energy, minimization, and dynamics calculations. *J. Comput. Chem.* **1983**, *4*, 187-217.

[51] Cornell, W.D.; Cieplak, P.; Bayly, C.I.; Gould, I.R.; Merz, K.M.; Ferguson, D.M.; Spellmeyer, D.C.; Fox, T.; Caldwell, J.W.; Kollman, P.A. A second generation force field for the simulation of proteins, nucleic acids, and organic molecules. *J. Am. Chem. Soc.* **1996**, *118*, 2309.

[52] Lindahl, E.; Hess, B.; van der Spoel, D. GROMACS 3.0: A package for molecular simulation and trajectory analysis. *J. Mol. Model.* **2001**, *7*, 306–317.

[53] Nelson, M.T.; Humphrey, W.; Gursoy, A.; Dalke, A.; Kale, L.V.; Skeel, R.D.; Schulten, K. NAMD: A parallel, object oriented molecular dynamics program. *Int. J. Supercomput. Applic.* **1996**, *10*, 251–268.

[54] Knegtel, R.M.; Kuntz, I.D.; Oshiro, C.M. Molecular docking to ensembles of protein structures. *J. Mol. Biol.* **1997**, *266*, 424-440.

[55] Ferrari, A.M.; Wei, B.Q.; Costantino, L.; Shoichet, B.K. Soft docking and multiple receptor conformations in virtual screening. *J. Med. Chem.* **2004**, *47*, 5076-5084.

[56] Clark, M.; Guarnieri, F.; Shkurko, I.; Wiseman, J. Grand canonical Monte Carlo simulation of ligand-protein binding. *J. Chem. Inf. Model.* **2006**, *46*, 231-242.

[57] Metropolis, N.; Rosenbluth, A.W.; Rosenbluth, M.N.; Teller, A.H.; Teller, E. Equations of state calculations by fast

computing machines. *J. Chem. Phys.* **1953**, *21*, 1087-1092.

[58] Kawai, H.; Kikuchi, T.; Okamoto, Y. A prediction of tertiary structures of peptide by the Monte Carlo simulated annealing method. *Protein Eng.* **1989**, *3*, 85-94.

[59] Yue, S.Y. Distance-constrained molecular docking by simulated annealing. *Protein Eng.* **1990**, *4*, 177-184.

[60] Caflisch, A.; Fischer, S.; Karplus, M. Docking by Monte Carlo minimization with a solvation correction: Application to an FKBP-substrate complex. **1997**, *18*, 723-743.

[61] Taylor, R.D.; Jewsbury, P.J.; Essex, J.W. FDS: flexible ligand and receptor docking with a continuum solvent model and soft-core energy function. *J. Comput. Chem.* **2003**, *24*, 1637-1656.

[62] Davis, I.W.; Baker, D. RosettaLigand docking with full ligand and receptor flexibility. *J. Mol. Biol.* **2009**, *385*, 381-392.

[63] Vieth, M.; Hirst, J.D.; Kolinski, A.; Brooks, C.L. Assessing search strategies for flexible docking. *J. Comput. Chem.* **1998**, *16*, 1623-1631.

[64] Morris, G.; Goodsell, D.S.; Halliday, R.S.; Huey, R.; Hart, W.E.; Belew, R.K.; Olson, A.J. Automated docking using a Lamarckian genetic algorithm and an empirical binding free energy function. *J. Comput. Chem.* **1999**, *19*, 1639-1662.

[65] Jones, G.; Willett, P.; Glen, R.C.; Leach, A.R.; Taylor, R. Development and validation of a genetic algorithm for flexible docking. *J. Mol. Biol.* **1997**, *267*, 727-748.

[66] Morris, G.M.; Huey, R.; Lindstrom, W.; Sanner, M.F.; Belew, R.K.; Goodsell, D.S.; Olson, A.J. AutoDock4 and AutoDockTools4: Automated docking with selective receptor flexibility. *J. Comput. Chem.* **2009**, *30*, 2785-2791.

[67] Corbeil, C.R.; Englebienne, P.; Moitessier, N. Docking ligands into flexible and solvated macromolecules. 1. Development and validation of FITTED 1.0. *J. Chem. Inf. Model.* **2007**, *47*, 435-449.

[68] Cui, Q.; Bahar, I., *Normal mode analysis: theory and applications to biological and chemical systems*. Chapman & Hall, London: **2006**.

[69] Perahia, D.; Mouawad, L. Computation of low-frequency normal modes in macromolecules: improvements to the method of diagonalization in a mixed basis and application to hemoglobin. *Comput. Chem.* **1995**, *19*, 241-246.

[70] Cavasotto, C.N.; Kovacs, J.A.; Abagyan, R.A. Representing receptor flexibility in ligand docking through relevant normal modes. *J. Am. Chem. Soc.* **2005**, *127*, 9632-9640.

[71] Tatsumi, R.; Fukunishi, Y.; Nakamura, H. A hybrid method of molecular dynamics and harmonic dynamics for docking of flexible ligand to flexible receptor. *J. Comput. Chem.* **2004**, *25*, 1995-2005.

[72] Floquet, N.; M'Kadmi, C.; Perahia, D.; Gagne, D.; Berge, G.; Marie, J.; Baneres, J.L.; Galleyrand, J.C.; Fehrentz, J.A.; Martinez, J. Activation of the Ghrelin Receptor is Described by a Privileged Collective Motion: A Model for Constitutive and Agonist-induced Activation of a Sub-class A G-Protein Coupled Receptor (GPCR). *J. Mol. Biol.* **2010** *395*, 769-778.

[73] Sander, T.; Liljefors, T.; Balle, T. Prediction of the receptor conformation for iGluR2 agonist binding: QM/MM docking to an extensive conformational ensemble generated using normal mode analysis. *J. Mol. Graph. Model.* **2008**, *26*, 1259-1268.

[74] May, A.; Zacharias, M. Protein-ligand docking accounting for receptor side chain and global flexibility in normal modes: evaluation on kinase inhibitor cross docking. *J. Med. Chem.* **2008**, *51*, 3499-3506.

[75] Berendsen, H.J.C.; Hayward, S. Collective protein dynamics in relation to function. *Current Opinion in Structural Biology* **2000**, *10*, 165-169.

[76] Zacharias, M. Rapid protein-ligand docking using soft modes from molecular dynamics simulations to account for protein deformability: binding of FK506 to FKBP. *Proteins* **2004**, *54*, 759-767.

[77] Timmers, L.F.; Caceres, R.A.; Dias, R.; Basso, L.A.; Santos, D.S.; de Azevedo, W.F., Jr. Molecular modeling, dynamics and docking studies of purine nucleoside phosphorylase from Streptococcus pyogenes. *Biophys. Chem.* **2009**, *142*, 7-16.

[78] Chang, M.W.; Lindstrom, W.; Olson, A.J.; Belew, R.K. Analysis of HIV wild-type and mutant structures via in silico docking against diverse ligand libraries. *J. Chem. Inf. Model.* **2007**, *47*, 1258-1262.

[79] Caflisch, A.; Miranker, A.; Karplus, M. Multiple copy simultaneous search and construction of ligands in binding sites: application to inhibitors of HIV-1 aspartic proteinase. *J. Med. Chem.* **1993**, *36*, 2142-2167.

[80] Elber, R.; Karplusm, M. Enhanced sampling in molecular dynamics: use of the time-dependent Hartree approximation for a simulation of carbon monoxide diffusion through myoglobin. *J. Am. Chem. Soc.* **1990**, *112*, 9161–9175.

[81] Wu, G.; Robertson, D.H.; Brooks, C.L.I.; Vieth, M. Detailed analysis of grid-based molecular docking: A case study of CDOCKER-a CHARMM-based MD docking algorithm. *J. Comput. Chem.* **2003**, *24*, 1549–1562.

[82] Vieth, M.; Cummins, D.J. DoMCoSAR: a novel approach for establishing the docking mode that is consistent with the structure-activity relationship. Application to HIV-1 protease inhibitors and VEGF receptor tyrosine kinase inhibitors. *J. Med. Chem.* **2000**, *43*, 3020-3032.

[83] Wu, G.; Vieth, M. SDOCKER: a method utilizing existing X-ray structures to improve docking accuracy. *J. Med. Chem.* **2004**, *47*, 3142-3148.

[84] Khandelwal, A.; Lukacova, V.; Comez, D.; Kroll, D. M.; Raha, S.; Balaz, S. A combination of docking, QM/MM methods,

and MD simulation for binding affinity estimation of metalloprotein ligands. *J. Med. Chem.* **2005**, *48*, 5437-5447.

[85]   Ferrara, P.; Curioni, A.; Vangrevelinghe, E.; Meyer, T.; Mordasini, T.; Andreoni, W.; Acklin, P.; Jacoby, E. New scoring functions for virtual screening from molecular dynamics simulations with a quantum-refined force-field (QRFF-MD). Application to cyclin-dependent kinase 2. *J. Chem. Inf. Model.* **2006**, *46*, 254-263.

[86]   Hoffmann, D.; Kramer, B.; Washio, T.; Steinmetzer, T.; Rarey, M.; Lengauer, T. Two-stage method for protein-ligand docking. *J. Med. Chem.* **1999**, *42*, 4422-4433.

[87]   Wang, J.; Kollman, P.A.; Kuntz, I.D. Flexible ligand docking: a multistep strategy approach. *Proteins* **1999**, *36*, 1-19.

[88]   Sherman, W.; Beard, H.S.; Farid, R. Use of an induced fit receptor structure in virtual screening. *Chem. Biol. Drug. Des.* **2006,** *67*, 83-84.

[89]   Nabuurs, S.B.; Wagener, M.; de Vlieg, J. A flexible approach to induced fit docking. *J. Med. Chem.* **2007**, *50*, 6507-6518.

[90]   Carlson, H.A.; Masukawa, K.M.; Rubins, K.; Bushman, F.D.; Jorgensen, W. L.; Lins, R.D.; Briggs, J.M.; McCammon, J.A. Developing a dynamic pharmacophore model for HIV-1 integrase. *J. Med. Chem.* **2000**, *43*, 2100-2114.

[91]   Kairys, V.; Gilson, M.K. Enhanced docking with the mining minima optimizer: acceleration and side-chain flexibility. *J. Comput. Chem.* **2002**, *23*, 1656-1670.

[92]   Leach, A.R. Ligand docking to proteins with discrete side-chain flexibility. *J. Mol. Biol.* **1994,** *235*, 345-356.

[93]   Frimurer, T.M.; Peters, G.H.; Iversen, L.F.; Andersen, H.S.; Moller, N.P.; Olsen, O.H. Ligand-induced conformational changes: improved predictions of ligand binding conformations and affinities. *Biophys. J.* **2003**, *84*, 2273-2281.

[94]   Hartmann, C.; Antes, I.; Lengauer, T. Docking and scoring with alternative side-chain conformations. *Proteins* **2009**, *74*, 712-726.

[95]   Schnecke, V.; Swanson, C.A.; Getzoff, E.D.; Tainer, J.A.; Kuhn, L.A. Screening a peptidyl database for potential ligands to proteins with side-chain flexibility. *Proteins* **1998**, *33*, 74-87.

[96]   Zavodszky, M.I.; Kuhn, L.A. Side-chain flexibility in protein-ligand binding: the minimal rotation hypothesis. *Protein Sci.* **2005**, *14*, 1104-1114.

[97]   Andrusier, N.; Mashiach, E.; Nussinov, R.; Wolfson, H.J. Principles of flexible protein-protein docking. *Proteins* **2008**, *73*, 271-289.

[98]   Jacobs, D.J.; Rader, A.J.; Kuhn, L.A.; Thorpe, M.F. Protein flexibility predictions using graph theory. *Proteins* **2001**, *44*, 150-165.

[99]   Zavodszky, M.I.; Lei, M.; Thorpe, M.F.; Day, A.R.; Kuhn, L.A. Modeling correlated main-chain motions in proteins for flexible molecular recognition. *Proteins* **2004**, *57*, 243-261.

[100]   Arora, K.; Brooks, C.L., 3rd. Large-scale allosteric conformational transitions of adenylate kinase appear to involve a population-shift mechanism. *Proc. Natl. Acad. Sci. U S A* **2007**, *104*, 18496-18501.

[101]   Lin, J.H.; Perryman, A.L.; Schames, J.R.; McCammon, J.A. The relaxed complex method: Accommodating receptor flexibility for drug design with an improved scoring scheme. *Biopolymers* **2003**, *68*, 47-62.

[102]   Amaro, R.E.; Baron, R.; McCammon, J.A. An improved relaxed complex scheme for receptor flexibility in computer-aided drug design. *J. Comput. Aided Mol. Des.* **2008**, *22*, 693-705.

[103]   Bowman, A.L.; Nikolovska-Coleska, Z.; Zhong, H.; Wang, S.; Carlson, H.A. Small molecule inhibitors of the MDM2-p53 interaction discovered by ensemble-based receptor models. *J. Am. Chem. Soc.* **2007**, *129*, 12809-12814.

[104]   Damm, K.L.; Carlson, H.A. Exploring experimental sources of multiple protein conformations in structure-based drug design. *J. Am. Chem. Soc.* **2007**, *129*, 8225-8235.

[105]   Wei, B.Q.; Weaver, L.H.; Ferrari, A.M.; Matthews, B.W.; Shoichet, B.K. Testing a flexible-receptor docking algorithm in a model binding site. *J. Mol. Biol.* **2004**, *337*, 1161-1182.

[106]   Huang, S.Y.; Zou, X. Ensemble docking of multiple protein structures: considering protein structural variations in molecular docking. *Proteins* **2007**, *66*, 399-421.

[107]   Huang, S.Y.; Zou, X. Efficient molecular docking of NMR structures: application to HIV-1 protease. *Protein Sci.* **2007**, *16*, 43-51.

[108]   Bolstad, E.S.; Anderson, A.C. In pursuit of virtual lead optimization: the role of the receptor structure and ensembles in accurate docking. *Proteins* **2008**, *73*, 566-580.

[109]   Durdagi, S.; Mavromoustakos, T.; Chronakis, N.; Papadopoulos, M. G. Computational design of novel fullerene analogues as potential HIV-1 PR inhibitors: Analysis of the binding interactions between fullerene inhibitors and HIV-1 PR residues using 3D QSAR, molecular docking and molecular dynamics simulations. *Bioorg. Med. Chem.* **2008**, *16*, 9957-9974.

[110]   Broughton, H.B. A method for including protein flexibility in protein-ligand docking: improving tools for database mining and virtual screening. *J. Mol. Graph. Model.* **2000**, *18*, 247-257.

[111]   Osterberg, F.; Morris, G.M.; Sanner, M.F.; Olson, A.J.; Goodsell, D.S. Automated docking to multiple target structures: incorporation of protein mobility and structural water heterogeneity in AutoDock. *Proteins* **2002**, *46*, 34-40.

[112] Bolstad, E.S.; Anderson, A.C. In pursuit of virtual lead optimization: pruning ensembles of receptor structures for increased efficiency and accuracy during docking. *Proteins* **2009**, *75*, 62-74.

[113] Claussen, H.; Buning, C.; Rarey, M.; Lengauer, T. FlexE: efficient molecular docking considering protein structure variations. *J. Mol. Biol.* **2001**, *308*, 377-395.

[114] Bottegoni, G.; Kufareva, I.; Totrov, M.; Abagyan, R. Four-dimensional docking: a fast and accurate account of discrete receptor flexibility in ligand docking. *J. Med. Chem.* **2009**, *52*, 397-406.

[115] Thorsteinsdottir, H.B.; Schwede, T.; Zoete, V.; Meuwly, M. How inaccuracies in protein structure models affect estimates of protein-ligand interactions: computational analysis of HIV-I protease inhibitor binding. *Proteins* **2006**, *65*, 407-423.

[116] de Azevedo, W.F., Jr.; Dias, R. Experimental approaches to evaluate the thermodynamics of protein-drug interactions. *Curr. Drug Targets* **2008**, *9*, 1071-1076.

[117] Liu, T.; Lin, Y.; Wen, X.; Jorissen, R.N.; Gilson, M.K. BindingDB: a web-accessible database of experimentally determined protein-ligand binding affinities. *Nucleic Acids Res.* **2007**, *35*, D198-201.

[118] Guvench, O.; MacKerell, A.D., Jr. Computational evaluation of protein-small molecule binding. *Curr. Opin. Struct. Biol.* **2009**, *19*, 56-61.

[119] Mobley, D.L.; Dill, K.A. Binding of small-molecule ligands to proteins: "what you see" is not always "what you get". *Structure* **2009**, *17*, 489-498.

[120] Gilson, M.K.; Zhou, H.X. Calculation of protein-ligand binding affinities. *Annu. Rev. Biophys. Biomol. Struct.* **2007**, *36*, 21-42.

[121] Woo, H.J.; Roux, B. Calculation of absolute protein-ligand binding free energy from computer simulations. *Proc. Natl. Acad. Sci. U S A* **2005**, *102*, 6825-6830.

[122] Doudou, S.; Burton, N.A.; Henchman, R.H. Standard free energy of binding from a one-dimensional potential of mean force. *J. Chem. Theory Comput.* **2009**, *5*, 909–918.

[123] Jorgensen, W.L.; Severance, D.L. Aromatic-aromatic interactions: free-energy profiles for the benzene dimer in water, chloroform, and liquid benzene. *J. Am. Chem. Soc.* **1990**, *112*, 4768–4774.

[124] Roux, B. The calculation of the potential of mean force using computer simulations. *Comp. Phys. Comm.* **1995**, *91*, 275–282.

[125] Ytreberg, F.M. Absolute FKBP binding affinities obtained via nonequilibrium unbinding simulations. *J. Chem. Phys.* **2009**, *130*, 164906.

[126] Hajjar, E.; Perahia, D.; Debat, H.; Nespoulous, C.; Robert, C.H. Odorant binding and conformational dynamics in the odorant-binding protein. *J. Biol. Chem.* **2006**, *281*, 29929-29937.

[127] Hummer, G.; Szabo, A. Free energy surfaces from single-molecule force spectroscopy. *Acc. Chem. Res.* **2005**, *38*, 504-513.

[128] Brandsdal, B.O.; Osterberg, F.; Almlof, M.; Feierberg, I.; Luzhkov, V.B.; Aqvist, J. Free energy calculations and ligand binding. *Adv. Protein Chem.* **2003**, *66*, 123-158.

[129] Shirts, M.R.; Mobley, D.L.; Chodera, J.D. Alchemical free energy computations: Ready for prime time?. *Annual Reports in Computational Chemistry* **2007**, *3*, 41-59.

[130] Knight, J.L.; Brooks, C.L. Validating charmm parameters and exploring charge distribution rules in structure-based drug design. *Journal of Chemical Theory and Computation* **2009**, *5*, 1680–1691.

[131] Kim, J.T.; Hamilton, A.D.; Bailey, C.M.; Domaoal, R.A.; Wang, L.; Anderson, K.S.; Jorgensen, W L. FEP-guided selection of bicyclic heterocycles in lead optimization for non-nucleoside inhibitors of HIV-1 reverse transcriptase. *J. Am. Chem. Soc.* **2006**, *128*, 15372-15373.

[132] Michel, J.; Essex, J.W. Hit identification and binding mode predictions by rigorous free energy simulations. *J. Med. Chem.* **2008**, *51*, 6654-6664.

[133] Blondel, A. Ensemble variance in free energy calculations by thermodynamic integration: theory, optimal "Alchemical" path, and practical solutions. *J. Comput. Chem.* **2004**, *25*, 985-993.

[134] Shyu, C.; Ytreberg, F.M. Reducing the bias and uncertainty of free energy estimates by using regression to fit thermodynamic integration data. *J. Comput. Chem.* **2009**, *30*, 2297-2304.

[135] Mobley, D.L.; Graves, A.P.; Chodera, J.D.; McReynolds, A.C.; Shoichet, B.K.; Dill, K.A. Predicting absolute ligand binding free energies to a simple model site. *J. Mol. Biol.* **2007**, *371*, 1118-1134.

[136] Boyce, S.E.; Mobley, D.L.; Rocklin, G.J.; Graves, A.P.; Dill, K.A.; Shoichet, B.K. Predicting Ligand Binding Affinity with Alchemical Free Energy Methods in a Polar Model Binding Site. *J. Mol. Biol.* **2009**, *394,* 747-763.

[137] Aqvist, J.; Luzhkov, V.B.; Brandsdal, B.O. Ligand binding affinities from MD simulations. *Acc. Chem. Res.* **2002**, *35*, 358-365.

[138] Almlöf, M.; Brandsdal, B.O.; Aqvist, J. Binding affinity prediction with different force fields: examination of the linear interaction energy method. *J. Comput. Chem.* **2004**, *25*, 1242–1254.

[139] Alam, M.A.; Naik, P.K. Applying linear interaction energy method for binding affinity calculations of podophyllotoxin analogues with tubulin using continuum solvent model and prediction of cytotoxic activity. *Journal of Molecular Graphics & Modelling* **2009**, *27*, 930–943.

[140] Orrling, K.M.; Marzahn, M.R.; Gutierrez-de-Teran, H.; Aqvist, J.; Dunn, B.M.; Larhed, M. alpha-Substituted norstatines as the transition-state mimic in inhibitors of multiple digestive vacuole malaria aspartic proteases. *Bioorg. Med. Chem.* **2009**, *17*, 5933-5949.

[141] Srinivasan, J.; Cheatham, T.E.; Cieplak, P.; Kollman, P.A.; Case, D.A. Continuum solvent studies of the stability of DNA, RNA, and phosphoramidate - DNA helices. *J. Am. Chem. Soc.* **1998**, *120*, 9401-9409.

[142] Luo, H.; Sharp, K. On the calculation of absolute macromolecular binding free energies. *Proc. Natl. Acad. Sci. U S A* **2002**, *99*, 10399-10404.

[143] Wong, S.; Amaro, R.E.; McCammon, J.A. MM-PBSA Captures Key Role of Intercalating Water Molecules at a Protein-Protein Interface. *J. Chem. Theory Comput.* **2009**, *5*, 422-429.

[144] Chang, C.E.; Chen, W.; Gilson, M.K. Ligand configurational entropy and protein binding. *Proc. Natl. Acad. Sci. U S A* **2007**, *104*, 1534-1539.

[145] Swanson, J.M.; Henchman, R.H.; McCammon, J.A. Revisiting free energy calculations: a theoretical connection to MM/PBSA and direct calculation of the association free energy. *Biophys. J.* **2004**, *86*, 67-74.

[146] Gohlke, H.; Case, D.A. Converging free energy estimates: MM-PB(GB)SA studies on the protein-protein complex Ras-Raf. *J. Comput. Chem.* **2004**, *25*, 238-250.

[147] Michel, J.; Tirado-Rives, J.; Jorgensen, W.L. Energetics of Displacing Water Molecules from Protein Binding Sites: Consequences for Ligand Optimization. *J. Am. Chem. Soc.* **2009**, *131*, 15403 -15411.

[148] Deng, Y.; Roux, B. Computation of binding free energy with molecular dynamics and grand canonical Monte Carlo simulations. *J. Chem. Phys.* **2008**, *128*, 115103.

[149] ten Brink, T.; Exner, T.E. Influence of protonation, tautomeric, and stereoisomeric states on protein-ligand docking results. *J. Chem. Inf. Model.* **2009**, *49*, 1535-1546.

[150] Ritschel, T.; Hoertner, S.; Heine, A.; Diederich, F.; Klebe, G. Crystal structure analysis and in silico pKa calculations suggest strong pKa shifts of ligands as driving force for high-affinity binding to TGT. *Chembiochem* **2009**, *10*, 716-727.

[151] Jiao, D.; Golubkov, P.A.; Darden, T.A.; Ren, P. Calculation of protein-ligand binding free energy by using a polarizable potential. *Proc. Natl. Acad. Sci. U S A* **2008**, *105*, 6290-6295.

[152] Khoruzhii, O.; Donchev, A.G.; Galkin, N.; Illarionov, A.; Olevanov, M.; Ozrin, V.; Queen, C.; Tarasov, V. Application of a polarizable force field to calculations of relative protein-ligand binding affinities. *Proc. Natl. Acad. Sci. U S A* **2008**, *105*, 10378-10383.

[153] Schames, J.R.; Henchman, R.H.; Siegel, J.S.; Sotriffer, C.A.; Ni, H.; McCammon, J.A. Discovery of a novel binding trench in HIV integrase. *J. Med. Chem.* **2004**, *47*, 1879-1881.

[154] Lin, J.H.; Perryman, A.L.; Schames, J.R.; McCammon, J.A. Computational drug design accommodating receptor flexibility: the relaxed complex scheme. *J. Am. Chem. Soc.* **2002**, *124*, 5632-5633.

[155] Tang, J.; Park, J.G.; Millard, C.B.; Schmidt, J.J.; Pang, Y.P. Computer-aided lead optimization: improved small-molecule inhibitor of the zinc endopeptidase of botulinum neurotoxin serotype A. *PLoS One* **2007**, *2*, e761.

[156] Hu, X.; Prehna, G.; Stebbins, C.E. Targeting plague virulence factors: a combined machine learning method and multiple conformational virtual screening for the discovery of Yersinia protein kinase A inhibitors. *J. Med. Chem.* **2007**, *50*, 3980-3983.

[157] Degliesposti, G.; Kasam, V.; Da Costa, A.; Kang, H.K.; Kim, N.; Kim, D.W.; Breton, V.; Kim, D.; Rastelli, G. Design and discovery of plasmepsin II inhibitors using an automated workflow on large-scale grids. *ChemMedChem* **2009**, *4*, 1164-1173.

[158] Hritz, J.; de Ruiter, A.; Oostenbrink, C. Impact of plasticity and flexibility on docking results for cytochrome P450 2D6: a combined approach of molecular dynamics and ligand docking. *J. Med. Chem.* **2008**, *51*, 7469-7477.

CHAPTER 7

Protein-Protein Interaction Inhibition (2P2I): Mixed Methodologies for the Acceleration of Lead Discovery

Philippe Roche and Xavier Morelli

Interactions et Modulateurs de Réponses (UPR3243) CNRS, 13402 Marseille Cedex 20, France ; Email: proche@ifr88.cnrs-mrs.fr or morelli@ifr88.cnrs-mrs.fr

Abstract: Protein–Protein Interactions (PPIs) constitute a promising class of targets for drug discovery. Inhibitors of these original interactions are certainly the next generation of highly innovative drugs that will reach the market in the next decade. However, the *in silico* design of such compounds still remains challenging. This review describes this particular protein-protein interaction chemical space and the main biophysical reasons that make them challenging targets for the drug discovery process. A state of the art of protein databases and servers dedicated to PPIs, their analysis and inhibition is also surveyed. It then presents some innovative methodologies that led to the development of new inhibitors. Finally, different families of protein-protein interactions for which an inhibitor is known are briefly introduced and potential tracks for the future are proposed such as a new classification based on protein-protein interfaces with known inhibitors into specific families, with a subsequent notion of focused databases dedicated to each specific class.

Key words: protein-protein interactions, protein-protein interfaces, PPIs, 2P2I, drug design, inhibitor, small molecule, structural databases, druggable target.

INTRODUCTION

A recent survey by John P. Overington *et al.* of the small molecules pipeline hold by pharmaceutical companies and current drugs per drug substance has pin-pointed that the majority of FDA-approved drugs derives from research targeting Rhodopsin-like GPCRs, nuclear receptors, ligand-gated ion channels, voltage-gated ion channels, and penicillin-binding protein. Moreover, beyond the ten most common drugged families, there were a further 120 domain families or singletons for which only a few drugs were successfully launched [1].

However, in the last decade, the inhibition of Protein-Protein Interaction (PPIs) has emerged from both academic and private research as a new way to modulate the activity of proteins [2-8]. In parallel to this new-born field, large scale genomics and proteomics programs have permitted the identification of entire protein networks interactomes - at the cellular level which led to major breakthrough in understanding biological pathways, host-pathogen interactions and cancer development. It is now commonly accepted that protein-protein complexes are an important constituent of the therapeutic targets [9]. They can be involved in a network of complex interactions that play a central role in numerous events in the cell. These interactions control processes involved in both normal and pathological actions ranging from signal transduction to cell adhesion through cell proliferation, growth, differentiation, viral self-assembly, programmed cell death and cytoskeleton structure (for a review see [10]). The modulation of these networks of interactions represents a promising therapeutic strategy and the growing tools of small molecules Protein-Protein Interaction Inhibitors (2P2Is) are thus certainly the next generation of highly innovative drugs that will reach the market in the next decade.

Despite the growing number of publications and reviews describing the inhibition of PPIs with small molecules, the field is still a matter of debates since successes in targeting these protein-protein 'hot spots' interfaces have been thwarted by a number of problems from hit discovery to pharmacokinetic and clinical studies. Another highly debatable question concern whether useful design and docking can be performed in all cases. Yet again, an analysis of the FDA approved drugs demonstrated that there is a high degree of structural characterization of drug targets, with more than 100 drug targets themselves and over 90% (>300) of drug targets being similar to known proteins in the Protein Data Bank (PDB) [1]. Taking into account the increasing number of PPIs complexes deposited in the

PDB every year, it makes no doubt that in a near future, the quality of PPIs inhibitors will increase and thus consequently, FDA-approved drug compounds.

The long-standing notion that these interfaces are excessively complicated and thus "poorly druggable" was founded on the inherent three-dimensional and biophysical complexity of this targeted space. This review describes this particular PPI chemical space and the main biophysical reasons making them challenging targets for the drug discovery process.

PROPERTIES OF PROTEIN-PROTEIN INTERACTIONS

"PPIs are large, flat and featureless and therefore are unsuitable as drug targets". This statement is unfortunately still a widely held belief in the scientific community. Wells and McClendon have been among the first ones to address this scientific dogma and their study has been confirmed more recently by numerous groups all over the world [2,11]. The aim of this present review is not to present another survey on protein-protein interfaces and their statistical analyses; however, to understand the subtle differences in the numerous available servers and databases, one might first know and understand the biophysical properties of this very specific chemical space. We will briefly focus on the most relevant parameters that permit to differentiate heterodimeric complexes from enzymes (and also from homodimeric and non specific complexes) and try to differentiate the true statements from the wrong interpretation that has conducted many scientists in the past (and still nowadays in some pharmaceutical companies and academic laboratories) to reject this "high hanging fruit" described by Wells and McClendon.

Protein–protein Interaction Sites are Large – True Statement -

The average size of an heterodimeric protein-protein interface is comprised between ~1200-2000 Å² [12], which is a "standard size" interface according to Lo Conte *et al.,* [13,14], but large if compared with those involved in protein-small molecule interactions (300-1000A²) [15,16]. The lack of interfaces with a size below 800 A² indicates that the formation of a stable and specific complex between two proteins requires a sufficient number of contacts and removing water at least from part of the protein interface. Many cellular functions are mediated through transient PPIs with short lifetimes. However, the structure determination by x-ray crystallography requires stable complex structures that can form well ordered crystals. The present analysis is thus restricted to sufficiently stable protein-protein complexes; the recognition principles for such stable interactions might differ from interfaces formed during transient interactions with a short lifetime.

Protein–protein Interaction Sites are Flat – True Statement -

Jones and Thornton derived a parameter called 'planarity' to assess whether a protein-protein interface is flat or curved [13]. The 'planarity' of the surfaces between two components of a complex is analyzed by calculating the root mean square deviation of all the interface atoms from the least-squares plane through the atoms. If all atoms would exactly fit to a plane, the planarity index would be zero. The average value of the planarity is 2.8 ± 0.9 Å for heterocomplexes [17]. This result indicates that protein-protein interfaces are generally flat in shape. Another parameter to characterize the shape of protein binding sites is the 'circularity' of the interface [13]. The circularity is the ratio of the lengths of the principal axes of the least-squares plane through the atoms in the interface. A ratio near 1.0 indicates that an interface is approximately circular meaning that the atoms that form the buried interface are all within a circular plane. On average, the interfaces of heterocomplexes are not perfectly circular (mean value 0.73 ± 0.05) but are rather oblong [17].

Protein-protein Interaction Sites are Featureless – Wrong Interpretation

Typical heterodimeric complexes are non-covalent assemblies of proteins that fold separately to carry out independent functions and associate under certain physiological conditions. These types of interactions are found in protease-inhibitor, antigen-antibody, enzyme complexes, and many signal transduction and cell cycle proteins [13,14]. Although their recognition sites share some common properties, protein hetero-complexes are extremely diverse in their cellular functions. These functional biological differences must be intrinsically related to their inherent properties - in term of composition in amino acid at the surface of the free protein and also to the three dimensional organization of these surfaces that must allow sufficient structural stability in the free form (too many hydrophobic residues in a same region would lead to an aggregation of the unbound protein).

Non-Contiguous Patches

Interface atoms may form one contiguous patch or they can be distributed in several patches throughout the interface. Bahadur described patches as clusters of residues or atoms in contact with the partner protein separated from other patches by surface residues or atoms that are not in contact with the protein-binding partner. In general, most of the single patch interfaces in homodimers and in heterodimers are standard size interfaces, and large interfaces contain more than one recognition patch [12]. Interface residues can be divided into two distinct regions, the 'core' and the 'rim' region, based on their accessibility to solvent. The 'core' region contains residues that have at least one fully buried interface atom surrounded by the 'rim' region, which contains residues having accessible atoms only. The two regions differ in their amino acid composition: The rim is very similar to the remainder of the protein surface, whereas the core region depicts the protein interior. On average the 'core' constitutes 72 to 77% of the interface area of homodimers and heterodimeric complexes [12].

Original Chemical Composition

The composition of amino acid residues at the specific interfaces differs from the rest of the protein surface. Relative to the solvent-accessible protein surface, homodimer and heterodimer complex interfaces are enriched in aliphatic (Leu, Val, Ile, Met) and aromatic (His, Phe, Tyr, Trp) residues, and depleted in charged residues (Asp, Glu, Lys) other than arginine [13,18-20].

Hot Spots Notion

A major breakthrough came from the discovery that only a limited number of key interactions contribute to binding affinity [11]. These key residues, called "hot spots", are critical in understanding the principles of protein interactions [2] and their dimensions are comparable to the size of a small ligand molecule which opens the door for the development of inhibitory compounds. Detailed analysis of amino acid composition of hot spots has revealed some critical residues such as Arg, Met, Phe, Trp and Tyr [21,22].

Proteins can undergo local structural rearrangements upon binding to their partner (see § conformational changes below). However, different studies have shown that small molecule inhibitors target virtually the same critical hot spots indicating that local adaptation is not necessarily an obstacle for the discovery of PPI disruptors [18,23,24].

Several methods have been developed to predict hot spots based on energy calculations [25,26], molecular dynamics simulations [27,28], solvent accessibility and statistical pairwise residue potentials of the interface residues [29] or machine learning and energy-based methods [30].

Cavities and Atomic Packing

Sonavane *et al.* have presented a comparative analysis of cavities enclosed in a tertiary structure of proteins and interfaces formed by the interaction of two protein subunits in obligate and non-obligate categories [31]. The total volume of cavities increases with the size of the protein (or the interface), though the exact relationship may vary in different cases. The larger cavities tend to be less spherical, solvated, and the interfaces are enriched in these. Relative to the frequency of occurrence in the whole structure (or interface), residues in beta-strands are found more often lining the cavities, and those in turn and loop the least.

Several algorithms have been developed to identify cavities on protein surfaces in relation to drug design (for a review, please refer to [32]). However, in the case of PPIs the small cavities are more difficult to detect. In a recent survey, Fuller *et al.* have observed that striking differences can be observed between the protein-ligand interfaces and protein–protein binding site pockets in a case study using Q-SiteFinder: the average active volume of the top ranked pockets for protein–ligand receptors (524 Å^3) is twice that of the protein–protein monomers (261 Å^3). Fifty-eight percent of the occupied protein–ligand binding pockets are the top predicted pocket, which has an average active volume of 471 Å³, as opposed to only 15% in PPI monomers (active volume: 234 Å³). PPIs are characterized typically by six (±3) small occupied pockets, with volumes that are the same as the average for all surface pockets (55 Å³). This contrasts with protein-ligand interfaces where the top 1 or 2 ranked pockets make by far the largest contribution to binding and the average occupied pocket volume (260 Å³) is over four times the average occupied volume for PPIs. Protein-ligand interfaces tend to occur in one or two disproportionately large pockets relative to the average pocket size. By contrast, PPIs tend to occur in several average-sized pockets that have a similar active volume to that of the average surface pocket. PPIs inhibitors target between 3 and 5 pockets. This is akin to the way in which PPIs also target several pockets rather than one large pocket as seen in protein-ligand interfaces. These later ones have one or two disproportionately large pockets that are generally occupied by their ligands; otherwise they show similar pocket size characteristics to the surfaces of proteins involved in PPIs [33] Another important information concerns the work by Eyrisch and Helms [34,35], Brown and Hajduk [36], and Lee and Craik [37] which suggest that there is often an advantage in looking at pockets using a dynamic model of protein structure, particularly where conformational selection is taking place, because many transient pocket openings happen within the picosecond–nanosecond timescale.

Conformational Changes Occur upon Complex Formation- It Depends -

Local or large scale rearrangements can occur upon complex formation but this phenomenon is mostly observed when the conformational changes mediate signaling events or trigger allosteric effects. In that case, lock and key/induced fit or equilibrium models or combination of both are observed [38]. Essentially, most structurally characterized complexes with interfaces area larger than 2000 Å² undergo large changes upon association.

However, many protein-protein complexes display a root mean square deviation (rmsd) between Cα atoms of less than 2Å [14,39]. It is, however, relevant to ask to what extent pockets at the interface are present in the unbound protein conformations, and to what extent does lock and key, induced fit, or equilibrium model play a role. Eyrisch and Helms have demonstrated that virtual screening methods allowing for picoseconds to nanosecond receptor flexibility is a very promising approach in cases where rigid body docking was unsuccessful [34]. They studied three PPI inhibitors (IL-2/IL-2Ra, MDM2/p53 and Bcl-X$_L$/BH3) using molecular dynamics, and a geometric pocket finding algorithm (PASS) [40] to determine potential binding pockets on the protein surface. The authors noted that in the case of the apo IL-2, PASS did not identify the native binding pocket, which was detectable in the inhibitor-bound structures of IL-2 with the inhibitors removed [34]. Similarly, it has been noted that Bcl-X$_L$ has a rather flat surface in the apo state [2]. However, transient pockets are observed in molecular dynamics simulations in the absence of small molecules in both Bcl-X$_L$ and IL-2 [36,34], showing that pockets on the protein surfaces frequently open and close on timescales ranging from picoseconds to hundreds of picoseconds. Indeed, there are limitations in describing potential small-molecule binding pockets when using static structures of either the free protein or the protein–protein complex. Fuller *et al.* have analyzed the free protein structures and the PPI inhibitor bound complexes using Q-SiteFinder [33]. Visual analysis shows that pocket detection is most successful when applied to the bound structure, as might be expected, but that key interaction pockets are often or preformed in the free protein or the protein–protein complex.

DATABASES RELATED TO PROTEIN-PROTEIN INTERACTIONS AND DRUG DESIGN

There is a growing number of structural databases dedicated to PPIs [41-43]. In this review we will focus on the structural databases and servers that depict the interaction at the atomic level. A summary of some available databases is given in Table **1**. All the datasets are derived from the Brookhaven Protein Data Bank unless mentioned (http://www.rcsb.org). Detailed analyses of 3D structures from protein-protein complexes have allowed the classification of the interfaces in different families according to different criteria. Most structure-based databases classify interfaces between domains and rely on domain definitions extracted from secondary databases Pfam, SCOP or CATH.

- ***iPfam*** [44] is a web resource that describes protein interactions using Pfam family of domains with a known 3D structure. It includes PPIs as well as interactions between domains in a single protein. For a given protein, the interaction network with its partners is displayed at the domain level giving information about which domains in a structure are interacting directly. In addition, the PDB structures can be visualized with the interacting residues highlighted and the interaction data at the level of amino acid residue can be accessed.

- **SCOPPI** [45] (Structural Classification of Protein-Protein Interfaces) is a comprehensive database that classifies and annotates domain interactions derived from all known protein 3D structures. Domains are defined according to the SCOP database. SCOPPI contains a total of 102084 domain-domain interactions and 14422 at 90% non-redundancy level which represents 3358 SCOP family-family interactions. These protein-protein interfaces are classified into 8850 types using distance criterion to determine inter-domain interfaces.

- **SNAPPI-DB** [46] provides a useful resource for the analysis on domain-domain and domain-ligand interactions Structural data are retrieved from the Macromolecular Database (MDS) developed at the European Bioinformatics Institute (EBI). The classification of domain–domain interactions is determined by the relative orientation of the interacting pair using the iRMSD (interaction root-mean square deviation) method described by Aloy *et al.*, [47]. Pairs of multiple structure alignments are generated for each partner of the interaction.

- **PIBASE** [48] is a structural database composed of binary interfaces extracted from the PDB. PIBASE was developed to cover all domain interfaces in the PDB as well as the computation and analysis of various descriptors of domain interaction. A diverse set of geometric, physiochemical and topologic properties are calculated for each complex, its domains, interfaces and binding sites. A subset of the interface properties are used to remove interface redundancy within PDB entries. The complexes are grouped into topological classes based on their patterns of domain-domain contacts. Pibase currently contains 269,821 interacting domains pairs between 212,071 domains.

- **3did** [49] is a structural database which stores information about PPIs as well as transient peptide-mediated interactions. Currently, 3did contains 115,559 domain–domain interactions of known 3D structure that were classified into 4887 unique interaction types. Unlike other databases, 3did also contains information about protein-peptide interactions. It can be searched by domain name, Pfam accession number or motif name. 3did is particularly helpful to display networks of interactions as for a given protein or domain, it displays all known domains, or peptides, that do physically interact with it and for which the 3D structure is known. The display of the network of interacting domains is interactive which makes it very convenient to use. It is possible to select a particular interaction to retrieve the specific details stored in 3did such as PDB codes, topologies, chains or residues involved in the interaction and to visualize it using molecular graphics program Rasmol.

- **3D complex** [50] proposes an original hierarchical classification of protein complexes with known x-ray structures based on representing their fundamental structural features and topology (identities, homology, residues in contact) as a graph. It addresses the question of domain organization at the level of quaternary structures (QS) and finds that they form a small number of arrangements compared with all theoretically possible ones. The database contains about 30,000 complexes which are classified into about 200 different QS topologies. Most complexes contain four subunits or less, and the large majority are homomeric. Through the web interface it is easy to retrieve a specific subset of the whole database that can be useful for *in silico* drug design such as all known heterodimers.

- **Bordner and Gorin** [51] have developed a method to predict biologically relevant PPI using high-resolution x-ray structures of protein complexes. Protein-protein interfaces were classified into different clusters using a machine learning method that utilizes a set of interface properties (interface area, number of intermolecular hydrogen bonds contacting pair residue counts, residue propensity ...). The whole dataset can be accessed through the Protein Interface Server (PInS) which can also be used to retrieve structures of predicted protein complexes. It is extremely useful to build datasets for protein docking benchmarks or to perform statistical analyses of PPIs.

- **The PROTCOM** database [52] is a collection of protein-protein homo- and hetero-complexes as well as domain-domain structures developed by Alexov (see chapter analysis of binding pockets in this book). It contains a collection of homo and hetero dimers. The database contains 17.024 entries of which 1350 are two-chain protein hetero-complexes, 7773 homodimers and 1589 are one-chain proteins parsed into two domains. The database is searchable with respect to different parameters (sequence, PDB ID, protein name, number of residues, number of residues on the interface, list of interfacial residues, number of helices and strands, absolute and relative interface areas). A detailed description of the interface with all above parameters can be retrieved. The whole database can be easily downloaded.

- **JAIL** [53] is a recently developed structure-based interface library for macromolecules. It combines data about PPIs as well as intra-chains contacts. It contains 157,000 protein-protein interfaces divided into two subsets: 81,000 between domains classified with SCOP; 76,000 between different protein chains. It also contains 19,000 interfaces which were calculated based on the assumed biological units and 8,000 interfaces between proteins and nucleic acids. It offers a web interface which allows a visual inspection of the interface through the interactive protein visualizer Jmol (http://www.jmol.org). User defined data sets with different levels of diversity can be downloaded from the entire library.

- *The Protein-Protein Interaction Database* [54] (PPIDB) contains PPIs with interfaces residues annotated on a per-residue basis. It has been design to allow large-scale analyses of interface residue properties as well as the training of machine learning classifiers to predict PPI sites. The current version of PPIDB contains 38,815 PDB structures consisting of 71,486 inter-chain protein-protein interfaces. Pair-wise sequence identity between pairs of chains is greater than 90% for 47,604 and less than 30% for 21,567 of these interfaces.

- *PRINT* [55] (Dataset of PRotein Protein INTerfaces) contains 49512 interfaces corresponding to 34817 structures extracted from PDB (65% homodimers and 35% heterodimers). An algorithm based on structure similarity and evolutionary conservation has been used to classify interfaces into 8205 different clusters. More than 50% of the interactions are verified in two publicly available interaction databases (DIP and BIND). Users may browse through the non-redundant dataset of representative interfaces, retrieve the list of structures similar to these interfaces, or retrieve general parameters of interaction for a given protein-protein complex. A link to the HotSprint database is provided to define putative hot spots.

All the databases described contain pre-calculated information about PPIs that can be searched using general criteria. Several servers are available to analyze and visualize PPIs at the atomic level and to predict putative binding sites.

- *The ProtorP server* [56] is a tool to analyze interfaces between protein chains in protein-protein associations. It calculates a series of structural and chemical properties of the protein-protein interface. These parameters include size using accessible surface area; shape defined in terms of planarity, length, breadth and numerical eccentricity. The number of intermolecular hydrogen bonds, bridging water, intermolecular salt bridges and disulphide bonds are also provided. The amino acid composition of residues defined in the interface is given as a percentage value of those classified as polar, non-polar and charged as described previously [13]. Residues at the interface are classified according to the secondary structure elements (α-helices, β-strands or coil). The analysis can be done using a single PDB code, a list of PDB codes or on a user defined PDB file. The ProtorP server is also useful to compare the properties of a given interface with observed parameters in non-homologous datasets and to obtain general statistics about protein-protein associations.

- *The InterProSurf server* [57] was originally developed to predict interacting sites on protein surfaces. It contains features to analyze interfaces in protein complex. It calculate the residues present at the interface, polar and apolar areas, the number of surface atoms, number of buried atoms and total area of the molecule in protein complex as well in its monomer form. Users can provide a valid PDB code or upload a local complex and submit the query to the server. The server automatically detects the different subunit and produces the results for all chain present in the protein complex.

- *The PPIDD* [58] (Protein-Protein Interaction Description Database) server allows the extraction and visualization of biological PPIs from an annotated subset of 501 representative crystallographic structures of high resolution. This tool was designed to distinguish between biological and crystallographic interfaces. The database contains 3428 different interfaces which were annotated and among them, 686 are considered as functional. The database can be searched through a web server using a large variety of interface properties. Specific descriptors of the protein-protein interface can then be identified and statistics computed for each subset. The influence of residue polarity on the interface, the frequency of aromatic and aliphatic residues at the interface, the hydrogen bonds and the salt bridges can be evaluated. PPDID is particularly useful to construct training sets and benchmarks for protein docking.

- *SCOWLP* [59] (Structural Characterization Of Water, Ligands and Proteins) is another database for detailed characterization and visualization of protein interfaces. It includes proteins, peptidic-ligands and interface water molecules as descriptors of protein interfaces. It contains currently 74,907 protein interfaces and 2,093,976 residue-residue interactions formed by 60,664 structural units and their interacting solvent. The database can be searched through a web-interface using PDB codes or SCOP domain definition. The interface can be displayed and analyzed interactively with Jmol java applet.

General and atomic descriptions of the interface are available (ASA, number of residues involved, nature of this residues, side-chain or main-chains interaction, as well as wet spots. Water molecules play a key role in protein interfaces that is often neglected; they can complement direct interactions contributing significantly to molecular recognition, function, and stability of protein association [60]. Thus, information about water molecules present at the interface is essential to achieve more accurate descriptions of protein interfaces. The water-bridged residues that interact through a water molecule, called wet spots, represent 14.5% of interfacial residues.

As mentioned in the paragraph about properties of PPIs, the notion of hot spots is essential to understand PPIs and their inhibition. Experimentally determined hot spots from alanine scanning mutagenesis experiments have been deposited in the Alanine Scanning Energetics Database (ASEdb, [61]). Binding Interface Database (BID) presents experimentally verified hot spots at interfaces collected from literature [62]. More recently, Tuncbag *et al.,* have developed the HotSprint database to define hot spots at protein interface [63].

- *HotSprint* [63] focuses on highly conserved interfacial residues and is regularly updated. A given interaction can be retrieved using PDB code of the complex or a set of interfaces can be accessed using advanced search criteria such as number of hot spots residues, number of conserved residues, average conservation score, conserved residue propensity, and buried accessible surface area. Overall characteristics and detailed information of the interface are available (interfacial residues, number of Hot Spots, conserved residues, buried ASA). More details about the type of interface can be obtained through the link to the PRINT database. Residues at the interface can be displayed and visualized interactively using Jmol viewer.

Table 1: Structural databases dedicated to PPIs.

Name	URL	Ref
iPFAM	*http://ipfam.sanger.ac.uk/*	[44]
SCOPPI	*http://www.scoppi.org/*	[45]
SNAPPI DB	*http://www.compbio.dundee.ac.uk/SNAPPI*	[46]
PIBASE	*http://salilab.org/pibase/*	[48]
3did	*http://gatealoy.pcb.ub.es/3did/*	[49]
3D Complex	*http://www.3Dcomplex.org*	[50]
PInS	*http://pins.ornl.gov/*	[51]
PROTCOM	*http://www.ces.clemson.edu/compbio/protcom/*	[52]
JAIL	*http://Bioinformaticscharite.de/jail*	[53]
PPIDB	*http://ppidb.cs.iastate.edu/ppidb*	[54]
PRINT	*http://prism.ccbb.ku.edu.tr/interface/*	[55]
ProtorP	*http://www.Bioinformaticssussex.ac.uk/protorp*	[56]
InterProSurf	*http://curie.utmb.edu/prosurf.html*	[57]
PPIDD	*http://ppidd.cgm.cnrs-gif.fr/ppidd/*	[58]
Pisite	*http://pisite.hgc.jp/*	[64]
Interpare	*http://interpare.kobic.re.kr/index.html*	[65]
Proface	*http://resources.boseinst.ernet.in/resources/bioinfo/interface*	[66]
SCOWLP	*http://www.scowlp.org/*	[59]
HotSprint	*http://prism.ccbb.ku.edu.tr/hotsprint/*	[63]

As discussed previously, targetable binding pockets of PPIs are significantly different from those of protein–ligand interfaces. The later use fewer, larger pockets, whereas PPIs and PPI inhibitor interactions use a greater number of smaller pockets instead. Interestingly, the pockets used by PPIs and PPI inhibitors are similar in size to those found anywhere on the surfaces of proteins, and the analysis shows that engaging multiple pockets is likely to be a productive strategy in inhibiting PPIs [33]. The detection of pockets on the protein surface is a great challenge for drug design and PPIs inhibition and several tools are available to predict such cavities : PocketPicker [67], Q-SiteFinder, PockerFinder [68], LigSite [69] and SURFNET [70]. Recently, some of these methods have been combined into the meta server metaPocket to improve the prediction success [71].

Table 2: Servers to find pocket and cavities at the surface of proteins.

Name	URL	Ref
PocketPicker	*http://gecco.org.chemie.uni-frankfurt.de/pocketpicker/*	[67]
PocketFinder	*http://www.modelling.leeds.ac.uk/pocketfinder/*	[68]
Q-SiteFinder	*http://www.modelling.leeds.ac.uk/qsitefinder/*	[68]
LigSite	*http://projects.biotec.tu-dresden.de/cgi-bin/index.php*	[69]
SurfNet	*http://www.biochem.ucl.ac.uk/~roman/surfnet/surfnet.html*	[70]
metaPocket	*http://metapocket.eml.org/*	[71]

MIXED METHODOLOGIES LEADING TO PPIS DRUG DISCOVERY

Although targeting PPIs inhibition has proven challenging, more than 100 small molecule inhibitors have been described in the literature. Here we decided to illustrate some original approaches that led to the discovery of new PPI inhibitors.

2P2I

For several years, our group has defended an innovative strategy that thwart most of the limitations encountered in drug design projects when there are no previously known inhibitors. This strategy combines virtual and experimental screening in few stages and allows a rapid tuning and adaptation of the screening protocol to the target [72]. The Protein-Protein Interaction Inhibition approach (2P2I approach) suffers from the same constraints, and our strategy appeared promising for this purpose. If there are no known inhibitors for the target, or if this target is totally original in the drug discovery perspective, it is impossible to apply with confidence virtual screening approaches. Indeed suitable scoring functions for the system cannot be selected. The same problem appears again for experimental screening approaches which cannot be developed without any positive control compound. At least, few low affinity compounds are needed to circumvent this problem, and this circumvention is the goal of the first part of our 2P2I approach: *"Find a Reference Compound"* (Fig. **3**, in [73]).

We used our consensus scoring function GFscore [74] to eliminate a maximum of non relevant molecules (true negatives) among a small subset of diverse compounds. Because of its generalist purpose, GFscore allowed us to avoid making a blind choice of a specific scoring function. The application on a small diverse library allows a quick exploration of many chemical scaffolds of the largest available chemical space. Once compounds are computationally selected, they must be confirmed *in vitro*. Again, NMR experiments like HSQC or Waterlogsy are well suited for this purpose since they allow the detection of low affinity binders (micromolar or even millimolar ones). These first compounds are exploited in the second part of our 2P2I approach: *"Predefine a Chemical Space"* (Fig. **3**, in [73]).

The first selected "hit" compounds can help as blank and validation to define, develop and tune, *in vitro* or *in cellulo* experimental screening assays, which are deployed on the entire diversity library. This part aims at defining the first small compounds that can inhibit the interaction between the two protein partners. The third 2P2I part explores large compound databases using computational strategies: *"Explore the Chemical Space"* (Fig. **3**, in [73]).

These secondary inhibitors are used to validate docking and three dimensional similarity search experiments. These computational strategies are applied on large databases of millions of compounds to select a few high potential compounds. These molecules are then validated by experimental assays and the resulting best "Hit compound" is sent to combinatorial chemistry to start a Quantitative Structure Activity Relationship program (QSAR) before being patented or published as "Lead Compound".

We have recently adapted our theoretical 2P2I approach to search for inhibitors of the Nef/SH3 protein protein complex as a "Proof of Concept". Several promising inhibitors were proposed [72]. In the *"Find a Reference Compound"* step, we selected one active molecule from the "Diversity" database (NCI, USA), using GFscore as consensus scoring function and the FlexX^{TM} program for HTD. This compound henceforth referred to as D1 and validated by NMR experiments was used to develop a CheckMate™ cellular-based screening assay on the entire "Diversity" database. Seven compounds were selected in this second step -"*Predefine a Chemical Space*"- comprising the new original reference compound, D1. Unfortunately none of them presented a better affinity than this compound. D1 was finally, in the last step "*Explore the Chemical Space*", used as template for a large scale similarity search in the Chembridge EXPRESS-PICK™ Database of 435,000 compounds. Among the 70 compounds selected, one compound presented a better affinity and a better solubility than D1. This "proof of concept" has demonstrated the high potential of this method for its application to other high potential therapeutic protein/protein complexes.

Pharmacophore-based Discovery of Inhibitors of PPIs

The protein–protein interactions (PPIs) between HIV-1 integrase and its cellular cofactor LEDGF/p75 was targeted for the development of new anti-HIV drugs using a structure-based pharmacophoric approach with the LigandScout software tool that allows to rapidly derive 3D pharmacophores from structural data ([75]. The cellular cofactor LEDGF/p75 plays a unique role during HIV integration and the identification of small molecules that interfere with HIV-1 IN–LEDGF/p75 interactions and block HIV replication might prove an enormous driving force in the field of antiretroviral therapeutics. LEDGF/p75 binds HIV-1 integrase via a small binding domain (~80 residues) which is both necessary and sufficient for an efficient interaction. The three dimensional crystal structure of full-length integrase or LEDGF/p75 is not available however, the complex between the interacting domains of integrase and its partner has been solved and was used as a starting point for the *in silico* drug design study. First, the authors defined residues Ile365, Asp366 and F406 as hot spots residues based on the detailed analysis of the complex interface and mutation data from the literature. A 3D pharmacophore model containing 14 features was generated. The *in house* CHIME library which consists of 3055 small molecules was screened *in silico* with Catalyst 4.1 and filtered with the pharmacophore model. This search returned 936 compounds among them 234 structures with a Catalyst fitness score above 2.0. The best compound CHIBA-3002 was tested *in vitro* and found to inhibit the integrase/LEDGFp75 interaction by up to 46% at a concentration of 100μM. However, visual inspection of the binding mode of CHIBA-3002 suggested suitable chemical modifications to improve this compound. Four CHIBA-3002 analogues were rationally designed, synthesized and tested *in vitro*. Two compounds showed a better inhibition activity than the original CHIBA-3002 with percentage of inhibition at 100 μM of 56 and 71 respectively and IC_{50} of 76 and 35 μM respectively. Although, more potent inhibitors would be necessary as therapeutic drugs, this study suggests that structure-based drug design can be used as a general approach in PPI inhibition strategies.

Dynamic Target-based Pharmacophoric Model

Glycoprotein (gp) 120 subunit is a protein from the HIV envelop recognized as a major target for drug discovery. It interacts with the cellular receptor protein CD4 and disruption of this interaction with small molecule inhibitors is a major challenge in the development of new treatment against HIV. Virtual screening of a commercially available library and a dynamic target-based pharmacophoric model were used to find inhibitors of gp120-CD4 PPI [76]. There is a large amount of structural data concerning the interaction between gp120 and CD4 [77]. The hydrophobic pocket of gp120 which interacts with highly conserved Phe43 of CD4 was chosen as the target for computational simulations. Molecular dynamics (MD) simulations in explicit water with AMBER were performed on the gp120 core structure to account for the flexibility of unbound protein in the protocol. Two thousand protein snapshots were saved and clustered on the basis of the root mean square deviation of gp120 residues within 10 Å from CD4 residues. Five frames, representing the ensemble of conformations of the key residues were selected. For each frame, molecular interaction fields (MIFs) with different probes were then computed with Peter Goodford's program GRID to identify favorable binding sites [78]. The results for the five frames were combined and used to generate a Pharmacophoric model that was then used as a three-dimensional query to perform a virtual screening of the Asinex database containing about 200,000 commercially available compounds. A total number of 729 compounds were selected and ranked using Chemscore and Goldsore scoring functions in a consensus docking protocol. The top 10% were visually inspected to discard compounds not binding into the Phe43 cavity. Finally, five entries were purchased and evaluated for their ability to inhibit the HIV-induced cytopathic effect. The activity and toxicity were compared to those of compound #155, a known inhibitor discovered through a cell-based screening assay [79]. Two compounds showed micromolar specific inhibition of HIV-1 replication in cells attributable to an interference with the entry step of infection, by direct interaction with gp120. This approach led to the identification of two novel chemical scaffolds, different from all known gp120–CD4 inhibitors that may serve as starting point for the development of more potent inhibitors.

SAR by NMR Approach

Proteins from the Bcl-2 or Bcl-X$_L$ families are overexpressed in many cancers and the development of potential anti-cancer therapeutics involves disrupting the interaction with their protein partners. Oltersdorf *et al.,* have used NMR-based screening, parallel synthesis and structure based design, to identify ABT-737, a small-molecule inhibitor of the anti-apoptotic proteins Bcl-2, Bcl-XL and Bcl-w [80], with an affinity two to three orders of magnitude more potent than previously reported compounds [81-83]. They used a method called 'SAR by NMR' [84] to screen a chemical library. In this method small organic molecules that bind to proximal subsites of a protein are identified, optimized, and linked together to produce high-affinity ligands. Two compounds that bind to the groove of Bcl-X$_L$ were first identified with Kds of 0.3 and 4.3 mM respectively. These small molecules bind to two distinct but proximal subsites within this cleft. In the next step, they used site directed synthesis to design new ligands by linking the two fragments together; the main challenge being to maintain the correct positioning of the key interacting atoms. A compound with high affinity to Bcl-X$_L$ (inhibitory constant Ki = 36 ± 1.6 nM) was discovered within the set of designed molecules. A structure-based approach was used to improve the selectivity of this compound. The lead compound, ABT-737 binds with high affinity (Ki < 1 nM) to Bcl-X$_L$, Bcl-2 and Bcl-w, but not to the less homologous proteins Bcl-B, Mcl-1 and A1 (Ki = 0.46±0.11 mM, >1μM and >1μM, respectively). A mammalian two-hybrid system was used to validate that ABT-737 can disrupt an intracellular Bcl-2 family protein–protein interaction. Following this work, ABT-737 has been modified into ABT-263 to make the drug orally available; this new drug is in clinical trials [18].

Inhibition of PPIs with Peptide Mimetics

Many PPIs are mediated by a short peptide region from one of the two partners. This interaction usually involves an alpha-helix or a beta strand secondary structure element at the interface and can be inhibited by small peptides, peptidomimetics or small organic molecules. The development of peptides as modulators of PPIs has been extensively reviewed recently [85]. The most common strategy to search for peptides that disrupt PPIs is to mimic the sequence of one of the interacting partners. This strategy has been applied successfully to inhibit the interaction between the catalytic domain of protein kinases and its regulatory domains [86]. Another popular strategy is to use peptidomimetics compounds that can mimic a natural peptide and retain the ability to interact with the biological target [87]. Peptides are flexible compared to small organic compounds and the main challenge in rational drug design is to account for this flexibility.

Alpha-helix Mimetics

The development of small organic molecules that can mimic the characteristics of alpha-helices constitutes an attractive drug discovery approach in targeting disease areas such as cancer, antimicrobials, and AIDS.

The challenge in designing organic compounds that mimic alpha-helices is to find scaffolds able to project functionality in a similar spatial arrangement as the peptidic side chain. A large number of helix mimetics have been designed using biphenyl [88,89] or terphenyl scaffolds ([90-92]). The aromatic core reproduces the backbone of the natural peptide whereas specific substitutions on key positions of the aromatic rings mimic the orientation of side chains. The strategy has been applied with success in the design of PPIs inhibitors of different complex families such as smMLCK/calmodulin [90], gp120/gp41, Bcl-X$_L$/Bak [93], p53/hDM2 [94]. More recently, new scaffolds have been developed in which nitrogen or oxygen heteroatoms are introduced in the general terphenyl scaffold to reproduce the hydrogen-bonding propensity of the peptide [95].

Stapled Peptides

Strategies based upon the development of peptides have historically failed because peptides lack the ability to enter cells, and peptides are inherently unstable within the body and are rapidly metabolized and filtered from the blood stream by the kidneys within minutes. One possibility to circumvent this issue is based on the fact that proteases can only recognize and break down peptides when they are unraveled, so if they are locked into certain folded shapes they are protected. Walensky *et al.* proposed to 'staple' peptides into an alpha-helical shape with an optimized cross-linking chemistry, mimicking the structure found at the interface of many PPIs [96-98]. The resulting Stapled Peptide drugs are endowed with unique properties, including efficient cell penetration, high affinity binding to large target protein surfaces, and excellent stability within the body. In preclinical studies, Stapled Peptide drugs have been shown to possess remarkable potency, *in vivo* stability and cell permeability. At the 2008 American Association for Clinical Cancer Research (AACR) annual meeting, AILERON presented data showing excellent *in vitro* and *in vivo* efficacies for Stapled Peptide drugs derived from the BH3 alpha-helical domains of the BID and BIM proteins in multiple cancer models. These Stapled Peptides targeted the rate-limiting step in programmed cell death, and this mechanism of action provides efficacy across a very broad range of cancers and also avoids known resistance pathways. These Stapled Peptides also possess *in vivo* half-lives greater than three hours and have displayed no major toxicities or other safety issues.

Other computational tools are being developed to facilitate the search for peptide-mimicking molecules [99,100]. A set of peptides which have been characterized by NMR or X-ray crystallography in complex with a protein partner have been used to investigate whether short peptides regions from PPIs can be mimicked with small molecules inhibitors [101]. The authors have defined 41 pharmacophores representing 30 classes of peptides based on the known 3D structures and searched for possible mimetics by virtually screening against 1253 FDA approved drug compounds. They found 204 small molecule compounds that satisfy at least 90% of the pharmacophoric constraints. Although this method is promising for the *in silico* design of peptidomimetics it still needs further validation.

Table 3: Summary of the drug discovery process for the PPI targets described in the above paragraph. General strategy and screening library used. The best small molecule inhibitor found as a direct result of the study or through lead optimization is presented as well as percentage of inhibition when available. Binding affinity to the target is given in terms of dissociation constant (K_d), half maximal inhibitory concentration (IC50), effective concentration (EC50) or inhibition constant (K_i).

PPI Target	Strategy	Library	Best Compound	% inhibition	Affinity (μM)	Ref
Nef/SH3	2P2I approach	Chembridge	DLC-27	>75% at 20μM	0.98 (K_d)	[72]
IN/LEDGF-p75	3D pharmacophore	CHIME	CHIBA-3003	71% at 100μM	35 (IC$_{50}$)	[75]
gp120/CD4	MD and 3D pharmacophore	Asinex	0775946	na	9 (EC$_{50}$)	[76]
Bcl-X$_L$	SAR by NMR	*in house* chemical library	ABT-737	55% at 0.1μM	0.001 (K_i)	[80]

COMPLEXES WITH KNOWN INHIBITORS

PPI inhibitors have been developed for several classes of protein. In this paragraph, some of the major targets in cancer, immune response, virus and antibacterial therapies are briefly described. The different target families with known three dimensional structures of the protein-protein and protein-ligand complexes were used as a test case to study the specific properties of PPIs that can be disrupted by small organic molecules ([102], and Table **4**).

Cancer Therapy

MDM2/p53 (Family #2, Table 4)

Tumor suppressor protein 53 (p53) is an attractive therapeutic target because it can be functionally activated to eradicate tumors [94]. In many tumor types, p53 is deleted or mutated, but in others p53 is inactivated by overexpression or amplification of its negative regulator human double minute 2 (HDM2) or its mouse homologue (MDM2). Disruption of the HDM2/p53 interaction with small organic molecules may offer a strategy for cancer therapy [103]. The crystal structure of MDM2 in complex with p53 was a major breakthrough to identify the key residues involved in the interaction [104]. The interaction involves a 15-residue α-helical peptide from p53 binding to a small but deep hydrophobic groove of HDM2. Alanine-scanning mutational analysis of the 15-residue peptide identified three dominant hydrophobic amino acids in the centre of the interface: Phe-19, Trp-23 and Leu-26. Several classes of inhibitors have been discovered and the most potent is Nutlin-3 [105-107] which belongs to the family of cis-imidazoline analogues and is able to disrupt HDM2–p53 complexes with an IC50 of 90 nM. Other classes have been reported including analogs of spiro-oxindole [108,109], benzodiazepinedione [110,111], terphenyl [92], quilinol [112], isoquinolin [113], chalcone [114] and sulfonamide [115]. They were discovered using different approaches such as in experimental HTS [116], focused libraries [114], *in silico* HTS [117], pharmacophore search [112,115,118], scaffold hoping and NMR [113] or structure-based de novo design [108,109]. Three dimensional structures of some MDM2/p53 disruptors in complex with the MDM2 target are available [110,117,119]. Analogs of MI-219 and Nutlin-3 are in preclinical development or early-phase clinical trial and thus very promising therapeutic agents.

IL-2/IL-2Rα (Family #7, Table 4)

Interleukin-2 (IL-2) is a 133 amino acid glycoprotein that belongs to a class of soluble, regulatory proteins known as cytokines. It stimulates the proliferation, survival and cytokine production of effectors T cells. IL-2 can also promote the suppression of inflammatory responses through its binding to the corresponding cell surface receptor (IL2Ralpha). IL-2/IL-2R is considered to be an important therapeutic target for immune disease that could be addressed through the development of small-molecule antagonists of the protein-protein interface. IL-2 adaptive region contains at least two cooperative binding sites where the binding of a first ligand to one site promotes or antagonizes the binding of a second ligand to the second site [120]. Several small molecule inhibitors that prevent interaction between IL-2 and its receptor have been developed [121-123]. Using a fragment assembly approach, Raimundo and collaborators have identified a 60 nM inhibitor of the IL-2/IL-2Rα interaction [124]. Briefly, they started with two small fragments binding to different sites at the interface with low affinities and used a modular chemical approach linking the two fragments together to improve the affinity in the micromolar range. These compounds were optimized into a lead series with nanomolar affinity by incorporating fragments identified through tethering. A high-affinity small molecule, SP4206 was found to block IL-2/IL-2Ralpha interaction with a Kd of 60 nM. The structural characterization of the IL-2/IL-2Rα complex [23] provided rationale for the binding of small molecule inhibitors at the interface [24]. Although the surface of IL-2 undergoes local structural adaptations of the binding site upon binding of SP4206, this small molecule targets virtually the same critical "hot-spot" residues on IL-2 that drive binding of IL-2Rα. The protein and small molecule bind to different conformations of IL-2 which shows that precise structural mimics of receptors are not required for high-affinity binding of small molecules.

Bcl-XL/Bak (Family #1, Table 4)

To survive, cancer cells typically acquire changes enabling evasion of death signals. One way they do this is by increasing the expression of anti-apoptotic B-cell lymphoma-2 (Bcl-2) proteins. Their function is regulated by the binding of anti- or pro-apoptotic factors such as Bak, Bax or Bid [125]. The development of specific agents interfering between Bcl-2 and its partners would be valuable for cancer therapy [126,127]. Resolution of the crystal structures of Bcl-2 family proteins, in the free form or in complex with the interacting protein has opened the way to the development of the small molecule inhibitors. The interaction with Bcl-2 involves an alpha helical segment from Bak and alanine-scanning mutational analysis of the Bak-derived peptide identified several key residues that are crucial for binding. Several laboratories have developed small molecules binding Bcl2 and/or Bcl-X_L. Solution NMR structures of small molecule inhibitors in complex with the target Bcl-X_L protein have been characterized [80,128]. The most potent inhibitor, ABT-737 was developed by Abbot Laboratories using a fragment-based nuclear magnetic resonance (NMR) method known as SAR by NMR [80,129]. This compound has a Ki of 0.6 nM; its affinity is therefore comparable to that of the α-helix. ABT-737, exhibits monotherapy efficacy in xenograft models of small-cell lung cancer and lymphoma. The x-ray structure of this compound bound to Bcl-X$_L$ has been solved [130]. Interestingly, the small molecule ABT-737 does not closely mimic the atomic details of the peptide. Instead, the small molecule traps a slightly different conformation of Bcl-X$_L$. A derivative of compound ABT-737, ABT-263, has progressed to phase I/II clinical testing [18]. Another small-molecule inhibitor of Bcl-2 family proteins, TW-37, induces cell growth inhibition and induces apoptosis in pancreatic cancer [131,132].

HPV E1/E2

Human papilloma virus (HPV) is the causative agent of warts and the "high-risk" HPV types are associated with cancers [133]. At present, there is no small-molecule drug that can treat these conditions. The interaction between the viral transcription factor E2 and the viral helicase E1 is crucial for the viral DNA replication and thus is an important PPI target. It has been reported that a 15-amino acid peptide derived from HPV16 E2 was able to inhibit the E1-E2 interaction *in vitro* [134]. A class of indandiones that disrupts this interaction ($Kd \approx 20$ μM) was found using HTS [135]. These compounds were later optimized by SAR and chemical design to high potent inhibitors with IC50 values as low as 6 nM [136,137]. The crystal structure of the complex between E2 and an inhibitor was solved and revealed that binding of the inhibitor caused no significant alteration of the protein backbone. However, some local side chains rearrangement resulting in the formation of a deep hydrophobic pocket that accommodates the indandione moiety were observed [138]. The structure of E2 in complex with E1 was reported soon later [139] revealing that the indandione derivative is in contact with only 7 out of the 20 residues of E2 that are in contact with E1. Interestingly, the inhibitor binds a cavity that is not observed in the protein–protein interface.

TNFα

The cytokine tumour-necrosis factor (TNF) plays a key role in inflammatory responses. TNFα receptors also called "death receptors" are potent regulators of apoptosis and therefore represent an attractive therapeutic approach. A small molecule that inhibits TNFα binding to its receptor has been discovered by using fragment screening [140]. This compound (SP304) does not directly disrupt the interaction between TNFα and its receptors instead it promotes TNFα trimer disassembly. The crystal structure of SP304 bound to TNFα revealed that the binding of the small molecule induces formation of a TNF dimer whereas in the free form TNFα assembles as a trimer. Compound SP304 is composed of trifluoromethylphenyl indole and dimethyl chromone moieties linked by a dimethylamine spacer and showed an inhibitory concentration (IC50) of 22 μM for inhibiting *in vitro* TNF receptor 1 (TNFR1) binding to TNFα. Although this molecule possesses moderate affinity it demonstrates that even constitutive homo-oligomeric interfaces can be disrupted with small organic molecules.

XIAP (Families #3 and 4, Table 4)

Inhibitor of Apoptosis Proteins (IAPs) such as XIAP, cIAP1, and cIAP2 constitute a family of regulatory proteins that suppress cell death. Several compounds targeting (IAP) and inducing cell death in cancer cells have been developed [141]. The recent discovery of Smac/DIABLO (Second Mitochondria-derived Activator of Caspases/Direct IAp Binding with LOw pI) and the elucidation of its structure and function have led to the rapid development of Smac mimetics for use in cancer treatment. To date, at least one Smac mimetic has been advanced into clinical development. Several other Smac mimetics are in an advanced preclinical development stage and are expected to enter human clinical testing for the treatment of cancer in the near future [141]. The crystal structure of the Smac/XIAP complex was solved and served as a guide for small molecule inhibitor design [142]. Embelin, the first PPI inhibitor that binds to the XIAP BIR3 domain, was discovered through computational structure-based database screening of an in-house traditional herbal medicine three-dimensional structure database of 8221 individual natural products, followed by biochemical testing of selected candidate compounds [143]. More potent embelin analogues (Ki 180 nM binding to XIAP BIR3) were later designed as lead compounds [144]. On the basis of the crystal structure of Smac in complex with XIAP BIR3 domain, a series of conformationally constrained, bicyclic Smac mimetics were designed [145]. These compounds bind to XIAP BIR3 with high affinities (Ki 14, 26 and 67 nM) and antagonize XIAP BIR3 in cell-free functional assays. The determination of the crystal structure of the most potent compound in complex with XIAP BIR3 provided the structural basis for its high-affinity interaction with XIAP BIR3 and for further design and optimization.

Recently, a series of 4-substituted azabicyclo[5.3.0]alkanes mimicking the N-terminal tetrapeptide sequence of Smac/DIABLO have been designed and tested for their ability to disrupt the interaction with XIAP [146]. The crystal structure of the BIR3 domain from XIAP in complex with the three most promising compounds has been solved and confirmed their predicted mode of binding [147]. Novel Smac/IAPs inhibitors have been designed, synthesized and characterized. Computational models and structural studies (crystallography, NMR) have elucidated the SAR of this class of inhibitors, and have permitted further optimization of their properties [148].

PDZ (Peptide)

PDZ domains play a critical role in organization and function of cellular signaling pathways. Thus, small molecule inhibitors of PDZ domain association with their targets have wide potential applications as research and therapeutic agents. PDZ domains mediate crucial protein–protein interactions and progress in the development of small-molecule PDZ inhibitors has been reviewed recently [149]. Different approaches have been taken to identify and develop PDZ domain inhibitors, including NMR based screening and chemical synthesis. Inhibitors of the tumor suppressor PTEN with the second PDZ domain of MAGI have been found using rational structure based design [150].

Dishevelled (Dvl) is an essential protein in the Wnt signaling pathways; it uses its PDZ domain to transduce the Wnt signals from the membrane receptor Frizzled (Fz) to downstream components. Drug-like small molecule inhibitors of the PDZ domain of Dvl have been characterized through structure-based ligand screening and NMR spectroscopy. The best Dvl PDZ domain inhibitor has a binding affinity to Dvl, which is similar to that of the natural partner Fz [151].

Using a pharmacophore model based on known inhibitors Shan and Zheng have screened the ChemDiv database and identified fifteen new inhibitors [152]. By using NMR spectroscopy they showed that all fifteen compounds bound to the PDZ domain. The different small molecules were used to developed SAR models of PDZ domain ligands, which they intend to use for hit to lead optimization.

Immune Response Therapy

CD80/CD28

The CD80-CD28 costimulatory pathway plays a critical role in regulation of the immune response and thus constitutes an attractive target for therapeutic manipulation of autoimmune diseases and transplantation outcomes. Compounds that specifically blocked binding of CD80 to CD28 were identified using a strategy involving a cell-based scintillation proximity assay as the initial step. Secondary screening revealed that these compounds are highly selective CD80 binders. Screening of structurally related derivatives led to the identification of the chemical features required for inhibition of the CD80-CD28 interaction. In addition, the optimization process led to a 10-fold increase in binding affinity of the CD80 inhibitors. These compounds serve as promising starting points for further development of CD80 inhibitors as potential immunomodulatory drugs [153]. CTLY4Ig, a second-generation small molecule that blocks the interaction between CD80 and CD28 is currently in preclinical and clinical trials [154].

HIV Therapy

Different classes of drugs have been developed against HIV targeting key stage of the virus replication cycle. However, because of its short replication cycle and the high-error rate of reverse transcriptase, HIV mutates very quickly, resulting in the rapid emergence of drug resistant HIV strains. Successful strategies involve the use of "drug cocktails" to maximize potency, minimize toxicity and diminish the risk for resistance. New targets involving protein–protein interactions are currently investigated for their therapeutic potential.

gp120/CD4

The HIV-1 replication cycle is initiated by the attachment of the viral envelope glycoprotein gp120 to host cells through binding to the cellular CD4 receptor. A series of compounds that disrupt efficiently gp120/CD4 interaction have been developed [155,156]. Lead optimization using structure activity relationship (SAR) led to the discovery of an inhibitor which exhibited antiviral potency in a cell based assay in the subnanomolar range [157]. Using the same approach, a series of piperazine derivatives have been described [158]. BMS-378806 discovered through a cell-based screening assay [159] has been optimized using SAR. The new lead compound BMS-488043 has been tested in phase II clinical trials [155].

Integrase/LEDGF-p75

Integration of viral-DNA into host chromosome mediated by the viral protein HIV-1 integrase (IN) is crucial step in the HIV-1 life cycle. The interaction with the integrase binding protein lens epithelium-derived growth factor (LEDGF)/p75 is essential and therefore disruption of the LEDGF/p75 interaction with IN has provided a special interest for anti-HIV agent discovery [160,161]. A benzoic acid derivative, (D77) was shown to inhibit the IN-LEDGF/p75 interaction and to exhibit antiretroviral activity. The possible binding mode of this compound was elucidated through molecular docking, site-directed mutagenesis analysis and surface plasmon resonance [162]. More recently, a different family of compounds (CHIBA-3003) was discovered using Pharmacophoric models ([75] and paragraph 3 of this chapter). Interestingly, the two series of compounds do not bind the same IN-binding site.

Antibacterial Therapy

ZipA/FtsZ (family #5, Table 4)

The ZipA-FtsZ PPI involved in separation of bacterial cells during cell division is a potential target for antibacterial therapy. A high-resolution x-ray structure of the FtsZ/ZipA complex revealed that FtsZ binds to its partner through a 17-residue peptide. Analysis of the structure together with alanine-scanning mutagenesis pointed out four interfacial residues that constitute a hot spot and were in drug discovery. NMR-based fragment screening approach was used to screen a diverse set of 825 compounds that yielded 7 molecules that bound to ZipA at the same site as FtsZ. However, none of the molecules was able to disrupt the FtsZ/ZipA interaction at a therapeutically relevant level of potency [163]. Other weak inhibitors were found using combinatorial chemical synthesis [164,165]. A potentially interesting inhibitor (Ki of 12 µM) was identified using HTS of 250,000 compounds [166]. However, this compound had potential issues with toxicity as it showed rather strong, non-specific activity in both bacterial and yeast-based assays. High resolution x-ray structure of this lead with ZipA was used to guide further development. The shape comparison program ROCS (Rapid Overlay of Chemical Structures) was used to identify new scaffolds with similar shapes within a multi-conformer database [166]. The crystal structure of one of the new discovered scaffold was solved in complex with ZipA and confirmed that it binds to ZipA at the same site as the FtsZ peptide [166].

CONCLUSIONS AND PERSPECTIVES

The long-standing notion that PPIs interfaces are excessively complicated and thus poorly druggable should be considered as an 'old story' in the next decade. As successes in discovering protein–protein inhibitors accumulate, principles will begin emerging, making successive clinical attempts more effective.

We have described in this review the particularities of the PPI chemical space and the main biophysical reasons making them challenging targets for the drug discovery process. A state of the art of protein databases and servers dedicated to PPIs was presented, with a particular emphasis on their analysis and inhibition. We have then depicted current approaches and some innovative methodologies that led to the development of specific inhibitors for the most common targets. During this assessment, we have pinpointed that the rational design of specific drugs targeting PPIs is greatly facilitated by structural data of the protein complex through alanine scanning or three-dimensional characterization to identify interaction hot spots. The incorporation of dynamic parameters and key interfacial water molecules during the hit discovery process greatly enhances the chances of success.

To this critical question *"What makes a PPI an attractive target for the discovery and development of small molecules?"* we have portrayed the good PPI candidates in few points that can be summarized from existing published data as follows i) The target should be validated at least biologically (RNAi, knock-out gene etc...) or, even more importantly, clinically; and the inhibition of the PPI should not be associated with toxicity ii) key residues involved in the interaction and defining hot spots should be characterized through mutagenesis analyses, alanine scanning or using server prediction; iii) Structural information on the PPI complex and/or unbound forms should be available to accelerate the process; bridging water molecules should be taken into account when available and since the best conformation to target for PPI inhibition is not necessarily the one found in the complex, the natural conformational dynamics of the target should be investigated through molecular dynamic simulations [167], normal mode analysis [168], conformation ensemble predictions [35] or NMR [37]; iv) Three to five pockets should be available at the interface in the free form or in the simulated conformations of the target, for a total volume of at least 250 Å^3, [33].

A recent study concluded that small molecule PPI modulators are larger (average molecular weight > 400 Da), more hydrophobic (average alogP ~4), with more aromatic rings (~4 in average) and make fewer hydrogen bonds with the protein than average drugs [169-171]. However, a detailed description of the chemical and topological spaces of protein interfaces that can be disrupted by small drugs is not available. Such a description would lead to a better definition of a potentially successful PPI target.

To define the general parameters that characterize known PPIs targets, we analyzed 16 protein-protein hetero-complexes that have been structurally characterized by x-ray crystallography or NMR (Table **4**). For each complex family, a three dimensional structure of one of the two partners bound to a small molecule inhibitor was also available, representing a total of 39 different small molecules inhibitors ([102] and Fig. **1**).

Table 4: PPI partners available in $2P2I_{DB}$.

#	Family	PP[a]	PI[b]	SMI[c]	Ref
1	Bcl-X$_L$/Bak	1bxl	1ysn, 2o2m, 1ysi, 2yxj, 1ysg, 1ysg, 2o2n, 2o22	1, 2, 3, 4, 5, 6, 7, 8	[80,128,130]
2	MDM2/p53	1ycr	1t4e, 1rv1	9, 10	[110,117]
		1ycq	1ttv	11	[119]
3	XIAP BIR3/CASPASE 9	1nw9	1tfq, 1tft	12, 13	[172]
4	XIAP BIR3/SMAC	1g73	2jk7, 2opy, 3clx, 3cm7, 3cm2, 3eyl	14, 15, 16, 16, 17, 18	[145-147,173]
5	ZipA/FtsZ	1f47	1y2g, 1s1j, 1s1s, 1y2f	19, 20, 21, 22	[165,166]
6	Chagasin/Papain	3e1z	1cvz	23	[174]
7	IL-2/IL-2R	1z92	1m49, 1pw6, 1m48, 1py2, 1nbp, 1m4a, 1m4b, 1qvn	24, 25, 26, 27, 28, 29, 30, 31	[120,123,175]
8	MMP1/TIMP1	2j0t	966c	32	[176]
9	MMP3/TIMP1	1oo9	1usn	33	[177]
10	Subtilisin/Eglin C	1cse	1bh6	34	[178]
		1r0r	1bh6	34	[178]
		1to2	1bh6	34	[178]
11	FKBP12/TGFR	1b6c	1j4r	35	[179]
12	Thrombin/Heparin	3b9f	1jwt	36	[180]
13	Trypsin/Trypsin inhibitor	2uuy	1aq7, 1az8, 1auj	37, 38, 39	[181]

[a] PDB code of the protein/protein complex, [b] PDB code of the protein/inhibitor complex, [c] small molecule inhibitor (see Fig. **1**).

We have analyzed the protein/protein and protein/inhibitor interfaces in terms of geometrical parameters, atom and residue properties, and buried accessible surface [102]. The results of our study permit to expand the current knowledge with new data, focusing at the interface of protein/protein complexes with prior structural knowledge. We have created the 2P2I database ($2P2I_{DB}$) containing the dataset used in this survey. This interacting database is publicly available and can be accessed and interrogated online at http://2p2idb.cnrs-mrs.fr. The general properties of the interfaces found in $2P2I_{DB}$ were compared to those of representative datasets of heterodimers complexes.

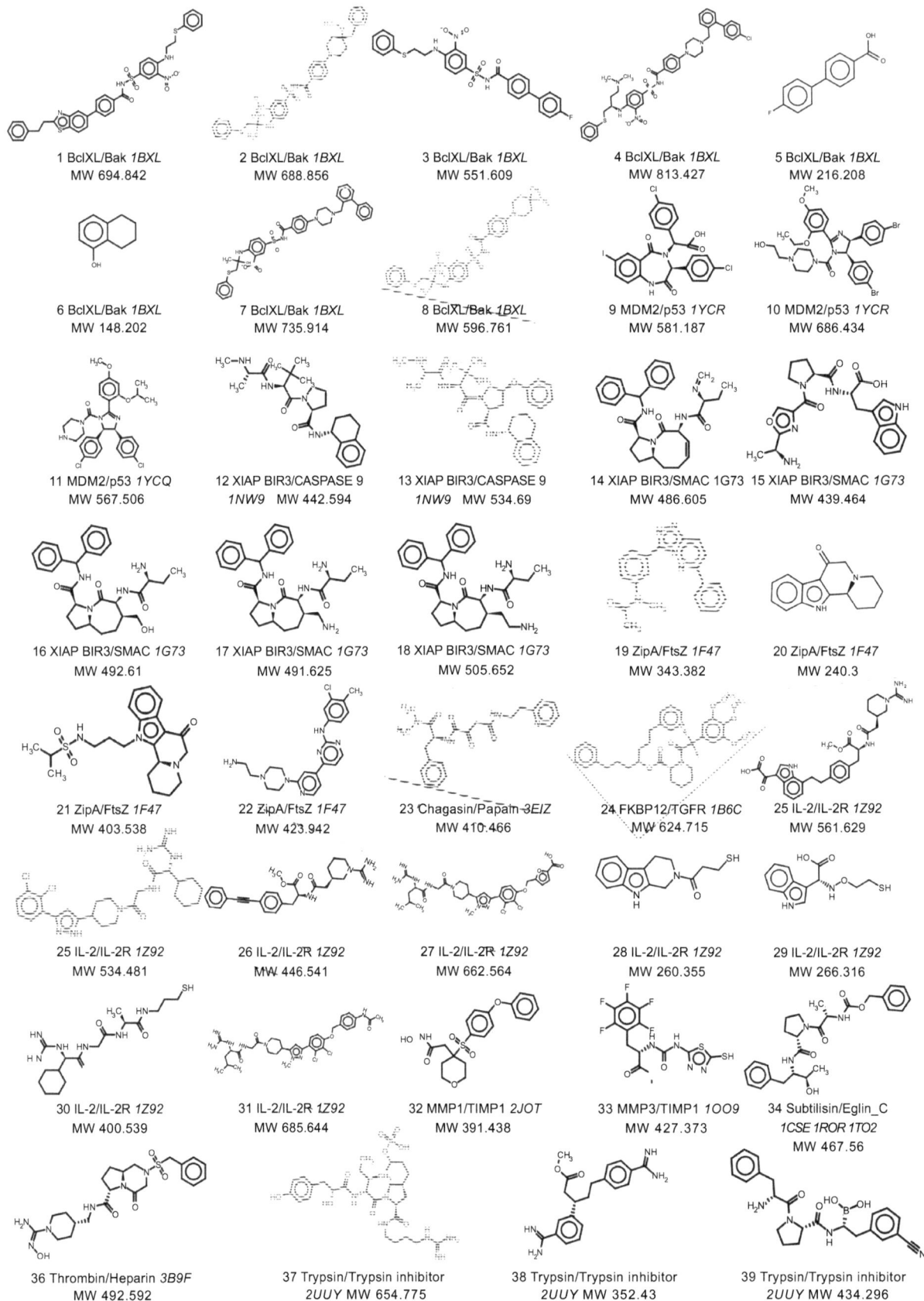

1 BclXL/Bak *1BXL*
MW 694.842

2 BclXL/Bak *1BXL*
MW 688.856

3 BclXL/Bak *1BXL*
MW 551.609

4 BclXL/Bak *1BXL*
MW 813.427

5 BclXL/Bak *1BXL*
MW 216.208

6 BclXL/Bak *1BXL*
MW 148.202

7 BclXL/Bak *1BXL*
MW 735.914

8 BclXL/Bak *1BXL*
MW 596.761

9 MDM2/p53 *1YCR*
MW 581.187

10 MDM2/p53 *1YCR*
MW 686.434

11 MDM2/p53 *1YCQ*
MW 567.506

12 XIAP BIR3/CASPASE 9
1NW9   MW 442.594

13 XIAP BIR3/CASPASE 9
1NW9   MW 534.69

14 XIAP BIR3/SMAC 1G73
MW 486.605

15 XIAP BIR3/SMAC *1G73*
MW 439.464

16 XIAP BIR3/SMAC *1G73*
MW 492.61

17 XIAP BIR3/SMAC *1G73*
MW 491.625

18 XIAP BIR3/SMAC *1G73*
MW 505.652

19 ZipA/FtsZ *1F47*
MW 343.382

20 ZipA/FtsZ *1F47*
MW 240.3

21 ZipA/FtsZ *1F47*
MW 403.538

22 ZipA/FtsZ *1F47*
MW 423.942

23 Chagasin/Papain *3EIZ*
MW 410.466

24 FKBP12/TGFR *1B6C*
MW 624.715

25 IL-2/IL-2R *1Z92*
MW 561.629

25 IL-2/IL-2R *1Z92*
MW 534.481

26 IL-2/IL-2R *1Z92*
MW 446.541

27 IL-2/IL-2R *1Z92*
MW 662.564

28 IL-2/IL-2R *1Z92*
MW 260.355

29 IL-2/IL-2R *1Z92*
MW 266.316

30 IL-2/IL-2R *1Z92*
MW 400.539

31 IL-2/IL-2R *1Z92*
MW 685.644

32 MMP1/TIMP1 *2JOT*
MW 391.438

33 MMP3/TIMP1 *1OO9*
MW 427.373

34 Subtilisin/Eglin_C
1CSE 1ROR 1TO2
MW 467.56

36 Thrombin/Heparin *3B9F*
MW 492.592

37 Trypsin/Trypsin inhibitor
2UUY MW 654.775

38 Trypsin/Trypsin inhibitor
2UUY MW 352.43

39 Trypsin/Trypsin inhibitor
2UUY MW 434.296

Figure 1: 2D structure of the 39 small molecule inhibitors available in 2P2I_{DB} database. For each compound, the protein partners, the PDB code of the protein/protein complex and the molecular weight are indicated.

From this analysis, we have proposed a new classification of protein-protein interfaces with known inhibitors into specific families [102]. Such a classification should lead to a better definition of a potentially successful PPI target and accelerate the process of designing new PPI drugs. As successes in discovering PPI inhibitors accumulate, the parameters will be refined and the classification updated. These optimized parameters will be used to retrieve putative targets from the whole PDB databank. Identification of new inhibitors for the best targets starting from our in-house library would provide ultimate proof-of-concept.

ACKNOWLEDGEMENTS

We would like to thank Marie Jeanne Basse and Raphaël Bourgeas for their help in the preparation of this review.

REFERENCES

[1] Overington, J.P.; Al-Lazikani, B.; Hopkins, A.L. How many drug targets are there? *Nat. Rev. Drug Discov.*, **2006**, *5*, 993-996.

[2] Wells, J.A.; McClendon, C.L. Reaching for high-hanging fruit in drug discovery at protein-protein interfaces. *Nature.*, **2007**, *450*, 1001-1009.

[3] Pagliaro, L.; Felding, J.; Audouze, K.; Nielsen, S.J.; Terry, R.B.; Krog-Jensen, C.; Butcher, S. Emerging classes of protein-protein interaction inhibitors and new tools for their development. *Curr. Opin. Chem. Biol.*, **2004**, *8*, 442-449.

[4] Arkin, M.R.; Whitty, A. The road less traveled: modulating signal transduction enzymes by inhibiting their protein-protein interactions. *Curr. Opin. Chem. Biol.*, **2009**, *13*, 284-290.

[5] Berg, T. Small-molecule inhibitors of protein-protein interactions. *Curr. Opin. Drug Discov. Devel.*, **2008**, *11*, 666-674.

[6] Fry, D.C. Drug-like inhibitors of protein-protein interactions: a structural examination of effective protein mimicry. *Curr. Protein Pept. Sci.*, **2008**, *9*, 240-247.

[7] Dömling, A. Small molecular weight protein-protein interaction antagonists: an insurmountable challenge? *Curr. Opin. Chem. Biol.*, **2008**, *12*, 281-291.

[8] Villoutreix, B.O.; Bastard, K.; Sperandio, O.; Fahraeus, R.; Poyet, J.; Calvo, F.; Déprez, B.; Miteva, M.A. In silico-in vitro screening of protein-protein interactions: towards the next generation of therapeutics. *Curr. Pharm. Biotechnol.*, **2008**, *9*, 103-122.

[9] Patel, S.; Player, M.R. Small-molecule inhibitors of the p53-hdm2 interaction for the treatment of cancer. *Expert Opin. Investig. Drugs.*, **2008**, *17*, 1865-1882.

[10] Toogood, P.L. Inhibition of protein-protein association by small molecules: approaches and progress. *J. Med. Chem.*, **2002**, *45*, 1543-1558.

[11] Wells, J.A. Structural and functional epitopes in the growth hormone receptor complex. *Biotechnology (N.Y.).*, **1995**, *13*, 647-651.

[12] Bahadur, R.P.; Zacharias, M. The interface of protein-protein complexes: analysis of contacts and prediction of interactions. *Cell. Mol. Life Sci.*, **2008**, *65*, 1059-1072.

[13] Jones, S.; Thornton, J.M. Principles of protein-protein interactions. *Proc. Natl. Acad. Sci. U.S.A.*, **1996**, *93*, 13-20.

[14] Lo Conte, L.; Chothia, C.; Janin, J. The atomic structure of protein-protein recognition sites. *J. Mol. Biol.*, **1999**, *285*, 2177-2198.

[15] Smith, R.D.; Hu, L.; Falkner, J.A.; Benson, M.L.; Nerothin, J.P.; Carlson, H.A. Exploring protein-ligand recognition with binding moad. *J. Mol. Graph. Model.*, **2006**, *24*, 414-425.

[16] Cheng, A.C.; Coleman, R.G.; Smyth, K.T.; Cao, Q.; Soulard, P.; Caffrey, D.R.; Salzberg, A.C.; Huang, E.S. Structure-based maximal affinity model predicts small-molecule druggability. *Nat. Biotechnol.*, **2007**, *25*, 71-75.

[17] Bahadur, R.P.; Chakrabarti, P.; Rodier, F.; Janin, J. A dissection of specific and non-specific protein-protein interfaces. *J. Mol. Biol.*, **2004**, *336*, 943-955.

[18] Tse, C.; Shoemaker, A.R.; Adickes, J.; Anderson, M.G.; Chen, J.; Jin, S.; Johnson, E.F.; Marsh, K.C.; Mitten, M.J.; Nimmer, P.; Roberts, L.; Tahir, S.K.; Xiao, Y.; Yang, X.; Zhang, H.; Fesik, S.; Rosenberg, S.H.; Elmore, S.W. Abt-263: a potent and orally bioavailable bcl-2 family inhibitor. *Cancer Res.*, **2008**, *68*, 3421-3428.

[19] Bahadur, R.P.; Chakrabarti, P.; Rodier, F.; Janin, J. Dissecting subunit interfaces in homodimeric proteins. *Proteins.*, **2003**, *53*, 708-719.

[20] Ofran, Y.; Rost, B. Protein-protein interaction hotspots carved into sequences. *PLoS Comput. Biol.*, **2007**, *3*, e119.

[21] Bogan, A.A.; Thorn, K.S. Anatomy of hot spots in protein interfaces. *J. Mol. Biol.*, **1998**, *280*, 1-9.

[22] Ma, B.; Nussinov, R. Trp/met/phe hot spots in protein-protein interactions: potential targets in drug design. *Curr. Top. Med. Chem.*, **2007**, *7*, 999-1005.

[23]   Rickert, M.; Wang, X.; Boulanger, M.J.; Goriatcheva, N.; Garcia, K.C. The structure of interleukin-2 complexed with its alpha receptor. *Science.*, **2005**, *308*, 1477-1480.

[24]   Thanos, C.D.; DeLano, W.L.; Wells, J.A. Hot-spot mimicry of a cytokine receptor by a small molecule. *Proc. Natl. Acad. Sci. U.S.A.*, **2006**, *103*, 15422-15427.

[25]   Kortemme, T.; Baker, D. A simple physical model for binding energy hot spots in protein-protein complexes. *Proc. Natl. Acad. Sci. U.S.A.*, **2002**, *99*, 14116-14121.

[26]   Gao, Y.; Wang, R.; Lai, L. Structure-based method for analyzing protein-protein interfaces. *J. Mol. Model.*, **2004**, *10*, 44-54.

[27]   Huo, S.; Massova, I.; Kollman, P.A. Computational alanine scanning of the 1:1 human growth hormone-receptor complex. *J. Comput. Chem.*, **2002**, *23*, 15-27.

[28]   González-Ruiz, D.; Gohlke, H. Targeting protein-protein interactions with small molecules: challenges and perspectives for computational binding epitope detection and ligand finding. *Curr. Med. Chem.*, **2006**, *13*, 2607-2625.

[29]   Tuncbag, N.; Gursoy, A.; Keskin, O. Identification of computational hot spots in protein interfaces: combining solvent accessibility and inter-residue potentials improves the accuracy. *Bioinformatics.*, **2009**, *25*, 1513-1520.

[30]   Lise, S.; Archambeau, C.; Pontil, M.; Jones, D. Prediction of hot spot residues at protein-protein interfaces by combining machine learning and energy-based methods. *BMC Bioinformatics.*, **2009**, *10*, 365.

[31]   Sonavane, S.; Chakrabarti, P. Cavities and atomic packing in protein structures and interfaces. *PLoS Comput. Biol.*, **2008**, *4*, e1000188.

[32]   Laurie, A.T.R.; Jackson, R.M. Methods for the prediction of protein-ligand binding sites for structure-based drug design and virtual ligand screening. *Curr. Protein Pept. Sci.*, **2006**, *7*, 395-406.

[33]   Fuller, J.C.; Burgoyne, N.J.; Jackson, R.M. Predicting druggable binding sites at the protein-protein interface. *Drug Discov. Today.*, **2009**, *14*, 155-161.

[34]   Eyrisch, S.; Helms, V. Transient pockets on protein surfaces involved in protein-protein interaction. *J. Med. Chem.*, **2007**, *50*, 3457-3464.

[35]   Eyrisch, S.; Helms, V. What induces pocket openings on protein surface patches involved in protein-protein interactions? *J. Comput. Aided Mol. Des.*, **2009**, *23*, 73-86.

[36]   Brown, S.P.; Hajduk, P.J. Effects of conformational dynamics on predicted protein druggability. *ChemMedChem.*, **2006**, *1*, 70-72.

[37]   Lee, G.M.; Craik, C.S. Trapping moving targets with small molecules. *Science.*, **2009**, *324*, 213-215.

[38]   Grünberg, R.; Leckner, J.; Nilges, M. Complementarity of structure ensembles in protein-protein binding. *Structure.*, **2004**, *12*, 2125-2136.

[39]   Smith, G.R.; Sternberg, M.J. E.; Bates, P.A. The relationship between the flexibility of proteins and their conformational states on forming protein-protein complexes with an application to protein-protein docking. *J. Mol. Biol.*, **2005**, *347*, 1077-1101.

[40]   Brady, G.P.; Stouten, P.F. Fast prediction and visualization of protein binding pockets with pass. *J. Comput. Aided Mol. Des.*, **2000**, *14*, 383-401.

[41]   Tuncbag, N.; Kar, G.; Keskin, O.; Gursoy, A.; Nussinov, R. A survey of available tools and web servers for analysis of protein-protein interactions and interfaces. *Brief. Bioinformatics.*, **2009**, *10*, 217-232.

[42]   Fuentes, G.; Oyarzabal, J.; Rojas, A.M. Databases of protein-protein interactions and their use in drug discovery. *Curr. Opin. Drug Discov. Devel.*, **2009**, *12*, 358-366.

[43]   Lehne, B.; Schlitt, T. Protein-protein interaction databases: keeping up with growing interactomes. *Hum. Genomics.*, **2009**, *3*, 291-297.

[44]   Finn, R.D.; Marshall, M.; Bateman, A. Ipfam: visualization of protein-protein interactions in pdb at domain and amino acid resolutions. *Bioinformatics.*, **2005**, *21*, 410-412.

[45]   Winter, C.; Henschel, A.; Kim, W.K.; Schroeder, M. Scoppi: a structural classification of protein-protein interfaces. *Nucleic Acids Res.*, **2006**, *34*, D310-314.

[46]   Jefferson, E.R.; Walsh, T.P.; Roberts, T.J.; Barton, G.J. Snappi-db: a database and api of structures, interfaces and alignments for protein-protein interactions. *Nucleic Acids Res.*, **2007**, *35*, D580-589.

[47]   Aloy, P.; Ceulemans, H.; Stark, A.; Russell, R.B. The relationship between sequence and interaction divergence in proteins. *J. Mol. Biol.*, **2003**, *332*, 989-998.

[48]   Davis, F.P.; Sali, A. Pibase: a comprehensive database of structurally defined protein interfaces. *Bioinformatics.*, **2005**, *21*, 1901-1907.

[49]   Stein, A.; Panjkovich, A.; Aloy, P. 3did update: domain-domain and peptide-mediated interactions of known 3d structure. *Nucleic Acids Res.*, **2009**, *37*, D300-304.

[50] Levy, E.D.; Pereira-Leal, J.B.; Chothia, C.; Teichmann, S.A. 3d complex: a structural classification of protein complexes. *PLoS Comput. Biol.*, **2006**, *2*, e155.

[51] Bordner, A.J.; Gorin, A.A. Comprehensive inventory of protein complexes in the protein data bank from consistent classification of interfaces. *BMC Bioinformatics.*, **2008**, *9*, 234.

[52] Kundrotas, P.J.; Alexov, E. Protcom: searchable database of protein complexes enhanced with domain-domain structures. *Nucleic Acids Res.*, **2007**, *35*, D575-579.

[53] Günther, S.; von Eichborn, J.; May, P.; Preissner, R. Jail: a structure-based interface library for macromolecules. *Nucleic Acids Res.*, **2009**, *37*, D338-341.

[54] Yan, C.; Wu, F.; Jernigan, R.L.; Dobbs, D.; Honavar, V. Characterization of protein-protein interfaces. *Protein J.*, **2008**, *27*, 59-70.

[55] Keskin, O.; Nussinov, R.; Gursoy, A. Prism: protein-protein interaction prediction by structural matching. *Methods Mol. Biol.*, **2008**, *484*, 505-521.

[56] Reynolds, C.; Damerell, D.; Jones, S. Protorp: a protein-protein interaction analysis server. *Bioinformatics.*, **2009**, *25*, 413-414.

[57] Negi, S.S.; Schein, C.H.; Oezguen, N.; Power, T.D.; Braun, W. Interprosurf: a web server for predicting interacting sites on protein surfaces. *Bioinformatics.*, **2007**, *23*, 3397-3399.

[58] Benoit, V.; Mucchielli-Giorgi, M.; Dumont, B.; Durosay, P.; Reymond, N.; Delacroix, H. Ppidd: an extraction and visualisation method of biological protein-protein interfaces. *Biochimie.*, **2008**, *90*, 640-647.

[59] Teyra, J.; Doms, A.; Schroeder, M.; Pisabarro, M.T. Scowlp: a web-based database for detailed characterization and visualization of protein interfaces. *BMC Bioinformatics.*, **2006**, *7*, 104.

[60] Teyra, J.; Pisabarro, M.T. Characterization of interfacial solvent in protein complexes and contribution of wet spots to the interface description. *Proteins.*, **2007**, *67*, 1087-1095.

[61] Thorn, K.S.; Bogan, A.A. Asedb: a database of alanine mutations and their effects on the free energy of binding in protein interactions. *Bioinformatics.*, **2001**, *17*, 284-285.

[62] Fischer, T.B.; Arunachalam, K.V.; Bailey, D.; Mangual, V.; Bakhru, S.; Russo, R.; Huang, D.; Paczkowski, M.; Lalchandani, V.; Ramachandra, C.; Ellison, B.; Galer, S.; Shapley, J.; Fuentes, E.; Tsai, J. The binding interface database (bid): a compilation of amino acid hot spots in protein interfaces. *Bioinformatics.*, **2003**, *19*, 1453-1454.

[63] Guney, E.; Tuncbag, N.; Keskin, O.; Gursoy, A. Hotsprint: database of computational hot spots in protein interfaces. *Nucleic Acids Res.*, **2008**, *36*, D662-666.

[64] Higurashi, M.; Ishida, T.; Kinoshita, K. Pisite: a database of protein interaction sites using multiple binding states in the pdb. *Nucleic Acids Res.*, **2009**, *37*, D360-364.

[65] Gong, S.; Park, C.; Choi, H.; Ko, J.; Jang, I.; Lee, J.; Bolser, D.M.; Oh, D.; Kim, D.; Bhak, J. A protein domain interaction interface database: interpare. *BMC Bioinformatics.*, **2005**, *6*, 207.

[66] Saha, R.P.; Bahadur, R.P.; Pal, A.; Mandal, S.; Chakrabarti, P. Proface: a server for the analysis of the physicochemical features of protein-protein interfaces. *BMC Struct. Biol.*, **2006**, *6*, 11.

[67] Weisel, M.; Proschak, E.; Schneider, G. Pocketpicker: analysis of ligand binding-sites with shape descriptors. *Chem. Cent. J.*, **2007**, *1*, 7.

[68] Laurie, A.T.R.; Jackson, R.M. Q-sitefinder: an energy-based method for the prediction of protein-ligand binding sites. *Bioinformatics.*, **2005**, *21*, 1908-1916.

[69] Huang, B.; Schroeder, M. Ligsitecsc: predicting ligand binding sites using the connolly surface and degree of conservation. *BMC Struct. Biol.*, **2006**, *6*, 19.

[70] Laskowski, R.A. Surfnet: a program for visualizing molecular surfaces, cavities, and intermolecular interactions. *Journal of Molecular Graphics.*, **1995**, *13*, 323-330.

[71] Huang, B. Metapocket: a meta approach to improve protein ligand binding site prediction. *OMICS.*, **2009**, *13*, 325-330.

[72] Betzi, S.; Restouin, A.; Opi, S.; Arold, S.T.; Parrot, I.; Guerlesquin, F.; Morelli, X.; Collette, Y. Protein protein interaction inhibition (2p2i) combining high throughput and virtual screening: application to the hiv-1 nef protein. *Proc. Natl. Acad. Sci. U.S.A.*, **2007**, *104*, 19256-19261.

[73] Betzi, S.; Guerlesquin, F.; Morelli, X. Protein protein interaction inhibition (2p2i):less and less undruggable target? *Comb. Chem. & HTS.*, **2009**, *12*, 968-983.

[74] Betzi, S.; Suhre, K.; Chétrit, B.; Guerlesquin, F.; Morelli, X. Gfscore: a general nonlinear consensus scoring function for high-throughput docking. *J. Chem. Inf. Model.*, **2006**, *46*, 1704-1712.

[75] De Luca, L.; Barreca, M.L.; Ferro, S.; Christ, F.; Iraci, N.; Gitto, R.; Monforte, A.M.; Debyser, Z.; Chimirri, A. Pharmacophore-based discovery of small-molecule inhibitors of protein-protein interactions between hiv-1 integrase and cellular cofactor ledgf/p75. *ChemMedChem.*, **2009**, *4*, 1311-1316.

[76]   Caporuscio, F.; Tafi, A.; González, E.; Manetti, F.; Esté, J.A.; Botta, M.A dynamic target-based pharmacophoric model mapping the cd4 binding site on hiv-1 gp120 to identify new inhibitors of gp120-cd4 protein-protein interactions. *Bioorg. Med. Chem. Lett.*, **2009**, *19*, 6087-6091.

[77]   Liu, S.; Liu, S.; Fu, Y. Molecular motions of human hiv-1 gp120 envelope glycoproteins. *J. Mol. Model.*, **2008**, *14*, 857-870.

[78]   Goodford, P.J. A computational procedure for determining energetically favorable binding sites on biologically important macromolecules. *J. Med. Chem.*, **1985**, *28*, 849-857.

[79]   Madani, N.; Perdigoto, A.L.; Srinivasan, K.; Cox, J.M.; Chruma, J.J.; LaLonde, J.; Head, M.; Smith, A.B.; Sodroski, J.G. Localized changes in the gp120 envelope glycoprotein confer resistance to human immunodeficiency virus entry inhibitors bms-806 and #155. *J. Virol.*, **2004**, *78*, 3742-3752.

[80]   Oltersdorf, T.; Elmore, S.W.; Shoemaker, A.R.; *et al.* An inhibitor of bcl-2 family proteins induces regression of solid tumours. *Nature.*, **2005**, *435*, 677-681.

[81]   Becattini, B.; Kitada, S.; Leone, M.; Monosov, E.; Chandler, S.; Zhai, D.; Kipps, T.J.; Reed, J.C.; Pellecchia, M. Rational design and real time, in-cell detection of the proapoptotic activity of a novel compound targeting bcl-x(l). *Chem. Biol.*, **2004**, *11*, 389-395.

[82]   Kitada, S.; Leone, M.; Sareth, S.; Zhai, D.; Reed, J.C.; Pellecchia, M. Discovery, characterization, and structure-activity relationships studies of proapoptotic polyphenols targeting b-cell lymphocyte/leukemia-2 proteins. *J. Med. Chem.*, **2003**, *46*, 4259-4264.

[83]   Wang, J.L.; Liu, D.; Zhang, Z.J.; Shan, S.; Han, X.; Srinivasula, S.M.; Croce, C.M.; Alnemri, E.S.; Huang, Z. Structure-based discovery of an organic compound that binds bcl-2 protein and induces apoptosis of tumor cells. *Proc. Natl. Acad. Sci. U.S.A.*, **2000**, *97*, 7124-7129.

[84]   Shuker, S.B.; Hajduk, P.J.; Meadows, R.P.; Fesik, S.W. Discovering high-affinity ligands for proteins: sar by nmr. *Science.*, **1996**, *274*, 1531-1534.

[85]   Rubinstein, M.; Niv, M.Y. Peptidic modulators of protein-protein interactions: progress and challenges in computational design. *Biopolymers.*, **2009**, *91*, 505-513.

[86]   Laudet, B.; Barette, C.; Dulery, V.; Renaudet, O.; Dumy, P.; Metz, A.; Prudent, R.; Deshiere, A.; Dideberg, O.; Filhol, O.; Cochet, C. Structure-based design of small peptide inhibitors of protein kinase ck2 subunit interaction. *Biochem. J.*, **2007**, *408*, 363-373.

[87]   Vagner, J.; Qu, H.; Hruby, V.J. Peptidomimetics, a synthetic tool of drug discovery. *Curr. Opin. Chem. Biol.*, **2008**, *12*, 292-296.

[88]   Rodriguez, J.M.; Nevola, L.; Ross, N.T.; Lee, G.; Hamilton, A.D. Synthetic inhibitors of extended helix-protein interactions based on a biphenyl 4,4'-dicarboxamide scaffold. *Chembiochem.*, **2009**, *10*, 829-833.

[89]   Jacoby, E. Biphenyls as potential mimetics of protein alpha-helix. *Bioorg. Med. Chem. Lett.*, **2002**, *12*, 891-893.

[90]   Orner, B.P.; Ernst, J.T.; Hamilton, A.D. Toward proteomimetics: terphenyl derivatives as structural and functional mimics of extended regions of an alpha-helix. *J. Am. Chem. Soc.*, **2001**, *123*, 5382-5383.

[91]   Yin, H.; Lee, G.; Sedey, K.A.; Kutzki, O.; Park, H.S.; Orner, B.P.; Ernst, J.T.; Wang, H.; Sebti, S.M.; Hamilton, A.D. Terphenyl-based bak bh3 alpha-helical proteomimetics as low-molecular-weight antagonists of bcl-xl. *J. Am. Chem. Soc.*, **2005**, *127*, 10191-10196.

[92]   Yin, H.; Lee, G.; Park, H.S.; Payne, G.A.; Rodriguez, J.M.; Sebti, S.M.; Hamilton, A.D. Terphenyl-based helical mimetics that disrupt the p53/hdm2 interaction. *Angew. Chem. Int. Ed. Engl.*, **2005**, *44*, 2704-2707.

[93]   Yin, H.; Lee, G.; Park, H.S.; Payne, G.A.; Rodriguez, J.M.; Sebti, S.M.; Hamilton, A.D. Terphenyl-based helical mimetics that disrupt the p53/hdm2 interaction. *Angew. Chem. Int. Ed. Engl.*, **2005**, *44*, 2704-2707.

[94]   Shangary, S.; Wang, S. Small-molecule inhibitors of the mdm2-p53 protein-protein interaction to reactivate p53 function: a novel approach for cancer therapy. *Annu. Rev. Pharmacol. Toxicol.*, **2009**, *49*, 223-241.

[95]   Saraogi, I.; Hamilton, A.D. Alpha-helix mimetics as inhibitors of protein-protein interactions. *Biochem. Soc. Trans.*, **2008**, *36*, 1414-1417.

[96]   Gavathiotis, E.; Suzuki, M.; Davis, M.L.; Pitter, K.; Bird, G.H.; Katz, S.G.; Tu, H.; Kim, H.; Cheng, E.H.; Tjandra, N.; Walensky, L.D. Bax activation is initiated at a novel interaction site. *Nature.*, **2008**, *455*, 1076-1081.

[97]   Walensky, L.D.; Kung, A.L.; Escher, I.; Malia, T.J.; Barbuto, S.; Wright, R.D.; Wagner, G.; Verdine, G.L.; Korsmeyer, S.J. Activation of apoptosis *in vivo* by a hydrocarbon-stapled bh3 helix. *Science.*, **2004**, *305*, 1466-1470.

[98]   Walensky, L.D.; Pitter, K.; Morash, J.; Oh, K.J.; Barbuto, S.; Fisher, J.; Smith, E.; Verdine, G.L.; Korsmeyer, S.J. A stapled bid bh3 helix directly binds and activates bax. *Mol. Cell.*, **2006**, *24*, 199-210.

[99]   Goede, A.; Michalsky, E.; Schmidt, U.; Preissner, R. Supermimic--fitting peptide mimetics into protein structures. *BMC Bioinformatics.*, **2006**, *7*, 11.

[100] Eckert, H.; Bajorath, J. Exploring peptide-likeness of active molecules using 2d fingerprint methods. *J. Chem. Inf. Model.*, **2007**, *47*, 1366-1378.

[101] Parthasarathi, L.; Casey, F.; Stein, A.; Aloy, P.; Shields, D.C. Approved drug mimics of short peptide ligands from protein interaction motifs. *J. Chem. Inf. Model.*, **2008**, *48*, 1943-1948.

[102] Bourgeas, R.; Basse, M.; Morelli, X.; Roche, P. Atomic analysis of protein-protein interfaces with known inhibitors: the 2p2i database. *PLoS ONE.* **2010**, *5*, e9598.

[103] Alsafadi, S.; Tourpin, S.; André, F.; Vassal, G.; Ahomadegbe, J. P53 family: at the crossroads in cancer therapy. *Curr. Med. Chem.*, **2009**, *16*, 4328-4344.

[104] Kussie, P.H.; Gorina, S.; Marechal, V.; Elenbaas, B.; Moreau, J.; Levine, A.J.; Pavletich, N.P. Structure of the mdm2 oncoprotein bound to the p53 tumor suppressor transactivation domain. *Science.*, **1996**, *274*, 948-953.

[105] Hasegawa, H.; Yamada, Y.; Iha, H.; Tsukasaki, K.; Nagai, K.; Atogami, S.; Sugahara, K.; Tsuruda, K.; Ishizaki, A.; Kamihira, S. Activation of p53 by nutlin-3a, an antagonist of mdm2, induces apoptosis and cellular senescence in adult t-cell leukemia cells. *Leukemia.*, **2009**, *23*, 2090-2101.

[106] Lee, Y.; Lim, J.; Chun, Y.; Moon, H.; Lee, M.K.; Huang, L.E.; Park, J. Nutlin-3, an hdm2 antagonist, inhibits tumor adaptation to hypoxia by stimulating the fih-mediated inactivation of hif-1alpha. *Carcinogenesis.*, **2009**, *30*, 1768-1775.

[107] Morselli, E.; Galluzzi, L.; Kepp, O.; Kroemer, G. Nutlin kills cancer cells via mitochondrial p53. *Cell Cycle.*, **2009**, *8*, 1647-1648.

[108] Ding, K.; Lu, Y.; Nikolovska-Coleska, Z.; Qiu, S.; Ding, Y.; Gao, W.; Stuckey, J.; Krajewski, K.; Roller, P.P.; Tomita, Y.; Parrish, D.A.; Deschamps, J.R.; Wang, S. Structure-based design of potent non-peptide mdm2 inhibitors. *J. Am. Chem. Soc.*, **2005**, *127*, 10130-10131.

[109] Ding, K.; Lu, Y.; Nikolovska-Coleska, Z.; Wang, G.; Qiu, S.; Shangary, S.; Gao, W.; Qin, D.; Stuckey, J.; Krajewski, K.; Roller, P.P.; Wang, S. Structure-based design of spiro-oxindoles as potent, specific small-molecule inhibitors of the mdm2-p53 interaction. *J. Med. Chem.*, **2006**, *49*, 3432-3435.

[110] Grasberger, B.L.; Lu, T.; Schubert, C.; Parks, D.J.; Carver, T.E.; Koblish, H.K.; Cummings, M.D.; LaFrance, L.V.; Milkiewicz, K.L.; Calvo, R.R.; Maguire, D.; Lattanze, J.; Franks, C.F.; Zhao, S.; Ramachandren, K.; Bylebyl, G.R.; Zhang, M.; Manthey, C.L.; Petrella, E.C.; Pantoliano, M.W.; Deckman, I.C.; Spurlino, J.C.; Maroney, A.C.; Tomczuk, B.E.; Molloy, C.J.; Bone, R.F. Discovery and cocrystal structure of benzodiazepinedione hdm2 antagonists that activate p53 in cells. *J. Med. Chem.*, **2005**, *48*, 909-912.

[111] Parks, D.J.; Lafrance, L.V.; Calvo, R.R.; Milkiewicz, K.L.; Gupta, V.; Lattanze, J.; Ramachandren, K.; Carver, T.E.; Petrella, E.C.; Cummings, M.D.; Maguire, D.; Grasberger, B.L.; Lu, T. 1,4-benzodiazepine-2,5-diones as small molecule antagonists of the hdm2-p53 interaction: discovery and sar. *Bioorg. Med. Chem. Lett.*, **2005**, *15*, 765-770.

[112] Lu, Y.; Nikolovska-Coleska, Z.; Fang, X.; Gao, W.; Shangary, S.; Qiu, S.; Qin, D.; Wang, S. Discovery of a nanomolar inhibitor of the human murine double minute 2 (mdm2)-p53 interaction through an integrated, virtual database screening strategy. *J. Med. Chem.*, **2006**, *49*, 3759-3762.

[113] Rothweiler, U.; Czarna, A.; Krajewski, M.; Ciombor, J.; Kalinski, C.; Khazak, V.; Ross, G.; Skobeleva, N.; Weber, L.; Holak, T.A. Isoquinolin-1-one inhibitors of the mdm2-p53 interaction. *ChemMedChem.*, **2008**, *3*, 1118-1128.

[114] Stoll, R.; Renner, C.; Hansen, S.; Palme, S.; Klein, C.; Belling, A.; Zeslawski, W.; Kamionka, M.; Rehm, T.; Mühlhahn, P.; Schumacher, R.; Hesse, F.; Kaluza, B.; Voelter, W.; Engh, R.A.; Holak, T.A. Chalcone derivatives antagonize interactions between the human oncoprotein mdm2 and p53. *Biochemistry.*, **2001**, *40*, 336-344.

[115] Galatin, P.S.; Abraham, D.J. A nonpeptidic sulfonamide inhibits the p53-mdm2 interaction and activates p53-dependent transcription in mdm2-overexpressing cells. *J. Med. Chem.*, **2004**, *47*, 4163-4165.

[116] Allen, J.G.; Bourbeau, M.P.; Wohlhieter, G.E.; Bartberger, M.D.; Michelsen, K.; Hungate, R.; Gadwood, R.C.; Gaston, R.D.; Evans, B.; Mann, L.W.; Matison, M. E.; Schneider, S.; Huang, X.; Yu, D.; Andrews, P.S.; Reichelt, A.; Long, A.M.; Yakowec, P.; Yang, E.Y.; Lee, T.A.; Oliner, J.D. Discovery and optimization of chromenotriazolopyrimidines as potent inhibitors of the mouse double minute 2-tumor protein 53 protein-protein interaction. *J. Med. Chem.*, **2009**, *52*, 7044-7053.

[117] Vassilev, L.T.; Vu, B.T.; Graves, B.; Carvajal, D.; Podlaski, F.; Filipovic, Z.; Kong, N.; Kammlott, U.; Lukacs, C.; Klein, C.; Fotouhi, N.; Liu, E.A. *In vivo* activation of the p53 pathway by small-molecule antagonists of mdm2. *Science.*, **2004**, *303*, 844-848.

[118] Bowman, A.L.; Nikolovska-Coleska, Z.; Zhong, H.; Wang, S.; Carlson, H.A. Small molecule inhibitors of the mdm2-p53 interaction discovered by ensemble-based receptor models. *J. Am. Chem. Soc.*, **2007**, *129*, 12809-12814.

[119] Fry, D.C.; Emerson, S.D.; Palme, S.; Vu, B. T.; Liu, C.; Podlaski, F. Nmr structure of a complex between mdm2 and a small molecule inhibitor. *J. Biomol. NMR.*, **2004**, *30*, 163-173.

[120] Hyde, J.; Braisted, A.C.; Randal, M.; Arkin, M.R. Discovery and characterization of cooperative ligand binding in the adaptive region of interleukin-2. *Biochemistry.*, **2003**, *42*, 6475-6483.

[121] Braisted, A.C.; Oslob, J.D.; Delano, W.L.; Hyde, J.; McDowell, R.S.; Waal, N.; Yu, C.; Arkin, M.R.; Raimundo, B.C. Discovery of a potent small molecule il-2 inhibitor through fragment assembly. *J. Am. Chem. Soc.*, **2003**, *125*, 3714-3715.

[122] Waal, N.D.; Yang, W.; Oslob, J.D.; Arkin, M.R.; Hyde, J.; Lu, W.; McDowell, R.S.; Yu, C.H.; Raimundo, B.C. Identification of nonpeptidic small-molecule inhibitors of interleukin-2. *Bioorg. Med. Chem. Lett.*, **2005**, *15*, 983-987.

[123] Thanos, C.D.; Randal, M.; Wells, J.A. Potent small-molecule binding to a dynamic hot spot on il-2. *J. Am. Chem. Soc.*, **2003**, *125*, 15280-15281.

[124] Raimundo, B.C.; Oslob, J.D.; Braisted, A.C.; Hyde, J.; McDowell, R.S.; Randal, M.; Waal, N.D.; Wilkinson, J.; Yu, C.H.; Arkin, M.R. Integrating fragment assembly and biophysical methods in the chemical advancement of small-molecule antagonists of il-2: an approach for inhibiting protein-protein interactions. *J. Med. Chem.*, **2004**, *47*, 3111-3130.

[125] Petros, A.M.; Olejniczak, E.T.; Fesik, S.W. Structural biology of the bcl-2 family of proteins. *Biochim. Biophys. Acta.*, **2004**, *1644*, 83-94.

[126] Kang, M.H.; Reynolds, C.P. Bcl-2 inhibitors: targeting mitochondrial apoptotic pathways in cancer therapy. *Clin. Cancer Res.*, **2009**, *15*, 1126-1132.

[127] Azmi, A.S.; Mohammad, R.M. Non-peptidic small molecule inhibitors against bcl-2 for cancer therapy. *J. Cell. Physiol.*, **2009**, *218*, 13-21.

[128] Bruncko, M.; Oost, T.K.; Belli, B.A.; Ding, H.; Joseph, M.K.; Kunzer, A.; Martineau, D.; McClellan, W.J.; Mitten, M.; Ng, S.; Nimmer, P.M.; Oltersdorf, T.; Park, C.; Petros, A.M.; Shoemaker, A.R.; Song, X.; Wang, X.; Wendt, M.D.; Zhang, H.; Fesik, S.W.; Rosenberg, S.H.; Elmore, S.W. Studies leading to potent, dual inhibitors of bcl-2 and bcl-xl. *J. Med. Chem.*, **2007**, *50*, 641-662.

[129] Stauffer, S.R. Small molecule inhibition of the bcl-x(l)-bh3 protein-protein interaction: proof-of-concept of an *in vivo* chemopotentiator abt-737. *Curr. Top. Med. Chem.*, **2007**, *7*, 961-965.

[130] Lee, E.F.; Czabotar, P.E.; Smith, B.J.; Deshayes, K.; Zobel, K.; Colman, P.M.; Fairlie, W.D. Crystal structure of abt-737 complexed with bcl-xl: implications for selectivity of antagonists of the bcl-2 family. *Cell Death Differ.*, **2007**, *14*, 1711-1713.

[131] Verhaegen, M.; Bauer, J.A.; Martín de la Vega, C.; Wang, G.; Wolter, K.G.; Brenner, J.C.; Nikolovska-Coleska, Z.; Bengtson, A.; Nair, R.; Elder, J.T.; Van Brocklin, M.; Carey, T.E.; Bradford, C.R.; Wang, S.; Soengas, M.S. A novel bh3 mimetic reveals a mitogen-activated protein kinase-dependent mechanism of melanoma cell death controlled by p53 and reactive oxygen species. *Cancer Res.*, **2006**, *66*, 11348-11359.

[132] Wang, Z.; Azmi, A.S.; Ahmad, A.; Banerjee, S.; Wang, S.; Sarkar, F.H.; Mohammad, R.M. Tw-37, a small-molecule inhibitor of bcl-2, inhibits cell growth and induces apoptosis in pancreatic cancer: involvement of notch-1 signaling pathway. *Cancer Res.*, **2009**, *69*, 2757-2765.

[133] Rezazadeh, A.; Laber, D.A.; Ghim, S.; Jenson, A.B.; Kloecker, G. The role of human papilloma virus in lung cancer: a review of the evidence. *Am. J. Med. Sci.*, **2009**, *338*, 64-67.

[134] Kasukawa, H.; Howley, P.M.; Benson, J.D. A fifteen-amino-acid peptide inhibits human papillomavirus e1-e2 interaction and human papillomavirus dna replication in vitro. *J. Virol.*, **1998**, *72*, 8166-8173.

[135] Yoakim, C.; Ogilvie, W.W.; Goudreau, N.; Naud, J.; Haché, B.; O'Meara, J.A.; Cordingley, M.G.; Archambault, J.; White, P.W. Discovery of the first series of inhibitors of human papillomavirus type 11: inhibition of the assembly of the e1-e2-origin dna complex. *Bioorg. Med. Chem. Lett.*, **2003**, *13*, 2539-2541.

[136] White, P.W.; Titolo, S.; Brault, K.; Thauvette, L.; Pelletier, A.; Welchner, E.; Bourgon, L.; Doyon, L.; Ogilvie, W.W.; Yoakim, C.; Cordingley, M.G.; Archambault, J. Inhibition of human papillomavirus dna replication by small molecule antagonists of the e1-e2 protein interaction. *J. Biol. Chem.*, **2003**, *278*, 26765-26772.

[137] Goudreau, N.; Cameron, D.R.; Déziel, R.; Haché, B.; Jakalian, A.; Malenfant, E.; Naud, J.; Ogilvie, W.W.; O'meara, J.; White, P.W.; Yoakim, C. Optimization and determination of the absolute configuration of a series of potent inhibitors of human papillomavirus type-11 e1-e2 protein-protein interaction: a combined medicinal chemistry, nmr and computational chemistry approach. *Bioorg. Med. Chem.*, **2007**, *15*, 2690-2700.

[138] Wang, Y.; Coulombe, R.; Cameron, D.R.; Thauvette, L.; Massariol, M.; Amon, L.M.; Fink, D.; Titolo, S.; Welchner, E.; Yoakim, C.; Archambault, J.; White, P.W. Crystal structure of the e2 transactivation domain of human papillomavirus type 11 bound to a protein interaction inhibitor. *J. Biol. Chem.*, **2004**, *279*, 6976-6985.

[139] Abbate, E.A.; Berger, J.M.; Botchan, M.R. The x-ray structure of the papillomavirus helicase in complex with its molecular matchmaker e2. *Genes Dev.*, **2004**, *18*, 1981-1996.

[140] He, M.M.; Smith, A.S.; Oslob, J.D.; Flanagan, W.M.; Braisted, A.C.; Whitty, A.; Cancilla, M.T.; Wang, J.; Lugovskoy, A.A.; Yoburn, J.C.; Fung, A.D.; Farrington, G.; Eldredge, J.K.; Day, E.S.; Cruz, L.A.; Cachero, T.G.; Miller, S.K.; Friedman, J.E.; Choong, I.C.; Cunningham, B.C. Small-molecule inhibition of tnf-alpha. *Science.*, **2005**, *310*, 1022-1025.

[141] Chen, D.J.; Huerta, S. Smac mimetics as new cancer therapeutics. *Anticancer. Drugs.*, **2009**, *20*, 646-658.

[142] Wu, G.; Chai, J.; Suber, T.L.; Wu, J.W.; Du, C.; Wang, X.; Shi, Y. Structural basis of iap recognition by smac/diablo. *Nature.*, **2000**, *408*, 1008-1012.

[143] Nikolovska-Coleska, Z.; Xu, L.; Hu, Z.; Tomita, Y.; Li, P.; Roller, P.P.; Wang, R.; Fang, X.; Guo, R.; Zhang, M.; Lippman, M.E.; Yang, D.; Wang, S. Discovery of embelin as a cell-permeable, small-molecular weight inhibitor of xiap through structure-based computational screening of a traditional herbal medicine three-dimensional structure database. *J. Med. Chem.*, **2004**, *47*, 2430-2440.

[144] Chen, J.; Nikolovska-Coleska, Z.; Wang, G.; Qiu, S.; Wang, S. Design, synthesis, and characterization of new embelin derivatives as potent inhibitors of x-linked inhibitor of apoptosis protein. *Bioorg. Med. Chem. Lett.*, **2006**, *16*, 5805-5808.

[145] Sun, H.; Stuckey, J.A.; Nikolovska-Coleska, Z.; Qin, D.; Meagher, J.L.; Qiu, S.; Lu, J.; Yang, C.; Saito, N.G.; Wang, S. Structure-based design, synthesis, evaluation, and crystallographic studies of conformationally constrained smac mimetics as inhibitors of the x-linked inhibitor of apoptosis protein (xiap). *J. Med. Chem.*, **2008**, *51*, 7169-7180.

[146] Mastrangelo, E.; Cossu, F.; Milani, M.; Sorrentino, G.; Lecis, D.; Delia, D.; Manzoni, L.; Drago, C.; Seneci, P.; Scolastico, C.; Rizzo, V.; Bolognesi, M. Targeting the x-linked inhibitor of apoptosis protein through 4-substituted azabicyclo[5.3.0]alkane smac mimetics. structure, activity, and recognition principles. *J. Mol. Biol.*, **2008**, *384*, 673-689.

[147] Cossu, F.; Mastrangelo, E.; Milani, M.; Sorrentino, G.; Lecis, D.; Delia, D.; Manzoni, L.; Seneci, P.; Scolastico, C.; Bolognesi, M. Designing smac-mimetics as antagonists of xiap, ciap1, and ciap2. *Biochem. Biophys. Res. Commun.*, **2009**, *378*, 162-167.

[148] Seneci, P.; Bianchi, A.; Battaglia, C.; Belvisi, L.; Bolognesi, M.; Caprini, A.; Cossu, F.; Franco, E.D.; Matteo, M.D.; Delia, D.; Drago, C.; Khaled, A.; Lecis, D.; Manzoni, L.; Marizzoni, M.; Mastrangelo, E.; Milani, M.; Motto, I.; Moroni, E.; Potenza, D.; Rizzo, V.; Servida, F.; Turlizzi, E.; Varrone, M.; Vasile, F.; Scolastico, C. Rational design, synthesis and characterization of potent, non-peptidic smac mimics/xiap inhibitors as proapoptotic agents for cancer therapy. *Bioorganic & Medicinal Chemistry.*, **2009**, *17*, 5834-5856.

[149] Wang, N.X.; Lee, H.; Zheng, J.J. Therapeutic use of pdz protein-protein interaction antagonism. *Drug News Perspect.*, **2008**, *21*, 137-141.

[150] Fujii, N.; Haresco, J.J.; Novak, K.A.P.; Gage, R.M.; Pedemonte, N.; Stokoe, D.; Kuntz, I.D.; Guy, R.K. Rational design of a nonpeptide general chemical scaffold for reversible inhibition of pdz domain interactions. *Bioorg. Med. Chem. Lett.*, **2007**, *17*, 549-552.

[151] Grandy, D.; Shan, J.; Zhang, X.; Rao, S.; Akunuru, S.; Li, H.; Zhang, Y.; Alpatov, I.; Zhang, X.A.; Lang, R.A.; Shi, D.; Zheng, J.J. Discovery and characterization of a small molecule inhibitor of the pdz domain of dishevelled. *J. Biol. Chem.*, **2009**, *284*, 16256-16263.

[152] Shan, J.; Zheng, J.J. Optimizing dvl pdz domain inhibitor by exploring chemical space. *J. Comput. Aided Mol. Des.*, **2009**, *23*, 37-47.

[153] Uvebrant, K.; da Graça Thrige, D.; Rosén, A.; Akesson, M.; Berg, H.; Walse, B.; Björk, P. Discovery of selective small-molecule cd80 inhibitors. *J. Biomol. Screen.*, **2007**, *12*, 464-472.

[154] Yabu, J.M.; Vincenti, F. Novel immunosuppression: small molecules and biologics. *Semin. Nephrol.*, **2007**, *27*, 479-486.

[155] Kadow, J.; Wang, H.H.; Lin, P. Small-molecule hiv-1 gp120 inhibitors to prevent hiv-1 entry: an emerging opportunity for drug development. *Curr. Opin. Investig. Drugs.*, **2006**, *7*, 721-726.

[156] Tilton, J.C.; Doms, R.W. Entry inhibitors in the treatment of hiv-1 infection. *Antiviral Res.*, **2010**, *85*, 91-100.

[157] Tran, T.; Adam, F.M.; Calo, F.; Fenwick, D.R.; Fok-Seang, J.; Gardner, I.; Hay, D.A.; Perros, M.; Rawal, J.; Middleton, D.S.; Parkinson, T.; Pickford, C.; Platts, M.; Randall, A.; Stephenson, P.T.; Vuong, H.; Williams, D.H. Design and optimisation of potent gp120-cd4 inhibitors. *Bioorg. Med. Chem. Lett.*, **2009**, *19*, 5250-5255.

[158] Williams, D.H.; Adam, F.; Fenwick, D.R.; Fok-Seang, J.; Gardner, I.; Hay, D.; Jaiessh, R.; Middleton, D.S.; Mowbray, C.E.; Parkinson, T.; Perros, M.; Pickford, C.; Platts, M.; Randall, A.; Siddle, D.; Stephenson, P.T.; Tran, T.; Vuong, H. Discovery of a small molecule inhibitor through interference with the gp120-cd4 interaction. *Bioorg. Med. Chem. Lett.*, **2009**, *19*, 5246-5249.

[159] Lin, P.; Blair, W.; Wang, T.; Spicer, T.; Guo, Q.; Zhou, N.; Gong, Y.; Wang, H.H.; Rose, R.; Yamanaka, G.; Robinson, B.; Li, C.; Fridell, R.; Deminie, C.; Demers, G.; Yang, Z.; Zadjura, L.; Meanwell, N.; Colonno, R. A small molecule hiv-1 inhibitor that targets the hiv-1 envelope and inhibits cd4 receptor binding. *Proc. Natl. Acad. Sci. U.S.A.*, **2003**, *100*, 11013-11018.

[160] Engelman, A. Mechanistic and pharmacological analyses of hiv-1 integration. *Methods.*, **2009**, *47*, 225-228.

[161] Marchand, C.; Maddali, K.; Métifiot, M.; Pommier, Y. Hiv-1 in inhibitors: 2010 update and perspectives. *Curr. Top. Med. Chem.*, **2009**, *9*, 1016-1037.

[162] Du, L.; Zhao, Y.; Chen, J.; Yang, L.; Zheng, Y.; Tang, Y.; Shen, X.; Jiang, H. D77, one benzoic acid derivative, functions as a novel anti-hiv-1 inhibitor targeting the interaction between integrase and cellular ledgf/p75. *Biochem. Biophys. Res. Commun.*, **2008**, *375*, 139-144.

[163] Tsao, D.H.H.; Sutherland, A.G.; Jennings, L.D.; Li, Y.; Rush, T.S.; Alvarez, J.C.; Ding, W.; Dushin, E.G.; Dushin, R.G.; Haney, S.A.; Kenny, C.H.; Malakian, A.K.; Nilakantan, R.; Mosyak, L. Discovery of novel inhibitors of the zipa/ftsz complex by nmr fragment screening coupled with structure-based design. *Bioorg. Med. Chem.*, **2006**, *14*, 7953-7961.

[164] Jennings, L.D.; Foreman, K.W.; Rush, T.S.; Tsao, D.H.H.; Mosyak, L.; Kincaid, S.L.; Sukhdeo, M.N.; Sutherland, A.G.; Ding, W.; Kenny, C.H.; Sabus, C.L.; Liu, H.; Dushin, E.G.; Moghazeh, S.L.; Labthavikul, P.; Petersen, P.J.; Tuckman, M.; Haney, S.A.; Ruzin, A.V. Combinatorial synthesis of substituted 3-(2-indolyl)piperidines and 2-phenyl indoles as inhibitors of zipa-ftsz interaction. *Bioorg. Med. Chem.*, **2004**, *12*, 5115-5131.

[165] Jennings, L.D.; Foreman, K.W.; Rush, T.S.; Tsao, D.H.H.; Mosyak, L.; Li, Y.; Sukhdeo, M.N.; Ding, W.; Dushin, E.G.; Kenny, C.H.; Moghazeh, S.L.; Petersen, P.J.; Ruzin, A.V.; Tuckman, M.; Sutherland, A.G. Design and synthesis of indolo[2,3-a]quinolizin-7-one inhibitors of the zipa-ftsz interaction. *Bioorg. Med. Chem. Lett.*, **2004**, *14*, 1427-1431.

[166] Rush, T.S.; Grant, J.A.; Mosyak, L.; Nicholls, A. A shape-based 3-d scaffold hopping method and its application to a bacterial protein-protein interaction. *J. Med. Chem.*, **2005**, *48*, 1489-1495.

[167] Novak, W.; Wang, H.; Krilov, G. Role of protein flexibility in the design of bcl-x(l) targeting agents: insight from molecular dynamics. *J. Comput. Aided Mol. Des.*, **2009**, *23*, 49-61.

[168] Tama, F.; Brooks, C.L. Symmetry, form, and shape: guiding principles for robustness in macromolecular machines. *Annu. Rev. Biophys. Biomol. Struct.*, **2006**, *35*, 115-133.

[169] Higueruelo, A.P.; Schreyer, A.; Bickerton, G.R.J.; Pitt, W.R.; Groom, C.R.; Blundell, T.L. Atomic interactions and profile of small molecules disrupting protein-protein interfaces: the timbal database. *Chem. Biol. Drug Des.*, **2009**, *74*, 457-467.

[170] Reynès, C.; Host, H.; Camproux, A.; Laconde, G.; Leroux, F.; Mazars, A.; Deprez, B.; Fahraeus, R.; Villoutreix, B.O.; Sperandio, O. Designing focused chemical libraries enriched in protein-protein interaction inhibitors using machine-learning methods. *PLoS Comput. Biol.*, **2010**, *6*, e1000695.

[171] Sperandio, O.; Reynès, C.H.; Camproux, A.; Villoutreix, B.O. Rationalizing the chemical space of protein-protein interaction inhibitors. *Drug Discov. Today.*, **2010**, *15*, 220-229.

[172] Oost, T.K.; Sun, C.; Armstrong, R.C.; Al-Assaad, A.; Betz, S.F.; Deckwerth, T. L.; Ding, H.; Elmore, S.W.; Meadows, R.P.; Olejniczak, E.T.; Oleksijew, A.; Oltersdorf, T.; Rosenberg, S.H.; Shoemaker, A.R.; Tomaselli, K.J.; Zou, H.; Fesik, S. W. Discovery of potent antagonists of the antiapoptotic protein xiap for the treatment of cancer. *J. Med. Chem.*, **2004**, *47*, 4417-4426.

[173] Wist, A.D.; Gu, L.; Riedl, S.J.; Shi, Y.; McLendon, G.L. Structure-activity based study of the smac-binding pocket within the bir3 domain of xiap. *Bioorg. Med. Chem.*, **2007**, *15*, 2935-2943.

[174] Tsuge, H.; Nishimura, T.; Tada, Y.; Asao, T.; Turk, D.; Turk, V.; Katunuma, N. Inhibition mechanism of cathepsin l-specific inhibitors based on the crystal structure of papain-clik148 complex. *Biochem. Biophys. Res. Commun.*, **1999**, *266*, 411-416.

[175] Arkin, M.R.; Randal, M.; DeLano, W.L.; Hyde, J.; Luong, T.N.; Oslob, J.D.; Raphael, D.R.; Taylor, L.; Wang, J.; McDowell, R.S.; Wells, J.A.; Braisted, A.C. Binding of small molecules to an adaptive protein-protein interface. *Proc. Natl. Acad. Sci. U.S.A.*, **2003**, *100*, 1603-1608.

[176] Lovejoy, B.; Welch, A.R.; Carr, S.; Luong, C.; Broka, C.; Hendricks, R.T.; Campbell, J.A.; Walker, K.A.; Martin, R.; Van Wart, H.; Browner, M.F. Crystal structures of mmp-1 and -13 reveal the structural basis for selectivity of collagenase inhibitors. *Nat. Struct. Biol.*, **1999**, *6*, 217-221.

[177] Finzel, B.C.; Baldwin, E.T.; Bryant, G.L.; Hess, G.F.; Wilks, J.W.; Trepod, C.M.; Mott, J.E.; Marshall, V.P.; Petzold, G.L.; Poorman, R.A.; O'Sullivan, T.J.; Schostarez, H.J.; Mitchell, M.A. Structural characterizations of nonpeptidic thiadiazole inhibitors of matrix metalloproteinases reveal the basis for stromelysin selectivity. *Protein Sci.*, **1998**, *7*, 2118-2126.

[178] Eschenburg, S.; Genov, N.; Peters, K.; Fittkau, S.; Stoeva, S.; Wilson, K.S.; Betzel, C. Crystal structure of subtilisin dy, a random mutant of subtilisin carlsberg. *Eur. J. Biochem.*, **1998**, *257*, 309-318.

[179] Dubowchik, G.M.; Vrudhula, V.M.; Dasgupta, B.; Ditta, J.; Chen, T.; Sheriff, S.; Sipman, K.; Witmer, M.; Tredup, J.; Vyas, D.M.; Verdoorn, T.A.; Bollini, S.; Vinitsky, A. 2-aryl-2,2-difluoroacetamide fkbp12 ligands: synthesis and x-ray structural studies. *Org. Lett.*, **2001**, *3*, 3987-3990.

[180] Lévesque, S.; St-Denis, Y.; Bachand, B.; Préville, P.; Leblond, L.; Winocour, P.D.; Edmunds, J.J.; Rubin, J.R.; Siddiqui, M.A. Novel bicyclic lactam inhibitors of thrombin: potency and selectivity optimization through p1 residues. *Bioorg. Med. Chem. Lett.*, **2001**, *11*, 3161-3164.

[181] Lee, S.L.; Alexander, R.S.; Smallwood, A.; Trievel, R.; Mersinger, L.; Weber, P.C.; Kettner, C. New inhibitors of thrombin and other trypsin-like proteases: hydrogen bonding of an aromatic cyano group with a backbone amide of the p1 binding site replaces binding of a basic side chain. *Biochemistry.*, **1997**, *36*, 13180-13186.

Application of *In Silico* Methods to Study ABC Transporters Involved in Multidrug Resistance

Ilza Pajeva[1],* and Michael Wiese[2],*

[1]*Centre of Biomedical Engineering, Bulgarian Academy of Sciences, Acad. G. Bonchev Str. Bl. 105, 1113 Sofia, Bulgaria, Fax +359 2 8723787, Tel. +359 2 979 3605, E-mail: pajeva@bio.bas.bg and* [2]*Pharmaceutical Institute, University of Bonn, An der Immenburg 4, 53121 Bonn, Germany, Fax: +49 228 737929, Tel: +49 228 735213, E-mail: mwiese@uni-bonn.de*

Abstract: ABC transporters are involved in variety of processes of physiological and clinical significance. Besides their function as natural physiological protectors of the living organisms against xenobiotics, they play a crucial role in drug pharmacokinetics and for the multidrug resistance in tumor cells. The *in silico* modeling of ABC transporters is extensively developing in the recent years making use of increasing data about 3D structures of transport proteins. The chapter describes the most recent achievements in the computational studies starting from ligand-based design approaches and classification algorithms to homology modeling and docking of ligands. Many of the models show a satisfying performance shading light on the structure-function relationships of the proteins and their substrates and inhibitors, as well as generating hypotheses of ligand-protein interactions and helping in design of further experimental studies. However, a number of problems related to reliability of the experimental data used for modeling and computational methodologies applied limit the applicability of these methods for virtual ligand screening. First attempts towards docking of ligands into transporter binding sites are discussed to illustrate these limitations.

Key words: ABC transporters, multidrug resistance, QSAR, pharmacophore modeling, homology models, docking

INTRODUCTION

The ATP-binding cassette (ABC) transporters form a large superfamily of proteins, which transport a number of molecules across the membrane bilayers in living organisms. Nowadays 48 members of the ABC superfamily have been found in humans and a much larger number of transport proteins have been identified in bacteria and parasites. Among them the eukaryotic ABC transporters, particularly the exporters, are in the main focus of research interest, because of their involvement in multidrug resistance (MDR) and drug transport. MDR has initially been associated with the ability of cancer cells to display resistance towards a wide spectrum of structurally and functionally diverse drugs. Later it has been related also to other living cells as bacteria, parasites, fungi. Over the past 30 years tremendous efforts have been made to find highly specific modulators of MDR in tumor cells, but so far all effective drugs have proven to be too toxic for human use [1]. Besides MDR, these transporters are also actively involved in absorption, distribution, metabolism and excretion (ADME) of drugs, thus influencing bioavailability and drug-drug interactions. Naturally occurring in a number of tissues and organs they serve also as natural physiological barriers protecting the organism against xenobiotics.

The 48 proteins of the ABC superfamily are phylogenetically highly conserved and according to their sequence similarity are classified into seven subfamilies (labeled from A to G) [2]. Three proteins from the B, C and G subfamilies have been primarily associated with MDR: P-glycoprotein, P-gp (ABCB1); the multidrug resistance-associated protein, MRP1 (ABCC1); and the breast cancer resistance protein, BCRP (ABCG2). In the recent years, however, other members of these families have become the subjects of research interest, mostly because of their clinical significance for cancer chemotherapy.

P-gp is the first ABC transporter discovered [3] and, undeniably, it is the most explored transport protein nowadays. Besides its involvement in the active efflux of chemotherapeutics from tumor cells [4], the protein plays also an essential role in ADME. A large and permanently increasing number of compounds are recognized as its substrates and inhibitors, including many cytotoxic anticancer drugs (anthracylines, *Vinca* alkaloids, taxanes, epipodophyllotoxins) and various MDR modulators. Well after P-gp, in 1994 the first MRP transporter MRP1 has been discovered [5] and later several more members of this subfamily have been identified with various tissue expression and physiological actions. The drugs interacting with MRPs vary depending on the necessity to be linked or not to glutathione (GSH) or glucuronic acid. MRP1 functions as a multispecific organic anion transporter but transports also neutral or weakly basic organic compounds [6]. Both, P-gp and MRP1 confer resistance to a similar but not identical spectrum of cytotoxic drugs [7, 8]. BCRP has been initially identified in cells that do not express P-gp and MRP1 [9-11] and also in cell lines selected for resistance to mitoxantrone that is a poor substrate of P-gp and MRP1 [12]. The protein displays also wide substrate recognition properties, including neutral, positively and negatively charged compounds. All three transporters are naturally expressed in many tissues with barrier functions. P-gp is expressed in the hepatocytes, the endothelial cells of the intracerebral blood vessels, the renal tubular and intestinal epithelial cells. MRP1 is present in almost all mammalian cells; and BCRP - in the intestine, the bile canalicular membrane, placenta and in cancer stem cells. The common and most striking feature of these and other similar MDR transporters remains the diversity of the recognized and transported substrates. They belong to various chemical classes and generally do not share structural homology.

The significant pharmacological and physiological role of these transporters explains the strong interest of researchers of different fields, reflected by the enormous number of papers and books published to date. In the last 20 years a number of modeling studies have also been conducted to elucidate structure-function relationships and ligand interactions of the MDR transporters, particularly P-gp, and more recently MRPs and BCRP. In this chapter we will focus on the most recent achievements in modeling of these transporters and will try to summarize current insights into the presentations of their interactions with substrates and inhibitors.

STRUCTURAL DATA, PRESENTATIONS OF FUNCTIONING AND BINDING SITES OF P-GLYCOPROTEIN AND RELATED MDR TRANSPORTERS

The eukaryotic ABC transporters are generally composed of four main structural units localized in a single polypeptide chain: two membrane domains (TMDs) and two nucleotide-binding domains (NBDs) (Fig. **1**). The NBDs are responsible for the binding and hydrolysis of ATP, and due to this common substrate they have highly conserved sequence motifs. The TMDs are involved in the binding and transport of substrates across the membrane, correspondingly they have variable sequences and this is reflected in the substrate diversity between the ABC transporters.

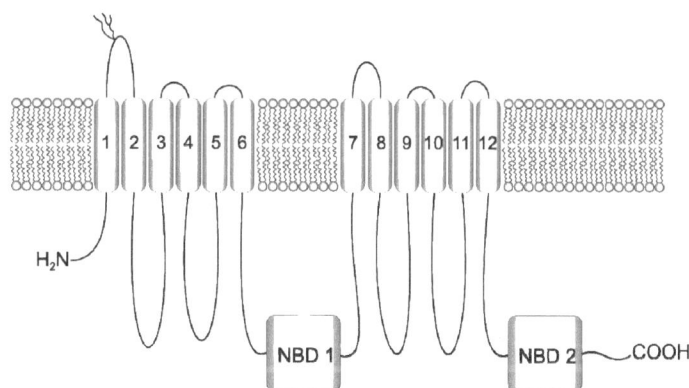

Figure 1: Schematic presentation of the topology of the ABC MDR transporter P-gp [13].

All experimental data agree that these transporters undergo large conformational changes during the transport cycle. Currently, two major mechanistic models of P-gp functioning are considered, differing in the presentation about the actual switch of the power stroke responsible for active drug export from the cell. According to the most popular ATP switch model, proposed by Higgins [14], the transport substrate binds to its high-affinity binding site on the TMDs from the inner leaflet of the membrane. This binding transmits a conformational change to the NBDs which increases affinity for ATP thus enhancing its binding. The bound ATP molecules generate a closed NBD dimer that itself induces conformational changes in TMDs in a way that the drug binding site is exposed to the extracellular side. The affinity decreases allowing drug release. During the next steps of the transport cycle the transporter restores its basal configuration. The ABC switch model is supported by the crystal structures of the bacterial MDR transporter Sav1866 [15] that has been interpreted as the ATP-bound outward-facing conformation allowing release of the transported drug from the low-affinity site into the extracellular space [16]. In contrast to the ATP switch model, other authors suggest that the ATP-induced dimerization of the NBDs alone is not sufficient to cause the high to low affinity transition because the occlusion of a nucleotide following the dimerization is required [17, 18]. No matter the ATP induced dimerization of the NBDs or the ATP hydrolysis provides the actual power stroke for the transition, the presence of binding sites of high and low affinity potentially corresponding to the inward- and outward-facing conformations of the transporter is of importance for the modeling studies.

Very recently the X-ray structure of mouse P-gp has been refined to 3.8 Å resolution [19]. The protein has 87% sequence identity to human P-gp. The apo and drug-bound structures have been obtained, both of inverted V-shape and open to the cytoplasm of the cell. They are thought to correspond to the inward-facing conformation of P-gp. According to the above discussed models of P-gp functioning, this conformation is considered to represent the initial stage of the transport cycle competent for drug binding. It has a large internal cavity, formed by the bundles of the TM helices, and can accommodate more than one compound simultaneously. The large internal cavity and possibility for binding large ligands or more than one ligand simultaneously has also been observed in other transporters [20, 21] thus confirming the relevance of the X-ray structure.

The large binding cavity observed in the conformations of the transporters raises the question about the number and location of drug binding sites. For P-gp the presentations vary from one to seven possible binding sites. A large common binding site was initially assumed and confirmed by the results from ATPase activity assays [22]. Later two binding sites within P-gp were proposed observing co-operative, competitive and noncompetitive interactions between compounds interacting with the protein [23-25]. Shapiro and Ling postulated two distinct binding sites, the rhodamine- and Hoechst 33342-site, which interacted in a positively cooperative manner [26]. Later a third, regulatory binding site was reported for progesterone and prazosin [27]. Martin *et al.* proposed at least four different binding sites able to allosterically communicate [28]. Seven binding sites were suggested by Safa summarizing data obtained by competition with photolabeling drugs [29].

Loo and coworkers are steadily reporting results on the interaction and transport characterization of P-gp substrates [30-38]. Differences in the cross-linking patterns observed upon binding of different drugs support the suggestion that P-gp could accommodate varying substrates through an induced-fit mechanism. Ecker *et al.* also studied ligand-binding sites using photoaffinity labeling and MALDI-TOF mass spectrometry [39]. They reported on particular TM helices of P-gp mostly labeled by the benzophenone-type photoaffinity ligands that suggests involvement of the interfaces between the two halves of the protein [40].

Considerable efforts have also been made to identify the binding sites of MRP1. Borst *et al.* [41] proposed two drug-binding sites, one, called GSH (G-site) with higher affinity for GSH-conjugated drugs, and the second one where neutral drugs bind (D-site). The G-site has been localized in the L0-loop and the D-site in the C-terminal half of the transporter [42]. Using photoaffinity experiments to find the MRP1 binding sites Daoud *et al.* concluded, that, independently of the differences observed in the binding sites of the transporters they certainly hold similar drug-binding regions [43]. No presentation about the BCRP binding sites has been reported so far, due to the complex behavior of the protein. Generally a homodimer connected via disulfide bonds is assumed to be the active form [44-47], but higher oligomers have also been proposed [48, 49].

In summary, the experimental studies reveal a complex behavior of the transport proteins related to changes in their conformation and drug binding affinity during the transport cycle. These facts, together with the extremely broad structural variety of the interacting compounds, require a careful selection of the data to be used in modeling and critical interpretation of the derived results.

IN SILICO MODELING OF ABC TRANSPORTERS

In Silico Methods for Modeling of ABC Transporters

The application of *in silico* approaches for modeling of transporters, particularly eukaryotic MDR exporters and their drug interactions, depends on the available data on the structure of the modeled protein. In the past years the studies were mostly restricted to the ligand-based methods, because of the lack of detailed information about the 3D structure of any ABC transporters. Mostly used were the classical QSAR analyses by Hansch [50] and Free-Wilson

[51]; and the 3D QSAR molecular field analyses CoMFA (comparative molecular field analysis) [52] and CoMSIA (comparative molecular similarity indices analysis) [53]. Besides the integral and local descriptors of chemical structures of different nature, descriptors generated from 3D-calculated GRID interaction energies [54, 55] or converted to 2D-form like the VolSurf descriptors [56] were applied. Different multivariate data techniques were explored in the correlations depending on the number and the type of the structural descriptors generated: multiple linear regression (MLR), principle component analysis, partial least squares (PLS) [57], genetic algorithms (GA) [58, 59], artificial neural networks [60, 61], support vector machine [62].

The publication of the first X-ray structure of the bacterial lipid A exporter MsbA in 2001 [63] followed by two structures with better resolution [64, 65] initiated a number of homology modeling studies of ABC MDR transporters. However, most of them had to be revised after the retraction of the template MsbA structures [66, 67]. In the last years a number of pharmacophore and homology models rapidly developed in parallel with more detailed information becoming available about drug binding sites and 3D structural data of other transport proteins. The first docking studies have also been performed making use of the recently solved structure of the mouse P-gp [68, 69].

Below the results of applications of these approaches are presented, grouped according to the methodology used. A summary of the earlier and the more recent results outlines the most significant achievements in modeling of the ABC MDR transporters obtained so far.

Ligand Based Studies

Over the past two decades a number of ligand-based studies have been performed. A large number of substrates and still higher number of inhibitors (MDR modulators) have been experimentally studied and their structure-activity relationships analyzed by different approaches. The most significant achievements in the ligand based studies of the MDR modulators could be related to the application of the 3D QSAR and pharmacophore modeling.

QSAR and 3D QSAR Studies

The QSAR and 3D QSAR studies of the first generation MDR modulators (compounds already used clinically for other therapeutic applications, like verapamil, cyclosporine A, quinidine, and others involved a variety of integral and local physiochemical structural descriptors for particular series of ligands. These studies have previously been discussed and we would like to refer to dedicated reviews for more details [70-74]. Generally, the empirical and QSAR analyses revealed several important structural characteristics of the drugs interacting with the transport proteins, in particular P-gp. They pointed to parameters related to the size (molecular weight, volume, solvent accessible surface area), lipophilicity (logP, logD, π-values of substituents), and electronic properties (quantum-chemical indices, hydrogen bond (HB) donor and acceptor capacities). Studies based on descriptors generated directly from the chemical structures (structural fragments), outlined the presence of a bulky aromatic ring system and tertiary basic nitrogen as obligatory structural elements. Wang *et al.*, for example, concluded that a highly effective P-gp modulator candidate should possess a log P value of 2.92 or higher, 18-atom-long or longer molecular axis, and a high E_{HOMO} value, as well as at least one tertiary basic nitrogen atom [75]. Neither of these analyses could outline any single set of descriptors of decisive importance for the interactions in all classes of studied compounds. Considering the suggested big binding pockets of the transporters and the possibility for polyspecific drug binding, such results are to be expected.

The development of the molecular field analyses, like CoMFA and CoMSIA, gave impetus to the application of 3D QSAR approaches. Perhaps, one of the most important outcomes from these studies, compared to the classical QSAR ones, is the understanding of the importance of hydrophobicity as a space distributed and directed molecular property rather than as an integral molecular property. Using HINT-derived hydrophobic fields to complement the CoMFA input matrix when modeling different series of MDR modulators of the 1st generation, Pajeva *et al.* observed a remarkable improvement in the quality of the 3D-QSAR models [76, 77]. Table 1 illustrates the superiority of the hydrophobic molecular fields over the partition coefficient logP for various classes of MDR modulators tested in different resistant cell lines.

Recently, the research interest has been directed to analysis of representatives of new, 3rd generation MDR modulators [78]. These compounds are composed of structural features preselected from structure-activity relationships analyses and then submitted to pharmacological screening [79]. In contrast to the previous generations, these transport inhibitors are aimed not to be cytochrome P450 3A4 substrates or inhibitors and not to influence significantly the pharmacokinetic profile of co-administered drugs [78, 80]. Their QSAR and 3D QSAR analyses revealed the important role of more than one molecular property field for the inhibitory potency of the compounds. In contrast to the 1st generation, the molecular fields contributing to the effects of the compounds demonstrate much larger variety; here steric, electrostatic, HB acceptor and donor fields have been identified. While for the tariquidar analogs, inhibitors of P-gp, the H-bond acceptor, steric, and hydrophobic fields were shown to be the top contributing properties [81-83]; for the BCRP inhibition – the HB donor fields were identified as the most important ones [84]. Recently, the electrostatic fields were found as the most significant for differentiating in the inhibition potency of compounds between P-gp and MRP1 [68]. Again, no correlation has been outlined in neither of the classes studied with logP (see refs in [83]).

Table 1: Lipophilicity (logP) versus 3D hydrophobicity (molecular hydrophobic field) as a predictor of MDR reversal activity.

Cell line / Chemical class (no. compounds)	Parameter			
	logP [*HINT*]		HINT field	
MCF-7/DOX [85, 86]	Q^2_{cv}	R^2	Q^2_{cv}	R^2
phenothiazines (21)	0.535	0.628	0.511	0.994
thioxanthenes (16)	0.333	0.462	0.789	0.935
phenothiazines (21) thioxanthenes (16)	0.607	0.647	0.789	0.916
phenothiazines (21) thioxanthenes (16) imipramines (2) acridines (1)	0.595	0.635	0.783	0.906
CCRF-CEMvcr1000 [87]	Q^2_{cv}	R^2	Q^2_{cv}	R^2
propafenones (13)	0.618	0.722	0.661	0.998
benzofuranes (6)	-1.06	0.357	-0.38	0.963
propafenones (13) benzofuranes (6)	0.090	0.373	0.758	0.983
KB-A1 [88]	logP [*ClogP*]		HINT field	
	Q^2_{cv}	R^2	Q^2_{cv}	R^2
triazines (30)	0.085	0.227	0.602	0.958
P388/ADR [89]	logP [*ACDLabs*]		CoMSIA# [90]	
	Q^2_{cv}	R^2	Q^2_{cv}	R^2
phenothiazines (18) thioxanthenes (3) imipramines (2) benzothiazines (2)	0.133	0.236	0.795	0.948

* calculated with MOLGEN [87] # CoMSIA hydrophobic indices

These results confirmed once more that logP, although an important lipophilicity characteristic, is not linearly related to the activities. In contrast, the hydrophobic molecular fields, showed high correlations; *e.g.* in [81] the hydrophobic field alone (in this case CoMSIA hydrophobic indices) produced models with $q^2=0.718$. It should be noticed, however, that, in contrast to the previous generations of modulators, relatively low predictive ability of the hydrophobic field alone has been recorded, thus illustrating that, although an important structural property, hydrophobicity is not the only determinant for the inhibitory potency of the new generations of MDR modulators.

A different group of MDR modulators being in the focus of research interest in the recent years is the class of flavonoids. Beside the antioxidant activity, these compounds have also been found to inhibit the function of some ABC transporters, like P-gp and BCRP. The flavonoid scaffold is able to perform an ATP-mimic function and thus, the flavonoids are presumed to interact with the highly conserved NBDs of the transporters. Among the most recent are studies employing a 3D linear solvation energy VolSurf model to quantify the affinity of flavonoid derivatives toward P-gp and BCRP. The shape parameters and hydrophobicity were outlined as the physicochemical parameters mostly responsible for the affinity of flavonoid derivatives towards P-gp; the HB capacities were defined as minor determinants of this activity [91]. For BCRP similar findings have been reported: shape and hydrophobic parameters were revealed to be the most important physicochemical parameters for the inhibition activity of the flavonoid derivatives and synthetic analogs towards the protein, whereas HB-donor capacity appeared highly unfavorable [92, 93].

Thus, independently of the differences in the binding sites between the different classes and generations of MDR modulators, the QSAR and 3D QSAR studies agree on the steric (shape) and hydrophobic properties as the major structural determinants for the transporter inhibition activity. No general tendency can be outlined for the effect of the type of the HB interactions (acceptor or donor). They seem to be dependent on the nature of the transporter and its binding sites and further studies in this direction could specify their role for the ligand-protein interactions.

Classification Studies

In classification studies rules are searched that usually discriminate substrates from non-substrates of the transporter, but occasionally discriminations between substrates and modulators have also been undertaken. In most cases medium sized data sets of structurally different compounds are used due to the limited amount of activity data available in the literature.

The validity of the classification heavily depends on the correctness of the group membership assigned. However, compared to classical enzyme inhibition assays, the measurement of transporter substrate properties is more difficult as the ratio of passive diffusion to active extrusion is decisive for the generation of a concentration gradient across the cell membrane. Thus activity data may depend on the assay, and it could happen that the same compound behaves as a non-substrate or substrate/inhibitor. Polli *et al.* have compared three different tests and found different combinations. Beside clear substrates and non-substrates, ambiguous compounds were found, showing interaction in one or two assays only [94]. Table **2** lists some of the compounds investigated by Polli *et al.* together with their assignments used in two classification studies. Obviously the classification of P-gp drugs as substrates and non-substrates is ambiguous. The cytostatics doxorubicin and daunorubicin exemplify the problem: they are well recognized P-gp substrates but only daunorubicin was classified as a transported substrate, while doxorubicin was recognized as an unambiguous non-substrate, because it showed no activity in either of the experimental tests. From inspection of Table **2** it becomes clear, that a significant fraction of compounds is inconsistently assigned in different classification studies, most probable due to different assays and interpretation of their results.

Table 2: Compounds assigned to different classes in classification studies and their behavior in different P-gp activity assays.

Compound	P-gp activity assays [31]			Classification studies			
	increase of calcein fluorescence	transport through Caco-2 cells	stimulation of ATPase activity	Cabrera *et al.* [Error! Bookmark not defined.]	Gombar *et al.* [Error! Bookmark not defined.]	Xue *et al.* [Error! Bookmark not defined.]	Seelig [Error! Bookmark not defined.]
Chlorpromazine	S	N	S	x	N	S	B
Diltiazem	S	S	S	S	N	S	S
Daunorubicin	N	S	N	S	S	S	S
Doxorubicin	N	N	N	S	N	S	S
GF120918	S	N	N	S		S	
Ketoconazole	S	N	S	x	N		
Mebendazole	S	N	S	x	N		
Midazolam	S	N	S	x	N	N	S
Nicardipine	S	N	S	x	N	S	S
Nifedipine	S	N	S	x	N	S	S
Nitrendipine	S	N	S	x	N		
Testosterone	S	N	N	N	N		
Verapamil	S	N	S	S	N	S	S
Yohimbine	N	N	N	S	N	S	S

N: non-substrate; S: substrate/modulator; B: borderline substrate; x: excluded from analysis

Several research groups have reported studies aiming to classify P-gp non-interacting and interacting drugs. In an early study a relatively small set of 44 compounds was investigated [95]. The authors formulated two simple rules to differentiate between transported substrates, inhibitors and non-interacting compounds. Compounds were classified as transport substrates if they fulfilled the following four criteria: at least one ≥ 6-membered ring system; molecular mass > 399 Da; lipophilicity < 2; and hydrogen bonding potential > 8. Compounds had to meet the following three criteria to be classified as inhibitors: at least one ≥ 6-membered ring system; molecular mass > 247 Da; a dipole moment ≥ 3.3. Compounds not fulfilling either of these rules were classified as non-interacting. Seelig [96, 97] compared 100 structurally diverse compounds and concluded that substrate recognition is connected to the presence of at least one of two specific hydrogen bond patterns. The type I and II pattern consists of two electron donor groups with a spatial separation of 2.5 ± 0.3 Å or 4.6 ± 0.6 Å, respectively. Bakken and Jurs studied a large data set of 609 compounds which multidrug resistance reversal activity had been measured using adriamycin-resistant P388 murine leukemia cells [98]. Two types of classification models were generated, a two-class and a three-class model, differentiating moderately active and actives additionally. Also the separation threshold between actives and inactives was varied. With a small threshold a model using nine topological descriptors reached a classification success of 83 percent on an external test set. With a larger separation of actives and inactives the correct classification rate raised to 92 % using a model with 6 topological descriptors. Gombar et al generated a model based on a set of 95 compounds that had been classified as transported substrates (63) and non-substrates (32) based on results from cell monolayer transport assays [99]. Using feature selection and linear discriminant analysis an excellent separation of substrates from non-substrates was achieved with 27 descriptors. Additionally a rule based only on the molecular electrotopological state (E-state) was developed. Compounds with an E-state > 110 were substrates and compounds with a value < 49 were non-substrates. While this rule correctly classified substrates and non-substrates of a test set, about two third of the compounds could not be classified, as they had intermediate E-state values in the interval from 49 to 110. Using very simple descriptors and atom counts and applying a stepwise classification method to a data set of more than 1000 substrates and non-substrates, Didziapetris *et al.* proposed a "rule of four" to discriminate substrates and non-substrates [100]. A compound is predicted to be a substrate if it has: molecular mass > 400 Da; the number of hydrogen-bond acceptors (O and N) ≥ 8 and most acidic pK_a > 4. Non-substrates were compounds with (N+O) ≤ 4, molecular mass < 400 Da and most basic pK_a < 8. More recently similar results were reported using an in house database of 2000 compounds with measured transport ratios [101]. With four simple descriptors (molecular mass, log P, positive or negative ionisation state) it was found that neutral or basic molecules with a molecular mass > 400 Da and log P > 4 are more likely substrates than zwitterionic or negatively charged compounds. Xue *et al.* used data of 201 compounds taken from the literature for classification together with a large set of 159 descriptors for shape, electrotopological states, and quantum chemical and geometrical properties. They compared the support vector machine (SVM) approach to other machine-learning algorithms. SVM with shape and molecular property descriptors was found to be superior to k-nearest neighbour, probabilistic neural networks and decision-tree classifier [102]. Cabrera *et al.* reported the classification of 163 compounds using topological descriptors and linear discriminant analysis with sensitivity and selectivity values of about 80 percent [103]. De Cerqueira Lima *et al.* applied an interesting combinatorial QSAR approach to the classification problem [104]. They investigated 195 diverse substrates and non-substrates of P-gp that had been used by Penzotti *et al.* for the generation of an ensemble pharmacophore model [105], applying different modelling techniques as k-nearest neighbours classification, decision tree, binary QSAR, and SVM. Each of the techniques was used with different descriptor sets comprising of molecular connectivity indices, atom pair and VolSurf descriptors. The authors found that all combinations yielded models with high accuracy for the training set, but only some had high classification accuracy for the test set. With this combinatorial QSAR approach, models were identified that showed higher predictivity than previously reported ones. Huang *et al.* also applied a SVM for classification of the same set of compounds as used by Cabrera *et al.* [103]. From an initial set of 929 descriptors calculated with the Dragon software, 79 were selected after preprocessing applying statistical criteria. For the selection of final descriptors a particle swarm algorithm was applied. The authors reported an increased accuracy compared to other classification studies, when applying this sophisticated combination [106].

More recently substrates of other transporters came into the focus. Pedersen *et al.* studied the inhibition of MRP2-mediated transport of estradiol-17beta-D-glucuronide. A classification model was generated, able to differentiate inhibitors and non-inhibitors with a predictive power of 86% for the training and 72% for the test set [107]. The same research group reported also a classification model for BCRP inhibitors and non-inhibitors using BCRP-mediated efflux of mitoxantrone [108]. Using only two descriptors, logD at pH 7.4 and molar refractivity, a correct classification rate of 93 % for the training set and 79 % for the test set was achieved. Recently a comparative classification study for P-gp, MRP2 and BCRP was reported [109] and a review was published summarizing the *in silico* predictions of substrate properties for ABC MDR transporters [110].

From the above analysis it can be concluded that the classification rules can be useful if the best candidates among a large number of potential ones undergo further experimental screening. The rules can also be applied as a preselection tool for *in silico* screening of substrates and inhibitors in combination with docking and virtual screening providing the level of knowledge about the drug-transporter interactions enables the use of such methods.

Pharmacophore Modeling

Elucidation of the pharmacophore patterns of compounds able to interact with ABC transporters is permanently in the focus of the research interest. The most studied is P-gp for which a number of pharmacophore models have been proposed [105, 111-117], but recently other transporters have also been studied [68,118 ,119]. In all models the binding of the compounds is predominantly driven by hydrophobic (aromatic and aliphatic), hydrogen bonding (HB) and van der Waals interactions. The patterns identified vary depending on the subset utilized and the methods applied (mostly the formalism used to define the functional groups and atoms involved in hydrophobic interactions, *e.g.* aliphatic and aromatic C-atoms, and, in a much lesser extent, the scoring function of the algorithm).

Independently of the various results reported, most of the P-gp studies agree on a pharmacophore pattern that combines at least one HB-acceptor and two hydrophobic elements, called "a conserved pharmacophore pattern"(a term firstly used by Pearce *et al.* in the earliest pharmacophore studies [120] of MDR modulators and adopted later by Wiese and Pajeva [73]). However, the spatial arrangement of these groups can be different and additional hydrophobic, HB-acceptor and also HB-donor points can be involved depending on the subset used. Such results suggest that there might be no general pharmacophore model for P-gp that could unify all pharmacophoric features identified and this certainly holds also for other similar MDR transporters. Further, this observation implies that the pharmacophore models proposed are dependent on and, thus, characteristic only for a particular subset and binding site, with which the compounds studied are interacting in the binding cavity of the transporter.

Fig. **2** illustrates the pharmacophore points proposed for the substrate Hoechst 33342 thought to interact with the H-site of P-gp [115]. Overlays of Hoechst 33342 with small and rigid compounds from the class of the so-called QB compounds [121] that act as P-gp activators in the H-site, pointed to the chain of hydrophobic centers H_1-H_3 and a minimum one HB-acceptor (A_1 or A_2) possibly involved in the interactions with the transporter. However, as it will be shown below in docking experiments, the HB-interactions could be performed by other functional groups too depending on the pose of the compound in the extremely large central cavity of the protein. Thus, the significance of the pharmacophore points identified has to be further proven.

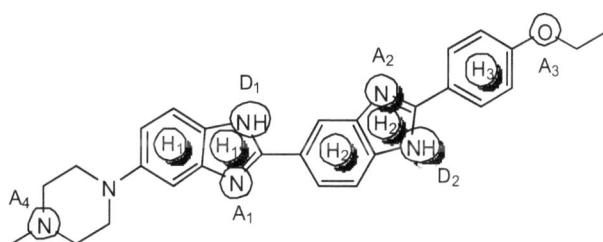

Figure 2: Structure of Hoechst 33342 with the pharmacophore points involved in the H-site of P-gp: H_1, H_1^1, H_2, H_2^1, H_3 hydrophobic centers; A_1, A_2, A_3, A_4 and D_1, D_2, D_4 are respectively HB acceptor and donor atoms; the functional groups and atoms involved in the "activator" pharmacophore model are shown by shaded circles.

The picture becomes even more complicated considering that the transporter-ligand interaction is certainly driven by an induced-fit mechanism presuming conformational changes upon binding in both, the protein and the substrate/inhibitor. Moreover, one and the same compound may have different binding poses with different amino acids involved in the interactions as illustrated by analysis of ligand binding in the recently obtained 3D structures of P-gp complexes with two stereoisomers of cyclic hexapeptide inhibitors QZ59 [69]. Besides, differences have been observed in the interactions of the same QZ59 isomer in both protein structures of the same crystal unit. It remains to be proven whether these differences are related to the experimental conditions under which the ligands have been co-crystallized or they really reflect different binding locations and orientations of the same ligand in the same protein environment. The docking of the QZ59 stereoisomers into the binding cavity of human P-gp homology model supports the latter suggestion. These results confirm the possibility for binding the same ligand in two different binding sites in the protein cavity, thus illustrating once again the complexity of the pharmacophore modeling of transport proteins.

Although sharing commonality in the interacting ligands, the binding sites of different MDR transporters possess also some differences as demonstrated recently for P-gp and MRP1 [68] and P-gp, MRP4 and MRP5 [122]; thus different pharmacophores can be expected for the same drug when it interacts with different transporters.

In summary, the definition of a pharmacophore model can be sound only for a particular transporter, binding site and subset of compounds, especially if the ligands have flexible structures as most of the drugs interacting with the transporters do. Every pharmacophore model generated has to be used with caution till strong evidence becomes available about the potential role of the pharmacophore features identified. An excellent example in this relation is the pharmacophore pattern recently reported for the subset of quinazolinone inhibitors with dual effect on MRP1 and P-gp transporters [68]. The pharmacophore model proposed for the P-gp binding of the compounds has been proven by docking into the binding cavity of the protein homology model, based on the solved 3D structure of mouse P-gp. The docking results confirmed the role of some and eliminated the participation of other pharmacophore points in the interactions of the ligands with the transporter.

Structure Based Studies

Homology Modeling

Elucidation of structures of several bacterial membrane transporters by X-ray crystallography in the beginning of this century strongly stimulated the homology modeling of human ABC transporters. As mentioned above, two homologous probable lipid A flippases, MsbA from *Escherichia coli* [63] and *Vibrio cholerae* [64, 65], as well as the *Escherichia coli* vitamin B12 transporter BtuCD [123], have been mostly used as templates. A number of studies reported homology models of P-gp [40, 115, 124, 125] and LmrA from *Lactococcus lactis* [126], the latter being the first bacterial homologue of P-gp described as a homodimeric MDR transporter. MsbA was mainly used as it was phylogenetically similar to and resembled most closely the topological organization of these transporters, in contrast to BtuCD that contained 20 TMDs and thus could not be easily used as a template. The homology models of the transporters based on these structures will not be discussed here as they have been developed before retracting the template MsbA structure [66, 67].

In the end of 2006 and beginning of 2007, however, high-resolution crystallographic structures of the bacterial ABCB1 homologue Sav1866 from *Staphylococcus aureus* have been reported [127, 128, which substantially differed from the previously published crystallographic structures of MsbA.

The Sav1866 structures have been interpreted to represent the ATP-bound state of the transporter, in which the protein TMDs form a large cavity exposed to the extracellular space (V-shaped conformation). This conformation is part of the drug translocation pathway, however, it is thought to correspond to the drug low affinity state and as such it is considered to be incompetent for drug binding. Despite of these facts the Sav1866 structural data reinitiated the homology modeling of ABC transporters and many studies have been published reporting models for P-gp [129, 130], LmrA [131], MRP1 [132], MRP4 [133], MRP5 [134, 135], and BCRP [136]. Other modeling studies made use of the corrected crystallographic data obtained for MsbA in a wide open inward-facing conformation [137] to model P-gp, MRP4 and MR5 [122]. Homology modeling of the human cystic fibrosis transmembrane conductance regulator CFTR has also been reported as an example to present a protein–protein docking approach for modeling dimeric structures without a priori structural knowledge about homologous protein dimers [138].

Some of the transporter homology models have been applied to elucidate the functional role of different segments of the proteins. Using the revised MsbA template Woebking *et al.* demonstrated the significance of the TM6 helix for the drug binding and transport. In a recent study Crowley *et al.* [139] performed *in silico* characterization of TM12 functionally perturbed mutants of P-gp using the homology model for the ATP-bound closed-state developed by

O'Mara and Tieleman [129]. The key role of TM12 in the progression of the ATP hydrolytic cycle in ABCB1 has been confirmed in agreement with the experimental observations.

Other homology modeling studies aimed at characterization of the drug binding sites and identification and localization of amino acids possibly involved in important protein-ligand interactions. Using the homology model of the bacterial LmrA [131] Federici *et al.* located the residue E314, important in the proton conduction by LmrA, at the interface between the two half-transporters in the Sav1866-based homodimer. Globisch *et al.* reported a homology model of human P-gp based on the X-ray structure of Sav1866 [140], incorporated and optimized into a membrane environment (phospholipid, water and ions). Three main binding regions have been identified: one at the interface between the membrane and cytosole and two others located in the TMDs. A big binding pocket inside the membrane cavity has also been observed (Fig. **3**). Analysis of the residues composing the binding regions and cavity revealed that most of them had been reported to influence the substrate specificity and protein function in experimental studies. Based on these results the authors concluded that the protein had multiple binding sites and may bind and/or release substrates in multiple pathways.

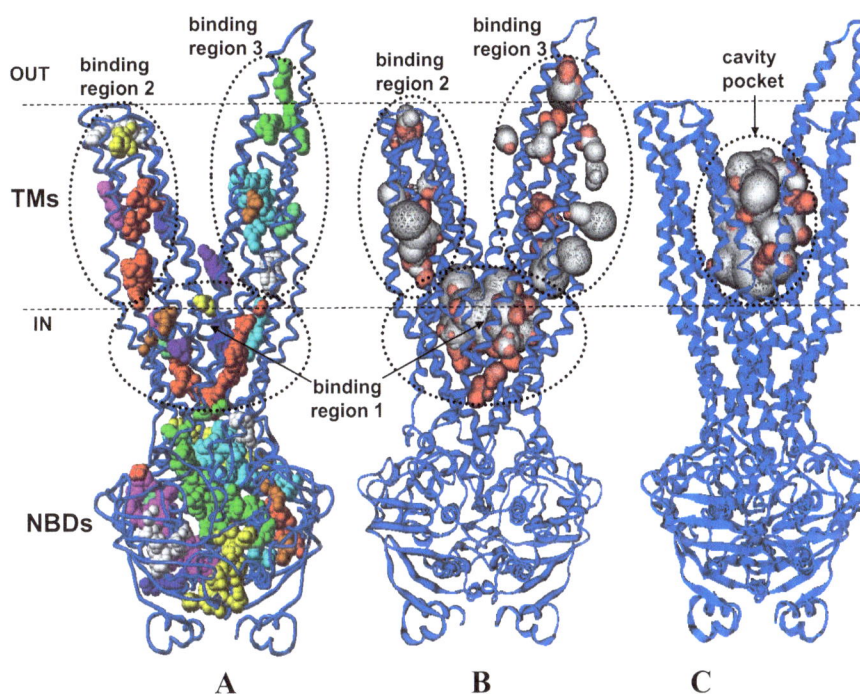

Figrue 3: Binding regions of P-gp: A) binding pockets identified by SiteID [141]; the pockets are filled with spheres set around the oxygen atoms of the water molecule; the same color is used for filling the spheres belonging to the same pocket; B) and C) binding pockets identified by SiteFinder [142]; the pockets are filled with alpha spheres; red spheres indicate hydrophilic atoms and white spheres, hydrophobic atoms; the protein backbone is shown as a blue tube; the TM related binding regions and the cavity pocket are outlined with dot lines; the horizontal dot lines show the approximate borders of the membrane. *Copyright Wiley-VCH Verlag GmbH & Co. KGaA. Reproduced from reference [140] with permission.*

Ravna *et al.* [122] investigated the differences in the electrostatic and molecular features of the drug binding sites in the molecular models of ABCB1, ABCC4 and ABCC5. Several amino acids, Ile306 (TM5), Ile340 (TM6), Phe343 (TM6), Phe728 (TM7), and Val982 (TM12), were found to form a putative substrate recognition site in the ABCB1 model, which is confirmed by both, the P-gp X-ray crystal structure [19] and site-directed mutagenesis studies. Distinct differences in the electrostatic properties of the drug recognition sites have been observed in the models of the three transporters.

Very recently Pajeva *et al.* [69] reported a homology model of human P-gp based on the recently solved X-ray structure of mouse P-gp [19]. The model corresponds to the open to inside or inward-facing conformation of the protein that is considered to be competent for drug binding. The authors compared this model to the previously developed one representing the outward-facing conformation [140]. Fig. **4** illustrates both conformations of the protein corresponding to two different functional states in the transport cycle. Using a similar approach the authors identified the residues involved in the binding cavity of the inward-open model and compared them to those identified previously in the outward-open conformation (Fig. **3**). The same amino acids have been found to face the cavity in both models thus suggesting that the interacting drugs may remain bound to the same residues during the transition of the protein from the inward- to outward-facing conformations.

Compared to other *in silico* methods, the homology modeling of the ABC transporters made the most significant progress in the recent years. This is mainly related to the fact that several 3D structures of ABC transport proteins have been obtained at reasonable resolution serving as appropriate templates for modeling, although there is no human ABC transporter structure determined so far. Combined with experimental studies, the homology models could undeniably be very effective in guiding and interpreting the experimental studies and elucidation of structure-function relationships of ABC transporters.

Docking Studies

In parallel with the homology modeling of the ABC transporters attempts have been undertaken to dock ligands into binding cavities of the proteins aiming at deeper understanding of the ligand-protein interactions. A few studies have been reported using the old MsbA templates to model the interactions with the TMDs [143, 144]. Homology models of the NBDs of the transporters have also been explored making use of the highly conserved structures of these domains. The NBD models were mostly applied for investigating compounds presumed to interact with the ATP-sites of the transporters. Badhan and Penny [145] performed docking into a homology model based on the high-resolution human TAP1 transporter structure to model interactions between NBD and flavonoids. They observed that the A-ring of an flavonoid was located within the NBD P-loop with the 5-hydroxyl group involved in HB; at the same time the B-phenylring was involved in hydrophobic stacking interactions, and the 3-hydroxyl group and carbonyl oxygen were extensively involved in HB interactions with amino acids within the NBD. Cassidy and Setzer [146] also reported docking of cytotoxic *Lonchocarpus* flavonoids, both non-prenylated and prenylated, into the ATP binding sites of the transporters LmrA, MRP1 and MJ0796. Consistent with the studies of Badhan and Penny they found that prenylation of flavonoids resulted in more stronger binding to human P-gp, which can be attributed to the positive contribution of hydrophobicity.

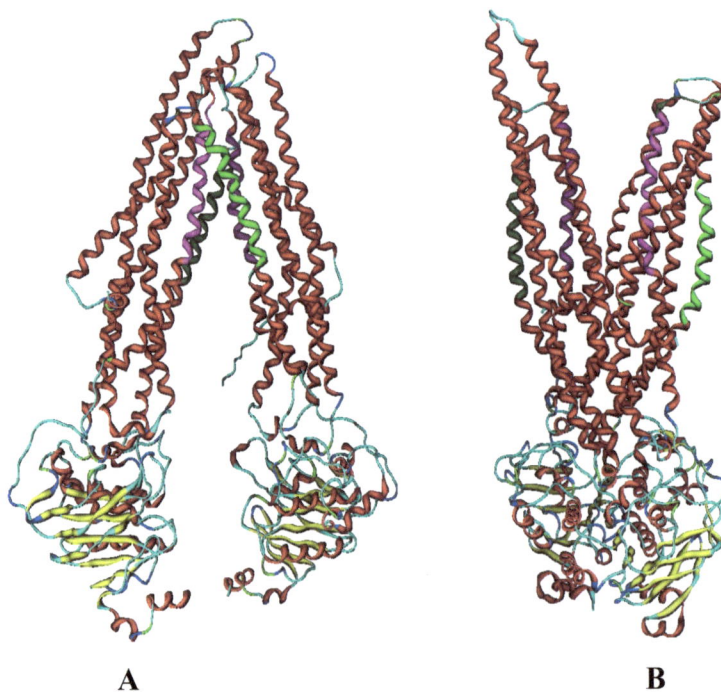

A   **B**

Figure 4: Homology models of the human P-gp in two different conformations: A) open to the inside (inward-open) [140]; B) open to outside of the membrane (outward-open) [69]. Four TMs are colored, forming two portals (TM4-TM6 and TM10-TM12) in the inward-open structure that give access of hydrophobic compounds for entering directly from the membrane to the protein cavity: TM4 (light green), TM6 (magenta); TM10 (dark green); TM12 (dark magenta). Clearly seen is that TM4 and TM10, being parts of the portals in the inward-open conformation do not face the cavity in the outward-open conformation.

Ravna *et al.* [134] docked the cyclic guanosine monophosphate, cGMP, which is known to act as a second messenger much like the cyclic AMP and shown to be transported by MRP5 with high affinity. A homology model of MRP5 was constructed based on the bacterial Sav1866 transporter in an outward-facing substrate releasing conformation. Two putative ligand binding sites were identified in the TMDs of MRP5. The docking of cGMP indicated that TM helices 1-3, 6, 11 and 12 were in contact with the ligands in binding site 1, while TM helices 1, 3, 5-8 were in contact with the ligands in binding site 2. Based on the results, the authors proposed 42 amino acids as possible candidates for single point mutations to direct further experimental studies of this transporter.

Recently Becker *et al.* [147] reported an extensive study on structure-function relationships of P-gp. The authors constructed four 3D models of two different catalytic states of the protein that were developed based on the crystal structures of the MDR transporters Sav1866 [127, 128] and MsbA [137]. Two potential pathways formed by a chain of interacting residues and involved in the propagation of a signal upon ATP binding from the catalytic site throughout to the TMDs have been proposed: one from adenine directly or through NBD aromatic residues to the intracellular loops 1 and 4, and the second pathway from the ATP phosphates directly to the Q-loop of the same NBD or through the ABC signature from the ATP of the facing NBD to the Q-loop and the intracellular loop 4. Next docking was carried out with four P-gp substrates (verapamil, rhodamine B, colchicine and vinblastine) into the binding cavity of the closed (inward-open) nucleotide-free model of P-gp. For each ligand, several positions were reported to involve interactions with residues identified to alter drug binding. In the model most of the residues, known from experimental studies to affect the specificity of individual substrates, faced the inside of the large pore and were located mainly at the level of the outer membrane leaflet. Compared to the residues identified in the study of Pajeva *et al.* [69] to form the binding pockets inside the P-gp cavity, some discrepancies can be observed between the amino acids reported in both studies. While in the study of Becker *et al.* [147] several residues belonging to TM4 (S222) and TM10 (I864, I867, I868, A871, G872) are reported to face the cavity; no amino acids belonging to these domains have been found by Pajeva *et al.* (see Table **1** in [69]). These differences could possibly be related to the templates used to build the models (bacterial MsbA in [147] and mouse P-gp in [19]) and the algorithms used to identify the amino acids involved in interaction with the ligands.

The 3D structure of the mouse P-gp solved in beginning of 2009 was the first that visualized ligand binding to the drug binding pocket of a multidrug ABC exporter [19]. Pajeva *et al.* [69] analyzed in detail the interaction of the co-crystallized cyclic peptides QZ59 with the protein in the X-ray complexes and also performed docking of these ligands into the binding cavity of a homology model of human P-gp derived from the mouse P-gp structure [19]. As discussed in the "Pharmacophore modeling" section, for the same QZ59 stereoisomer differences were recorded in the residues involved as well as in the specific interactions with the two protein molecules in the crystal unit cell. The results confirm the possibility for multiple binding sites and binding poses of the interacting compounds into the large binding cavity. This observation is further illustrated on the example of docking experiments with the P-gp substrate Hoechst 33342 into the binding cavity of the human P-gp homology model (Fig. **5**).

In the figure three out of the ten best poses of the P-gp substrate Hoechst 33342 docked into the binding cavity of the human P-gp homology model are presented. The GOLD [148] docking scores were very close each other and none of them could be given any preferences. The highest score pose (in orange) is located along the left part of the cavity and the following pharmacophore points (Fig. **3**) are involved in specific interactions: H_1 with Tyr307 (arene-arene interactions), H_3 with Phe770 (arene-arene interactions), and A_2 with Asn721 (HB). In the middle score pose (in pink) an arene-arene interaction with Phe343 (H_2) and a HB interaction with Tyr307 (A_2) were identified, and in the lowest score pose (green) – an arene-arene interaction with Phe978 (H_2') and a HB interaction with Tyr307 (A_1). Notably, Tyr307 is involved in all poses and this residue has also been identified in previous modeling studies of this protein [115]. In the different poses different pharmacophore points are involved, thus confirming the possibility for polyspecific binding of the P-gp substrates, similarly to those of the cocrystallized P-gp ligands QZ59 [69]. This example comes to show that docking of ligands for proteins like ABC tranporters has many uncertainties: the large binding cavities and the polyspecific way of binding for the same ligand make the task much more difficult compared to modeling of the enzymatic protein-ligand interactions.

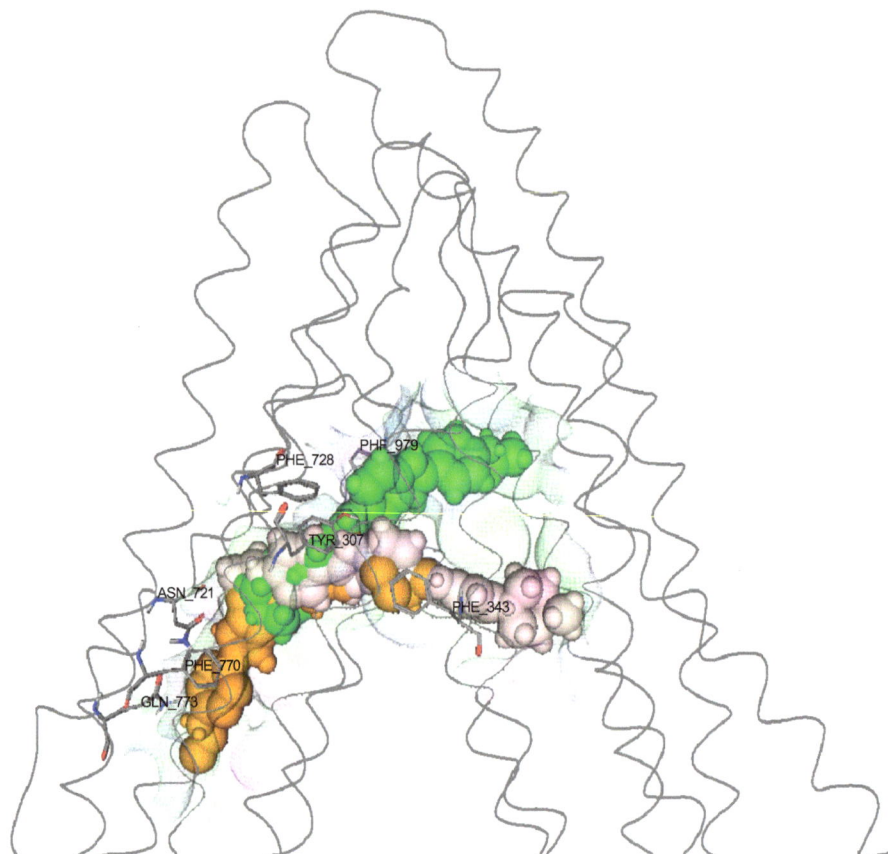

Figure 5: Front overview of the binding cavity of the human P-gp homology model with the substrate Hoechst 33342 (space filled) docked and the amino acids (sticks) involved in the specific interactions with the ligand; the binding site is outlined by the Gaussian contact surface: green – hydrophobic; magenta – hydrogen bonding; blue – mildly polar. The ligand structures are colored according to the score: orange (highest); pink (middle); green (lowest).

CONCLUSIONS

In the recent years there have been considerable modeling efforts to target promiscuous proteins. Compared to other proteins, the ABC transporters perhaps are one of the hardest objects for *in silico* modeling. The high conformational flexibility of the proteins, the large binding cavities composed of multiple binding sites and able to accommodate more than one ligand simultaneously; the induced-fit mechanisms governing the ligand-protein interactions, the change in the level of affinity of the binding sites during the transport cycle – all these facts render the modeling task extremely difficult and explain the absence of successful virtual ligand screening applications for these transporters.

It is important to remember that effectiveness and usefulness of any modeling study is a function of the data used for modeling. While experimental data necessary for applying ligand–based approaches are easier to produce and validate, a number of questions of crucial importance for gathering reliable and significant data for structure-based studies have still to be answered: there is a need of a clearer understanding of the conformational changes governing coordination between the TMDs and NBDs in the protein transport cycle; it is questionable how comparable are the physiological activities and cellular environments between the template and the modeled protein to make a direct relation between the structures and, correspondingly, the mechanism of both proteins; it is necessary to have evidences that the substrates that are transported by the template can also be transported by the modeled protein; the difference and similarities in the effect of the substrate – stimulated ATPase activity has to be also taken into account; the role of the lipid composition surrounding the protein may essentially affect the activity of some transporters, *e.g.* BCRP, and this fact should also be considered. Having these questions answered in parallel with application of more sophisticated modeling tools such as molecular dynamics (MD), is a promising direction for future research. MD simulations, for example, have been shown to detect incorrect packing of otherwise correctly predicted TM-helices in a membrane protein model [149].

Despite these limitations, the *in silico* modeling of ABC transporters has made a significant progress in the recent years and its impact on both, theoretical and application oriented studies, is permanently increasing. The *in silico* approaches to ABC transporters hold and will continue to play a significant role for generating hypotheses and designing further experimental studies. The theoretical and experimental investigations of this important family of proteins should complement each other in an iterating process towards the better understanding of the protein structure-function relationships and rational design of transporter inhibitors.

ACKNOWLEDGMENTS

We are grateful to our colleagues whose efforts contributed to our results and knowledge on the topic. We apologize to all whose studies have not been discussed in more detail and regret omitting any important contributions, which may have escaped our attention. We acknowledge also the financial support of our studies from the Alexander von Humboldt Foundation, DFG, and the National Science Fund of Bulgaria (grant TDK 02/58).

REFERENCES

[1]   Shukla, S.; Wu, C.P.; Ambudkar, S.V. Development of inhibitors of ATP-binding cassette drug transporters: present status and challenges. *Expert. Opin. Drug Metab. Toxicol.*, **2008**, *4*, 205–223.

[2]   Klein, I.; Sarkadi, B.; Varadi, A. An inventory of the human ABC proteins. *Biochim. Biophy.s Acta*, **1999**, *1461*, 237–262.

[3]   Juliano, R.L.; Ling, V. A surface glycoprotein modulating drug permeability in Chinese hamster ovary cell mutants. *Biochim. Biophys. Acta*, **1976**, *455*, 152-162.

[4]   Gottesman, M.M.; Ling, V. The molecular basis of multidrug resistance in cancer: the early years of P-glycoprotein research. *FEBS Lett.*, **2006**, *580*, 998–1009.

[5]   Cole, S.P.; Sparks, K.E.; Fraser, K.; Loe, D.W.; Grant, C.E.; Wilson, G.M.; Deeley, R.G. Pharmacological characterization of multidrug resistant MRP-transfected human tumor cells. *Cancer Res.*, **1994**, *54*, 5902-5910.

[6]   Boumendjel, A.; Baubichon-Cortay, H.; Trompier, D.; Perrotton, T.; Di Pietro, A. Anticancer multidrug resistance mediated by MRP1: recent advances in the discovery of reversal agents. *Med. Res. Rev.*, **2005**, *25*, 453-472.

[7]   Hipfner, D.R.; Deeley, R.G.; Cole, S.P.C. Structural, mechanistic and clinical aspects of MRP1. *Biochim. Biophys. Acta*, **1999**, *1461*, 359-376.

[8]   Leslie, E.M.; Deeley, R.G.; Cole, S.P.C. Toxicological relevance of the multidrug resistance protein 1, MRP1 (ABCC1) and related transporters. *Toxicology*, **2001**, 167, 3-23.

[9]   Kusuhara, H.; Sugiyama, Y. ATP-binding cassette subfamily G (ABCG family). *Pflugers Arch.*, **2007**, *453,* 735-744.

[10]   Doyle, L.A.; Yang, W.D.; Abruzzo, L.V.; Krogmann, T.; Gao, Y.; Rishi, A.K.; Ross, D.D. A multidrug resistance transporter from human MCF-7 breast cancer cells. *Proc. Natl. Acad. Sci. USA*, **1998**, *95*, 15665-15670.

[11]   Litman, T.; Brangi, M.; Hudson, E.; Fetsch, P.; Abati, A.; Ross, D.D.; Miyake, K.; Resau, J.H.; Bates, S.E. The multidrug-resistant phenotype associated with overexpression of the new ABC half-transporter, MXR (ABCG2). *J. Cell Sci.*, **2000**, *113*, 2011-2021.

[12]   Robey, R.W.; Polgar, O.; Deeken, J.; To, K.W.; Bates, S.E. ABCG2: determining its relevance in clinical drug resistance. *Cancer Metastasis Rev.*, **2007**, *26*, 39-57.

[13]   Globisch, C.T. PhD Thesis. Molecular-Modelling-Untersuchungen des ABC-Transporters P-Glykoprotein, University of Bonn, 2008.

[14]   Higgins, C.F.; Linton, K.J. The ATP switch model for ABC transporters. *Nat. Struct. Mol. Biol.*, **2004**, *11*, 918–926

[15]   Dawson, R.J.; Hollenstein, K.; Locher, K.P. Uptake or extrusion: crystal structures of full ABC transporters suggest a common mechanism. *Mol. Microbiol.*, **2007**, *65*, 250–257.

[16]   Dawson, R.J.; Locher, K.P. Structure of a bacterial multidrug ABC transporter. *Nature*, **2006**, *443*, 180–185.

[17]   Tombline, G.; Muharemagic, A.; White, L.B.; Senior, A.E. Involvement of the "occluded nucleotide conformation" of P-glycoprotein in the catalytic pathway. *Biochemistry*, **2005**, *44*, 12879–12886.

[18]   Sauna, Z.E.; Nandigama, K.; Ambudkar, S.V. Exploiting reaction intermediates of the ATPase reaction to elucidate the mechanism of transport by P-glycoprotein (ABCB1). *J. Biol. Chem.*, **2006**, *281*, 26501–26511.

[19]   Aller, S.G.; Yu, J.; Ward, A.; Weng, Y.; Chittaboina, S.; Zhuo, R.; Harrell, P.M.; Trinh, Y.T.; Zhang, Q.; Urbatsch, I.L.; Chang, G. Structure of P-Glycoprotein Reveals a Molecular Basis for Poly-Specific Drug Binding. *Science*, **2009**, *323*, 1718-1722.

[20]   Loo, T.W.; Bartlett, M.C.; Clarke, D.M. Simultaneous binding of two different drugs in the binding pocket of the human multidrug resistance P-glycoprotein. *J. Biol. Chem.*, **2003**, *278*, 39706-39710.

[21]   Lugo, M.R.; Sharom, F.J. Interaction of LDS-751 and Rhodamine 123 with P - Glycoprotein: Evidence for Simultaneous Binding of Both Drugs. *Biochemistry*, **2005**, *44*, 14020-14029.

[22]   Borgnia, M.J.; Eytan, G.D.; Assaraf, Y.G. Competition of hydrophobic peptides, cytotoxic drugs, and chemosensitizers on a common P-glycoprotein pharmacophore as revealed by its ATPase activity. *J. Biol. Chem.*, **1996**, *271*, 3163-3171.

[23]   Ayesh, S.; Shao, Y.M.; Stein, W.D. Co-operative, competitive and non-competitive interactions between modulators of P-glycoprotein. *Biochim. Biophys. Acta*, **1996**, *1316*, 8-18.

[24]   Dey, S.; Ramachandra, M.; Pastan, I.; Gottesman, M.M.; Ambudkar, S.V. Evidence for two nonidentical drug-interaction sites in the human P-glycoprotein. *Proc. Natl. Acad. Sci. USA*, **1997**, *94*, 10594-10599.

[25]   Scala S.; Akhmed N.; Rao, U. S.; Paull, K.; Lan, L.-B.; Dickstein, B.; Lee, J.-S.; Elgemeie, G.H.; Stein, W.D.; Bates, S.E. P-glycoprotein substrates and antagonists cluster into two distinct groups. *Mol. Pharmacol.*, **1997**, *51*, 1024-1033.

[26]   Shapiro, A.B.; Ling, V. Positively cooperative sites for drug transport by P-glycoprotein with distinct drug specificities. *Eur. J. Biochem.*, **1997**, *250*, 130-137.

[27]   Shapiro, A.B.; Fox, K.; Lam, P.; Ling, V. Stimulation of P-glycoprotein-mediated drug transport by prazosin and progesterone. Evidence for a third drug-binding site. *Eur. J. Biochem.*, **1999**, *259*, 841-850.

[28]   Martin, C.; Berridge, G.; Higgins, C.F.; Mistry, P.; Charlton, P.; Callaghan, R.. Communication between multiple drug binding sites on P-glycoprotein. *Mol. Pharmacol.*, **2000**, *58*, 624-632.

[29]   Safa, A.R. Identification and characterization of the binding sites of P-glycoprotein for multidrug resistance-related drugs and modulators. *Curr. Med. Chem. – Anti-Cancer Agents*, **2004**, *4*, 1-17.

[30]   Loo, T.W.; Clarke, D.M. Identification of residues in the drug-binding site of human P-glycoprotein using a thiol-reactive substrate. *J. Biol. Chem.*, **1997**, *272*, 31945-31948.

[31]   Loo, T.W.; Clarke, D.M. Identification of residues within the drug-binding domain of the human multidrug resistance P-glycoprotein by cysteine-scanning mutagenesis and reaction with dibromobimane. *J. Biol. Chem.*, **2000**, *275*, 39272-39278.

[32]   Loo, T.W.; Clarke, D.M. Defining the drug-binding site in the human multidrug resistance P-glycoprotein using a methanethiosulfonate analog of verapamil, MTS-verapamil. *J. Biol. Chem.*, **2001**, *276*, 14972-14979.

[33]   Loo, T.W.; Clarke, D.M. Location of the rhodamine-binding site in the human multidrug resistance P-glycoprotein. *J. Biol. Chem.*, **2002**, *277*, 44332-44338.

[34]   Loo, T.W.; Bartlett, M.C.; Clarke, D.M. Methanethiosulfonate derivatives of rhodamine and verapamil activate human P-glycoprotein at different sites. *J. Biol. Chem.*, **2003**, *278*, 50136-50141.

[35]   Loo, T.W.; Bartlett, M.C.; Clarke, D.M. Disulfide cross-linking analysis shows that transmembrane segments 5 and 8 of human P-glycoprotein are close together on the cytoplasmic side of the membrane. *J. Biol. Chem.*, **2004**, *279*, 7692-7697.

[36]   Loo, T.W.; Bartlett, M.C.; Clarke, D.M. Val133 and Cys137 in transmembrane segment 2 are close to Arg935 and Gly939 in transmembrane segment 11 of human P-glycoprotein. *J. Biol. Chem.*, **2004**, *279*, 18232-18238.

[37]   Loo, T.W.; Bartlett, M.C.; Clarke, D.M. Transmembrane segment 7 of human P-glycoprotein forms part of the drug-binding pocket. *Biochem. J.,* **2006**, *399*, 351-359.

[38]   Loo, T.W.; Bartlett, M.C.; Clarke, D.M. Identification of residues in the drug-translocation pathway of the human multidrug resistance P-glycoprotein by arginine mutagenesis. *J. Biol. Chem.*, **2009**, *284*, 24074-24087.

[39]   Ecker, G.F.; Csaszar, E.; Kopp, S.; Plagens, B.; Holzer, W.; Ernst, W.; Chiba, P. Identification of ligand-binding regions of P-glycoprotein by activated-pharmacophore photoaffinity labeling and matrix-assisted laser desorption/ionization-time-of-flight mass spectrometry. *Mol. Pharmacol.*, **2002**, *61*, 637-648.

[40]   Pleban, K.; Kopp, S.; Csaszar, E.; Peer, M.; Hrebicek, T.; Rizzi, A.; Ecker, G. F.; Chiba, P. P-glycoprotein substrate binding domains are located at the transmembrane domain/transmembrane domain interfaces: a combined photoaffinity labeling-protein homology modeling approach. *Mol. Pharmacol.*, **2005**, *67*, 365-374.

[41]   Borst, P.; Evers, R.; Kool, M.; Wijnholds, J. The multidrug resistance protein family. *Biochem. Biophys. Acta*, **1999**, *1461*, 347-357

[42]   Ren, X.Q.; Furukawa, T.; Aoki, S.; Nakajima, T.; Sumizawa, T.; Haraguchi, M.; Chen, Z. S.; Kobayashi, M.; Akiyama, S. Glutathione-dependent binding of a photoaffinity analog of agosterol A to the C-terminal half of human multidrug resistance protein. *J. Biol. Chem.*, **2001**, *276*, 23197-23206.

[43]   Daoud, R.; Julien, M.; Gros, P.; Georges, E. Major photoaffinity drug binding sites in multidrug resistance protein 1 (MRP1) are within transmembrane domains 10-11 and 16-17. *J. Biol. Chem.*, **2001**, *276*, 12324-12330.

[44]   Henriksen, U.; Fog, J.U.; Litman, T.; Gether, U. Identification of intra- and intermolecular disulfide bridges in the multidrug resistance transporter ABCG2. *J. Biol. Chem.*, **2005**, *280*, 36926-36934.

[45]   Liu, Y.; Yang, Y.; Qi, J.; Peng, H.; Zhang, J. Effect of cysteine mutagenesis on the function and disulfide bond formation of human ABCG2. *J. Pharmacol. Exp. Ther.*, **2008**, *326*, 33-40.

[46]   Bhatia, A.; Schäfer, H.; Hrycyna, C.A. Oligomerization of the human ABC transporter ABCG2: evaluation of the native protein and chimeric dimmers. *Biochemistry*, **2005**, *44,* 10893-10904.

[47] Kage, K.; Fujita, T.; Sugimoto, Y. Role of Cys-603 in dimer/oligomer formation of the breast cancer resistance protein BCRP/ABCG2, *Cancer Sci.*, **2005**, *96*, 866-872.

[48] Xu, J.; Liu, Y.; Yang, Y.; Bates, S.; Zhang, J. Characterization of oligomeric human half-ABC transporter ATP-binding cassette G2. *J. Biol. Chem.*, **2004**, *279*, 19781-19789.

[49] McDevitt, C.A.; Collins, R.F.; Conway, M.; Modok, S.; Storm, J.; Kerr, I.D.; Ford, R.C.; Callaghan, R. Purification and 3D structural analysis of oligomeric human multidrug transporter ABCG2. *Structure*, **2006**, *14*, 1623-1632.

[50] Kubinyi, H. In: *QSAR. Hansch Analysis and Related Approaches*, Mannhold, R.; Kroogsgard-Larsen, P.; Timmerman, H., Eds.; VCH: Weinheim, **1993**, Vol *1*, pp. 57-85.

[51] Kubinyi, H. In: *Comprehensive Medicinal Chemistry*, Hansch, C., Ed.; Pergamon Press: Oxford, **1990**, Vol. *4*, pp. 589-643.

[52] Cramer III, R.D.; DePriest, S.A.; Patterson, D.E.; Hecht, P. In: *3D QSAR in drug design. Theory methods and applications*, Kubinyi, H., Ed.; Escom: Leiden, **1993**, pp. 443-486.

[53] Klebe, G. In: *3D QSAR in drug design. Recent advances*, Kubinyi, H.; Klebe, G.; Martin, Y., Eds.; Kluwer/Escom: Dordrecht, **1998**; Vol. *3*, pp. 87-104.

[54] Wade, R.C.; Goodford, P.J. The role of hydrogen-bonds in drug binding. *Prog. Clin. Biol. Res.*, **1989**, *289*, 433-444.

[55] Cruciani, G.; Pastor, M.; Clementi, S.; Clementi, S. In: *Rational Approaches to Drug Design, Proceedings of the 13th European Symposium on Quantitative Structure-Activity Relationships*, Hoeltje, H.-D.; Sippl, W., Eds.; Prous Science: Barcelona, **2001**; pp. 251-260.

[56] Cruciani, G.; Crivori, P.; Carrupt, P.-A.; Testa, B. Predicting blood-brain barrier permeation from three-dimensional molecular structure. *J. Mol. Struct. (Theochem)*, **2000**, *503*, 17–30.

[57] Leach, A.R.; Gillet, V.J. *An introduction to Chemoinformatics*, Kluwer Academic Publishers: Dordrecht, **2003**.

[58] Leardi, R. Genetic algorithms in chemometrics and chemistry: a review. *J. Chemom.*, **2001**, *15*, 559-569.

[59] Clark, D.E. In: *Rational Drug Design: Novel Methodology and Practical Applications*, ACS Symposium Series Vol. 719, Parrill, A.L., Reddy, M.R., Eds., American Chemical Society: Washington DC, **1999**; pp. 255-270.

[60] Zupan, J.; Gasteiger, J. *Neural Networks for Chemists: An Introduction*, 2nd ed.; Wiley-VCH: Weinheim, **1999**.

[61] Anzali, S.; Gasteiger, J.; Holzgrabe, U.; Polanski, J.; Sadowski, J.; Teckentrup, A.; Wagener M. In: *3D QSAR in Drug Design,* Kubinyi, H.; Folkers, G.; Martin Y. C. Eds.; Kluwer/ESCOM: Dordrecht, **1998**; Vol. *2*, pp. 273-299.

[62] Li, H.; Yap, C.W.; Ung, C.Y.; Xue, Y.; Li, Z.R.; Han, L.Y.; Lin, H.H.; Chen, Y.Z. Machine learning approaches for predicting compounds that interact with therapeutic and ADMET related proteins. *J. Pharm. Sci.*, **2007**, *96*, 2838-2860.

[63] Chang, G.; Roth, C.B. Structure of MsbA from E. coli: a homolog of the multidrug resistance ATP binding cassette (ABC) transporters. *Science*, **2001**, *293*, 1793-1800.

[64] Chang, G. Structure of MsbA from Vibrio cholera: a multidrug resistance ABC transporter homolog in a closed conformation. *J. Mol. Biol.*, **2003**, *330*, 419-430.

[65] Reyes, C.L.; Chang, G. Structure of the ABC transporter MsbA in complex with ADP.vanadate and lipopolysaccharide. *Science*, **2005**, *308*, 1028-1031.

[66] Chang, G.; Roth, C.B.; Reyes, C.L.; Pornillos, O.; Chen, Y.J.; Chen, A.P. Retraction. *Science*, **2006**, *314*, 1875.

[67] Chang G. Retraction of "Structure of MsbA from Vibrio cholera: a multidrug resistance ABC transporter homolog in a closed conformation" [J. Mol. Biol. (2003) 330 419-430]. *J. Mol. Biol.*, **2007**, *369*, 596.

[68] Pajeva, I.K.; Globisch, C.; Wiese, M. Combined pharmacophore, docking and 3D QSAR study of inhibitors of ABCB1 and ABCC1 transporters. *ChemMedChem*, **2009**, *11*, 1883-1896.

[69] Pajeva, I.K.; Globisch, C.; Wiese, M. Comparison of the inward- and outward-open homology models of human P-glycoprotein, *FEBS J.*, **2010**, *276*, 7016-7026.

[70] Wiese, M.; Pajeva, I.K. Structure-activity relationships of multidrug resistance reversers. *Curr. Med. Chem.*, **2001**, *8*, 685-713.

[71] Wiese, M.; Pajeva, I. In: *Virtual ADMET assessment in target selection and maturation*, Solvay Pharmaceuticals Conferences Series (Volume 6), Testa, B.; Turski, L. Eds., *IOS* Press: Amsterdam, **2006**, pp. 187-208.

[72] Ekins, S.; Ecker, G.F.; Chiba, P.; Swaan, P.W. Future directions for drug transporter modelling. *Xenobiotica*, **2007**, *37*, 1152-1170.

[73] Wiese, M.; Pajeva, I. In: *Comprehensive Medicinal Chemistry II*, Editors-in-Chief: Taylor, J.B.; Triggle, D.J.; Vol. *5*: *ADMET/Property based approaches*, Testa, B.; Waterbeemd, H. Eds., Elsevier: Oxford, **2007**, pp. 767-795.

[74] Ecker, G.F.; Stockner, T.; Chiba, P. Computational models for prediction of interactions with ABC-transporters. *Drug Discov. Today*, **2008**, *13*, 311-317.

[75] Wang, R.B.; Kuo, C.L.; Lien, L.L.; Lien, E.J. Structure–activity relationship: analyses of p-glycoprotein substrates and inhibitors. *J. Clin. Pharm. Ther.*, **2003**, *28*, 203–228.

[76] Pajeva, I.K.; Wiese, M. Molecular Modeling of Phenothiazines and Related Drugs As Multidrug Resistance Modifiers: A Comparative Molecular Field Analysis Study. *J. Med. Chem.*, **1998**, *41*, 1815-1826.

[77]   Pajeva, I.; Globisch, C.; Fleischer, R.; Tsakovska, I.; Wiese, M. Molecular Modeling of P-Glycoprotein and related Drugs. *Med. Chem. Res.*, **2005**, *14*, 106-117.

[78]   Thomas, H.; Coley, H.M. Overcoming multidrug resistance in cancer: an update on the clinical strategy of inhibiting p-glycoprotein. *Cancer Control*, **2003**, *10*, 159-165.

[79]   Robert, J.; Jarry, C. Multidrug resistance reversal agents. *J. Med. Chem.*, **2003**, *46*, 4805-4817.

[80]   Krishna, R.; Mayer, L.D. Multidrug resistance (MDR) in cancer. Mechanisms, reversal using modulators of MDR and the role of MDR modulators in influencing the pharmacokinetics of anticancer drugs. *Eur. J. Pharm. Sci.*, **2000**, *11*, 265-283.

[81]   Müller, H.; Pajeva, I.K.; Globisch, C.; Wiese, M. Functional assay and structure-activity relationships of new third-generation P-glycoprotein inhibitors. *Bioorg. Med. Chem.*, **2008**, *16*, 2448-2462.

[82]   Klinkhammer, W.; Müller, H.; Globisch, C.; Pajeva, I.K.; Wiese, M. Synthesis and biological evaluation of a small molecule library of 3rd generation multidrug resistance modulators. *Bioorg. Med. Chem.*, **2009**, *17*, 2524-2535.

[83]   Pajeva, I.K.; Wiese, M. Structure-Activity Relationships of Tariquidar Analogs as Multidrug Resistance Modulators. *AAPS J.*, **2009**, *11*, 435-444.

[84]   Pick, A.; Müller, H.; Wiese, M. Structure-activity relationships of new inhibitors of breast cancer resistance protein (ABCG2). *Bioorg. Med. Chem.*, **2008**, *16*, 8224-8236.

[85]   Ford, J.M.; Prozialeck, W.C.; Hait, W.N. Structural features determining activity of phenothiazines and related drugs for inhibition of cell growth and reversal of multidrug resistance. *Mol. Pharmacol.*, **1989**, *35*, 105-115.

[86]   Ford, J.M.; Bruggeman, E.P.; Pastan, I.; Gottesmann, M.; Hait, W.T. Cellular and biochemical characterization of thioxanthenes for reversal of multidrug resistance in human and murine cell lines. *Cancer Res.*, **1990**, *50*, 1748-1756.

[87]   Ecker, G.; Chiba, P.; Hitzler, M.; Schmid, D.; Visser, K.; Cordes, H.P.; Csöllei, J.; Seydel, J.K.; Schaper, K.-J. Structure-activity relationship studies on benzofuran analogs of propafenone-type modulators of tumor cell multidrug resistance. *J. Med. Chem.*, **1996**, *39*, 4767-4774.

[88]   Dhainaut, A.; Régnier, G.; Atassi, Gh., Pierré, A.; Léonce, St.; Kraus-Berthier, L.; Prost J.-F. New triazine derivatives as potent modulators of multidrug resistance. *J. Med. Chem.*, **1992**, *35*, 2481-2496.

[89]   Ramu, A.; Ramu, N. Reversal of multidrug resistance by bis(phenylalkyl)amines and structurally related compounds. *Cancer Chemother. Pharmacol.*, **1994**, *34*, 423-430.

[90]   Tsakovska, I.M. QSAR and 3D-QSAR of phenothiazine type multidrug resistance modulators in P388/ADR cells. *Bioorg. Med. Chem.*, **2003**, *11*, 2889-2899.

[91]   Boccard, J.; Bajot, F.; Di Pietro, A.; Rudaz, S.; Boumendjel, A.; Nicolle, E.; Carrupt, P.A. A 3D linear solvation energy model to quantify the affinity of flavonoid derivatives toward P-glycoprotein. *Eur. J. Pharm. Sci.*, **2009**, 36, 254-264.

[92]   Nicolle, E.; Boumendjel, A.; Macalou, S.; Genoux, E.; Ahmed-Belkacem, A.; Carrupt, P.A. Di Pietro, A. QSAR analysis and molecular modeling of ABCG2-specific inhibitors. *Adv. Drug Deliv. Rev.*, **2009**, 61, 34-46.

[93]   Nicolle, E.; Boccard, J.; Guilet, D.; Dijoux-Franca, M.G.; Zelefac, F.; Macalou, S.; Grosselin, J.; Schmidt, J.; Carrupt, P.A.; Di Pietro, A.; Boumendjel, A. Breast cancer resistance protein (BCRP/ABCG2): new inhibitors and QSAR studies by a 3D linear solvation energy approach. *Eur. J. Pharm. Sci.*, **2009**, *38*, 39-46.

[94]   Polli, J.W.; Wring, S.A.; Humphreys, J.E.; Huang, L.; Morgan, J.B.; Webszerr, L.O.; Serabjit-Singh, C. S. Rational use of *in vitro* P-glycoprotein assays in drug discovery. *J. Pharmacol. Exp. Ther.*, **2001**, *299*, 620-628.

[95]   Bain, L.; McLachlan, J.; LeBlanc, G. Structure-activity relationships for xenobiotic transport substrates and inhibitory ligands of P-glycoprotein. *Environ. Health Perspect.*, **1997**, *105*, 812-818.

[96]   Seelig, A. A general pattern for substrate recognition by P-glycoprotein. *Eur. J. Biochem.*, **1998**, *251*, 252-261.

[97]   Seelig, A.; Landwojtowicz, E. Structure-activity relationship of P-glycoprotein substrates and modifiers. *Eur. J. Pharm. Sci.*, **2000**, *12*, 31-40.

[98]   Bakken, G.A.; Jurs, P.C. Classification of multidrug-resistance reversal agents using structure-based descriptors and linear discriminant analysis. *J. Med. Chem.*, **2000**, *43*, 4534-4541.

[99]   Gombar, V.K.; Polli, J.W.; Humphreys, J.E.; Wring, S.A.; Serabjit-Singh, C.S. Predicting P-glycoprotein substrates by a quantitative structure-activity relationship model. *J. Pharm. Sci.*, **2004**, *93*, 957-968.

[100]   Didziapetris, R.; Japertas, P.; Avdeef, A.; Petrauskas, A. Classification analysis of P-glycoprotein substrate specificity. *J. Drug Target* **2003**, *11*, 391-406

[101]   Gleeson, M.P. Generation of a set of simple, interpretable ADMET rules of thumb. *J. Med. Chem.*, **2008**, *51*, 817-834

[102]   Xue, Y.; Yap, C.W.; Sun, L.Z.; Cao, Z.W.; Wang, J.F.; Chen, Y.Z. Prediction of P-glycoprotein substrates by a support vector machine approach. *J. Chem. Inf. Comput. Sci.*, **2004**, *44*, 1497-1505.

[103]   Cabrera, M.A.; González, I.; Fernández, C.; Navarro, C.; Bermejo, M. A topological substructural approach for the prediction of P-glycoprotein substrates. *J. Pharm. Sci.*, **2006**, *95*, 589-606.

[104]   de Cerqueira Lima, P.; Golbraikh, A.; Oloff, S.; Xiao, Y.; Tropsha, A. A combinatorial QSAR modeling of P-glycoprotein substrates. *J. Chem. Inf. Model.*, **2006**, *46*, 1245-1254.

[105]   Penzotti, J.E.; Lamb, M.L.; Evensen, E.; Grootenhuis, P.D. A computational ensemble pharmacophore model for identifying substrates of P-glycoprotein. *J. Med. Chem.*, **2002**, *45*, 1737–1740.

[106]   Huang, J.; Ma, G.; Muhammad, I.; Cheng Y. Identifying P-glycoprotein substrates using support vector machine optimized by a particle swarm. *J. Chem. Inf. Model.*, **2007**, *47*, 1638-1647.

[107]   Pedersen, J.M.; Matsson, P.; Bergström, C.A.; Norinder, U.; Hoogstraate, J.; Artursson, P. Prediction and identification of drug interactions with the human ATP-binding cassette transporter multidrug-resistance associated protein 2 (MRP2; ABCC2). *J. Med. Chem.*, **2008**, *51*, 3275-3287.

[108]   Matsson, P.; Englund, G.; Ahlin, G.; Bergström, C.A.; Norinder, U.; Artursson, P. A global drug inhibition pattern for the human ATP-binding cassette transporter breast cancer resistance protein (ABCG2). *J. Pharmacol. Exp. Ther.*, **2007**, *323*, 19-30.

[109]   Matsson, P.; Pedersen, J.M.; Norinder, U.; Bergström, C.A.; Artursson, P. Identification of novel specific and general inhibitors of the three major human ATP-binding cassette transporters P-gp, BCRP and MRP2 among registered drugs. *Pharm. Res.*, **2009**, *26*, 1816-1831.

[110]   Demel, M.A.; Schwaha, R.; Krämer, O.; Ettmayer, P.; Haaksma, E.E.; Ecker, G.F. *In silico* prediction of substrate properties for ABC-multidrug transporters. *Expert Opin. Drug Metab. Toxicol.*, **2008**, *4*, 1167–1180.

[111]   Ekins, S.; Kim, R.B.; Leake, B.F.; Dantzig, A.H.; Schuetz, E.G.; Lan, L.B.; Yasuda, K.; Shepard, R.L.; Winter, M.A.; Schuetz, J.D. Wikel, J.H.; Wrighton, S.S. Application of three-dimensional quantitative structure-activity relationships of P-glycoprotein inhibitors and substrates. *Mol. Pharmacol.*, **2002**, *61*, 964-973.

[112]   Pajeva, I.; Wiese, M. Pharmacophore model of drugs involved in P-glycoprotein multidrug resistance: explanation of structural variety (hypothesis). *J. Med. Chem.*, **2002**, *45*, 5671-5686.

[113]   Garrigues, A.; Loiseau, N.; Delaforge, M.; Ferte, J.; Garrigos, M.; Andre, F.; Orlowski, S. Characterization of two pharmacophores on the multidrug transporter P-glycoprotein. *Mol. Pharmacol.*, **2002**, *62*, 1288–1298.

[114]   Rebitzer, S.; Annibali, D.; Kopp, S.; Eder, M.; Langer, T.; Chiba, P.; Ecker, G.F.; Noe, C.R. *Farmaco*, **2003**, *58*, 185–191.

[115]   Pajeva, I.; Globisch, C.; Wiese, M. Structure-function relationships of multidrug resistance P-glycoprotein. *J. Med. Chem.* **2004**, *47*, 2523-2533.

[116]   Globisch, C.; Pajeva, I.; Wiese, M. Structure-activity relationships of a series of tariquidar analogs as multidrug resistance modulators. *Bioorg. Med. Chem.*, **2006,** *14*, 1588-1598.

[117]   Jain, S.; Abraham, I.; Carvalho, P.; Kuang, Y.H.; Shaala, L.A.; Youssef, D.T.; Avery, M.A.; Chen, Z S.; El Sayed, K.A. Sipholane Triterpenoids: Chemistry, Reversal of ABCB1/P-Glycoprotein-Mediated Multidrug Resistance, and Pharmacophore Modeling. *J. Nat. Prod.*, **2009**, *72*, 1291-1298.

[118]   Chang, C.; Ekins, S.; Bahadduri, P.; Swaan, P.W. Pharmacophore-based discovery of ligands for drug transporters. *Adv. Drug Deliv. Rev.*, **2006**, 58, 1431-1450.

[119]   Tawari, N.R.; Bag, S.; Degani, M.S. Pharmacophore mapping of a series of pyrrolopyrimidines, indolopyrimidines and their congeners as multidrug-resistance-associated protein (MRP1) modulators. *J. Mol. Model.*, **2008**, *14*, 911–921.

[120]   Pearce, H.L.; Safa, A.R.; Bach, N.J.; Winter, M.A.; Cirtain, M.C.; Beck, W.T. Essential features of the P-glycoprotein pharmacophore as defined by a series of reserpine analogs that modulate multidrug resistance. *Proc. Natl. Acad. Sci. USA*, **1989**, *86*, 5128–5132.

[121]   Kondratov, R.V.; Komarov, P.G.; Becker, Y.; Ewenson, A.; Gudkov, A.V. Small molecules that dramatically alter multidrug resistance phenotype by modulating the substrate specificity of P-glycoprotein. *Proc. Natl. Acad. Sci. USA*, **2001**, *98*, 14078–14083.

[122]   Ravna, A.W.; Sylte, I.; Sager, G. Binding site of ABC transporter homology models confirmed by ABCB1 crystal structure. *Theor. Biol. Med. Model.*, **2009**, *6*, 20.

[123]   Locher, K.P.; Lee, A.T.; Rees, D.C. The E. coli BtuCD structure: a framework for ABC transporter architecture and mechanism. *Science*, **2002**, *296*, 1091-1098.

[124]   Seigneuret, M.; Garnier-Suillerot, A. A structural model for the open conformation of the mdr1 P-glycoprotein based on the MsbA crystal structure. *J. Biol. Chem.*, **2003**, *278*, 30115-30124.

[125]   Stenham, D.R.; Campbell, J.D.; Sansom, M.S.; Higgins, C.F.; Kerr, I.D.; Linton, K. An atomic detail model for the human ATP binding cassette transporter P-glycoprotein derived from disulfide cross-linking and homology modeling. *FASEB J.*, **2003**, *17*, 2287-2289.

[126]   Pleban, K.; Macchiarulo, A.; Costantino, G.; Pellicciari, R.; Chiba P.; Ecker, G. F. Homology model of the multidrug transporter LmrA from Lactococcus lactis. *Bioorg. Med. Chem. Lett.*, **2004**, *14*, 5823-5826.

[127]   Dawson, R.J.; Locher, K.P. Structure of a bacterial multidrug ABC transporter. *Nature*, **2006**, *443*, 180–185.

[128]   Dawson, R.J.; Locher, K.P. Structure of the multidrug ABC transporter Sav1866 from *Staphylococcus aureus* in complex with AMP-PNP. *FEBS Lett.*, **2007**, *581*, 935–938.

[129] O'Mara, M.L.; Tieleman, D.P. P-glycoprotein models of the apo and ATP-bound states based on homology with Sav1866 and MalK. *FEBS Lett.*, **2007**, *581*, 4217-4222.

[130] Stockner, T.; de Vries, S.J.; Bonvin, A.M.; Ecker, G.F.; Chiba, P. Data-driven homology modelling of P-glycoprotein in the ATP-bound state indicates flexibility of the transmembrane domains. *FEBS J.*, **2009**, *276*, 964-972.

[131] Federici, B.; Woebking, S.; Velamakanni, R.A.; Shilling, B.; Luisi, H.W.; van Veen, H.W. New structure model for the ATP-binding cassette multidrug transporter LmrA, *Biochem. Pharmacol.*, **2007**, *74*, 672–678.

[132] DeGorter, M.K.; Conseil, G.; Deeley, R.G.; Campbell, R.L.; Cole, S.P. Molecular modeling of the human multidrug resistance protein 1 (MRP1/ABCC1). *Biochem. Biophys. Res. Commun.*, **2008**, *365*, 29-34.

[133] Ravna, A.W., Sager, G. Molecular model of the outward facing state of the human multidrug resistance protein 4 (MRP4/ABCC4). *Bioorg. Med. Chem. Lett.*, **2008**, *18*, 3481-3483.

[134] Ravna, A.W.; Sylte, I.; Sager, G. A molecular model of a putative substrate releasing conformation of multidrug resistance protein 5 (MRP5). *Eur. J. Med. Chem.*, **2008**, *43*, 2557-2567.

[135] Ravna, A.W.; Sylte, I.; Sager, G. Molecular model of the outward facing state of the human P-glycoprotein (ABCB1), and comparison to a model of the human MRP5 (ABCC5). *Theor. Biol. Med. Model.*, **2007**, *4*, 33.

[136] Hazai, E.; Bikádi, Z. Homology modeling of breast cancer resistance protein (ABCG2). *J. Struct. Biol.*, **2008**, *162*, 63-74.

[137] Ward, A.; Reyes, C.L.; Yu, J.; Roth, C.B.; Chang, G. Flexibility in the ABC transporter MsbA: alternating access with a twist. *Proc. Natl. Acad. Sci. USA*, **2007**, *104*, 19005–19010.

[138] Huang, S.Y.; Bolser, D.; Liu, H.Y.; Hwang, T.C.; Zou, X. Molecular modeling of the heterodimer of human CFTR's nucleotide-binding domains using a protein-protein docking approach. *J. Mol. Graph. Model.*, **2009**, *27*, 822-828.

[139] Crowley, E.; O'Mara, M.L.; Reynolds, C.; Tieleman, D.P.; Storm, J.; Kerr, I.D. Callaghan R. Transmembrane helix 12 modulates progression of the ATP catalytic cycle in ABCB1. *Biochemistry*, **2009**, *48*, 6249-6258.

[140] Globisch, C.; Pajeva, I.K.; Wiese M. Identification of putative binding sites of P-glycoprotein based on its homology model. *ChemMedChem.*, **2008**, *3*, 280-295.

[141] SYBYL 7.3, Tripos Inc., 1699 South Hanley Road, St. Louis, MO 63114-2917.

[142] MOE 2002.03 (Molecular Operating Environment), Chemical Computing Group, 1010 Sherbrooke Street West, Suite 910; Montreal, Que., Canada H3A 2R7.

[143] Vandevuer, S.; Van Bambeke, F.; Tulkens, P.M.; Prévost, M. Predicting the three-dimensional structure of human P-glycoprotein in absence of ATP by computational techniques embodying crosslinking data: insight into the mechanism of ligand migration and binding sites. *Proteins*, **2006**, *63*, 466-478.

[144] Omote, H.; Al-Shawi, M.K. Interaction of transported drugs with the lipid bilayer and P-glycoprotein through a solvation exchange mechanism. *Biophys. J.*, **2006**, *90*, 4046-4059.

[145] Badhan, R.; Penny, J. *In silico* modelling of the interaction of flavonoids with human P-glycoprotein nucleotide-binding domain. *Eur. J. Med. Chem.*, **2006**, *41*, 285-295.

[146] Cassidy, C.E.; Setzer, W.N. Cancer-relevant biochemical targets of cytotoxic Lonchocarpus flavonoids: A molecular docking analysis. *J. Mol. Model.*, **2010**, *16*, 311-326.

[147] Becker, J.P.; Depret, G.; Van Bambeke, F.; Tulkens, P.M.; Prévost, M. Molecular models of human P-glycoprotein in two different catalytic states. *BMC Struct. Biol.,* **2009**, *9*, 3.

[148] GOLD Version 4.0.1, The Cambridge Crystallographic Data Centre, 12 Union Road, Cambridge, CB2 1EZ, UK.

[149] Ivetac, A.; Sansom, M.S.P. Molecular dynamics simulations and membrane protein structure quality. *Eur. Biophys. J.,* **2008**, *37*, 403–409.

In Silico Lead Discovery, 2011, 163-175

CHAPTER 9

Successful Applications of *In Silico* Approaches for Lead/Drug Discovery

Andrea Bortolato[1], Francesca Perruccio[1] and Stefano Moro[2],*

[1]*Syngenta Crop Protection Research, Schaffhauserstrasse, CH- 4332 Stein, Switzerland and* [2]*Molecular Modeling Section, Dipartimento di Scienze Farmaceutiche, Università di Padova, Via Marzolo 5, 35131 Padova, Italy; Tel: +39-049-8275704, Fax: +39-049-8275366, E-mail: stefano.moro@unipd.it*

Abstract: Rational drug design approaches represent an important alternative strategy to the discovery of new therapeutics based on simply serendipity. Computational chemists have to understand the peculiar aspects linked to the biological system in study and on this basis choose the right collection of complementary *in silico* approaches to tackle in the better way the medicinal chemistry problem in study. In this chapter several examples of successful applications of *in silico* approaches for lead/drug discovery are presented. In the first paragraph are discussed interesting structure based methods to design inhibitors of protein kinases, while in the second both structure and ligand based techniques are exploited in a consensus fashion to develop novel Adenosine A_3 receptor antagonists. The last paragraph presents an overview of ligand based virtual screening and an example of a successful application of this approach to identify a small molecule agonist of the NTS1 receptor. Even if the cases reported cannot cover completely the complexity of the computational techniques available nowadays, they embody different interesting examples useful to understand the potentialities, limits and pitfalls of modern molecular modeling approaches.

INTRODUCTION

In the last years several efforts have been made to improve the results of rational drug design approaches as an alternative to the discovery of new therapeutics based on simply serendipity. In this contest, *in silico* approaches represent a valuable strategy to tackle the estimated 10^{60} small drug-like molecules present in the chemical space [1]. Databases of largest pharmaceutical/agrochemical companies have around 10^5-10^7 compounds, while all the virtual databases could reach 10^{20} molecules. On the opposite the biologically relevant part of this chemical space seems to be a small fraction, even smaller if we consider that not all the active compounds have the desired physicochemical properties to be a drug [2]. Pharmaceutical industry investments in research and development have grown in an exponential way, in particular in the period from 1993 to 2004 of the 147%, now exceeding 50 billion of US dollars annually. Nevertheless, drug approval applications to the US Food and Drug Administration have shown an increment remarkable minor of what expected, especially for drugs against new targets: an average of only two to three innovative agents from the entire industry each year for the past decade [3]. Such disappointing results constitute a continuous stimulus for the medicinal chemistry community [4] and in particular in recent years several efforts have been made for the improvement of rational drug design approaches and for the development of innovative computational techniques.

An emblematic example that can be helpful to understand the evolution of medicinal chemistry research can be the development of the prostaglandin-forming cyclooxygenase (COX) 2 inhibitors [5]. The discovery of the first non-steroidal anti-inflammatory COX1 inhibitor was probably based on serendipity and acute observation. Hippocrates already in the 400 BC used extracts from the willow tree (containing salicylic acid) as pain reliever. In 1889 Aspirin was introduced with the beneficial effect of a lower irritating action on the stomach thanks to its additional acetyl function. However we have to wait the 70s for the identification of its target, while only in the early 1990s a second isozyme revitalized the field and stimulated a hunt for new and selective isoform inhibitors. Rational structure-based drug design approaches, exploiting the X-ray diffraction crystal spectra of COX1 [6] and 2 [7] solved in the mid 90s, helped the development of rofecoxib (Vioxx; Merck) and celecoxib (Celebrex; Pfizer). The binding channel of the two enzymes was discovered to be different as consequence of the replacement of a bulky isoleucine in COX1 with a valine in COX2. The related conformational changes created a 'side pocket' that allowed the binding of COX2-specific inhibitors. Molecular modeling represented a part of the whole drug discovery program useful to support crystallography to prioritize the chemical synthesis of new derivatives and the design of new COX2-selective scaffolds [8]. Rofecoxib and celecoxib approved by FDA for osteoarthritis and rheumatoid arthritis in 1998/99 quickly became multi-billion dollar selling drugs. However in about five years they showed in a randomized placebo-controlled trial an increased risk of heart attack and stroke [9]. The consequence was the withdrawn from market of Vioxx [10], while Celebrex has now a label warning that its use could be associated with an increased risk of cardiovascular events, including heart attack or stroke [11]. This unexpected negative conclusion can help to highlight the limits and the many challenges that medicinal chemistry has still to face nowadays.

In this chapter several examples of successful applications of *in silico* approaches for lead/drug discovery are presented. In the first and second paragraph are discussed two of the most important drug targets for pharmaceutical companies: protein kinases and G protein-coupled receptors (GPCRs). Different structure based approaches are presented and in particular for the Adenosine A_3 receptor the results obtained are compared to ligand based studies. In the last paragraph the ligand based virtual screening approach is discussed and as example is reported its application to a corporate database to identify a small molecule agonist of the neurotensin receptor (NTS1R). Even if the cases discussed in this chapter cannot cover completely the complexity of the computational techniques available nowadays, they embody different interesting applications of promising strategies part of the portfolio of modern molecular modeling.

RATIONAL DESIGN OF PROTEIN KINASE INHIBITORS

Structural biology heavily contributed to the discovery of protein kinase (PK) inhibitors [12]. In particular X-ray diffraction crystallography data represented a solid starting point for the rational drug design supported by molecular modeling studies. Kinases represent one of the most intensively pursued classes of drug targets for the treatment of cancer, immunological, neurological, metabolic and infectious diseases. Approximately 30 distinct PKs are targets of drug candidates developed to the level of a Phase I clinical trial and approximately 80 inhibitors have been advanced to some stage of clinical evaluation [13]. This important protein family presents challenges related to its complex flexibility, difficulties to gain selective activity and to avoid or to overcome the development of mutations providing drug resistance. Computational studies helped not only the design of new inhibitors but also to understand how the dynamics of the protein is related to its regulation and in particular to better understand how mutations lead to activation of PKs and to drug resistance.

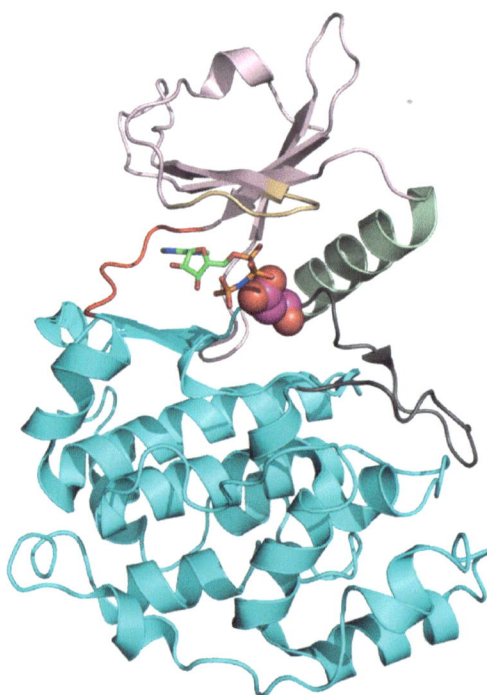

Figure 1: Ribbon diagram of a protein kinase domain in complex with the ATP analogue AMP. The N-terminal lobe is shown in pink with the αC-helix in green and the P-loop in orange. The C-terminal lobe (cyan) includes the activation loop in grey with the Asp residue of the DFG motif (magenta). The two lobes are linked by the hinge region shown in red.

Structure of the Protein Kinase Domain

PKs and protein phosphatases exert a tight and reversible control on protein phosphorylation that is strictly correlated to protein function and, consequently to the cell life. In particular PKs catalyze the transfer of the γ-phosphoryl group of an ATP or a GTP molecule to the peptide substrate. On the contrary, phosphatases are responsible for the removal of the phosphate group from the substrate. The kinase domain is very conserved in all PKs and responsible for the enzymatic activity. It is characterized by two lobes separated by the ATP binding cleft (Fig. **1**). The smaller N-terminal lobe is principally composed by β-sheets, with only one α-helix (C-helix), while the larger C-terminal has several α-helical secondary structures motifs. The two protein subdomains are linked by a small loop known as hinge region and responsible of anchoring the ATP molecule to the cavity via hydrogen bonds. This molecule creates further interactions of hydrophobic nature with the roof of this binding pocket, composed by the P-loop. An important flexible structural element of this domain is the activation loop (A-loop) that can have an open and a close conformation. The first is typical of the active state and it is generally stabilized by the phosphorylation of precise residues, like tyrosine, serine or threonine. The latter represents an inactive state composed by various conformations where the substrate site is not accessible. At the beginning of the activation loop is present a highly conserved element, *i.e.* the DFG motif (Asp-Phe-Gly), important for the interaction with the ATP molecule.

Rational Structure-Based Design of Protein Kinase CK2 Inhibitors

Different interesting applications of molecular modeling techniques have been applied to study CK2. This enzyme is a pleiotropic, ubiquitous, and constitutively active PK, with both cytosolic and nuclear localization in most mammalian cells. CK2 is believed to promote tumorigenesis [14], Alzheimer's disease, ischemia, chronic alcohol exposure and immunodeficiency virus HIV [15]. X-ray diffraction spectroscopy together with molecular modeling analysis pointed out how the common pharmacophore for CK2 activity is composed of an important hydrophobic/aromatic area in the centre of the pocket. This part is essential and present in all submicromolar inhibitors, and it is understandable on the basis of the many apolar bulky aminoacids characterizing the CK2 ATP-binding pocket. Notably two of them, *i.e.* Ile66 and Ile174, are unique or nearly unique to CK2, being replaced by either alanine or smaller residues in the other human kinases. This aspect may account for the exceptionally modest responsiveness to the not specific PK inhibitor staurosporine, whose large molecular mass may hamper accessibility to the active site. Furthermore the topology of the ATP-binding site of CK2 allows a specific inhibition action by relatively small molecules like the 4,5,6,7-tetrabromobenzotriazole (TBB, Fig. **2**) [16]. This molecule reaches a Ki of 0.4 μM and creates strong apolar contacts filling almost perfectly the pocket with a total buried surface, after the binding, of approximately 730 Å^2. Further optimized TBB derivates reached a highly selective action on CK2 on a panel of 80 PKs [17]. Their selectivity could be linked to the exploiting not only of the particular hydrophobic properties and size of the pocket but also of its atypical electrostatic features. Interesting on the opposite part of the active pocket, near the DFG motif, has been individuated a region with a positive electrostatic potential equal to 1.5 kcal/mol. In this region in the CK2 apo form are present three conserved waters molecules that can be substituted by a carboxyl acid moiety or a sufficient polar group of the inhibitor [18]. As an attempt to rationalize if the electrostatic tuning of the inhibitor can be exploited as general strategy to improve the selectivity of known CK2 ligands a total of 71 structures of different human kinase domains have been compared to the crystal structure of CK2. The molecular electrostatic potentials of the ATP-binding pockets of the different PKs have been evaluated quantitatively by calculating similarity indices exploiting the PIPSA (Protein Interaction Property Similarity Analysis) web server [19]. Interesting only Polo Like Kinase 1 resulted similar from an electrostatic point of view to CK2 at the contrary of the phylogenetically related CK1. Furthermore also PIM1 and PIM2, commonly affected by CK2 inhibitors, have been found in this analysis quite different suggesting the possibility to exploit electrostatic interaction to obtain a selective inhibitory effect. This analysis supported the hypothesis that the peculiarities of CK2 ATP-binding cleft are not only the small size and the hydrophobicity, but also of electrostatic nature. All these different aspects represent important features that can be exploited to obtain a selective inhibitory action [15].

The importance of the shape and size complementarity of the inhibitors to CK2 ATP-binding pocket has been further confirmed by quantitative binding free energy models based on the Linear Interaction Energy (LIE) method [20]. This method developed by Aqvist represents a plausible compromise between accuracy and computational speed in determining free energy of binding [21]. The LIE approach is based on the assumption that the inhibitor free energy of binding to a macromolecule is linearly correlated to several energy terms that can be calculated using a molecular mechanic force field. Using a training set of molecules with known activity, a semiempirical energy model has been built by fitting an electrostatic, a van der Waals and a hydrophobic energy term to the experimental free energy of binding. A first model has been built using a total 45 known benzoimidazole derivatives and structural information coming from four different X-ray diffraction crystals. The result confirmed that the crucial binding interactions for this class of inhibitors are principally due to hydrophobic and van der Waals contributions. These considerations are in agreement also with the crystallographic data of TBB in complex with the PK CDK2. Indeed in this case the loss of complementarity of the inhibitor for the wider CDK2 active site results in a different binding mode and less van der Waals contacts causing more than ten times decrease in activity. Afterward a second energy model based on coumarins derivatives has been based on the binding mode shown by the X-ray diffraction crystallographic data of DBC (3,8-dibromo-7-hydroxy-4-methylchromen-2-one) in complex with CK2 [22] (Fig. **2**). Derivatives satisfying a precise pharmacophore resulted to bind the enzyme in a comparable mode to DBC. Therefore they have been included with this particular binding orientation in the new LIE energy model that resulted in agreement with the first LIE analysis (Fig. **3**). For this scaffold the study pointed out the importance of an electron withdrawing substituent in orto position to the hydroxyl group in position 7 to enhance its polarization and create a stronger interaction with one of the water molecules conserved in the majority of the crystals of CK2 in complex with inhibitors. The estimated different energy contributions to the protein-ligand binding event together with a robust statistical validation, have been used to describe at a molecular level the possible difference in the CK2-inhibitor recognition processes.

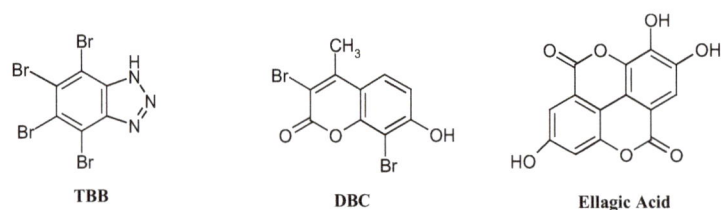

Figure 2: CK2 inhibitors TBB, DBC and ellagic acid.

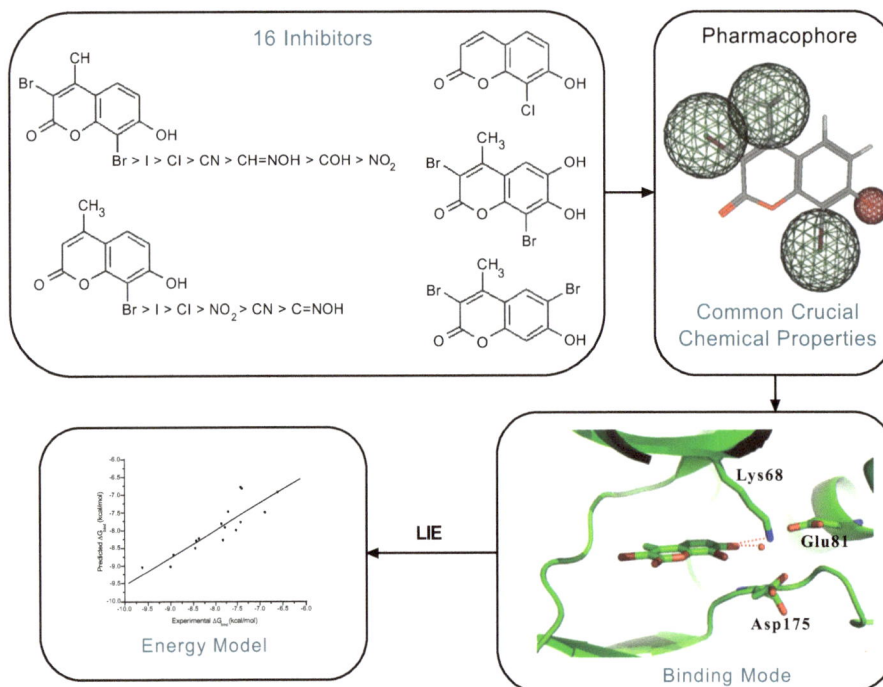

Figure 3: The16 inhibitors used to build the LIE energy model using the DBC binding mode as starting position for the computational study are shown in the upper left. The more the scaffold satisfied the base pharmacophore shown on the upper right (the green spheres represent hydrophobic molecular regions, the small red region stands for a strongly polarized hydrogen bonding donor), the higher resulted the probability that it possessed the chemical features to achieve a binding mode in the CK2 ATP binding pocket similar to DBC (bottom right) used for the LIE model (bottom left).

An interesting example of structure-based discovery of a potent CK2 inhibitor has been the identification by means of virtual screening of the ellagic acid a naturally occurring tannic acid derivative (Fig. **2**). This molecule has been demonstrated to be a very potent inhibitor (Ki=20 nM) specific for CK2 on a panel of 12 PKs [23]. A virtual screening experiment targeting the ATP binding site of CK2 has been performed assessing an in-house molecular database (defined as "MMS-database") with almost 2000 natural-occurring compounds. This virtual library comprised several families of polyphenolic compounds, including a large class of flavones, flavonols, isoflavones, catechins, anthraquinones, coumarins, and tannic acid derivatives. In the protocol a combination of high-throughput docking programs in tandem with a *consensus* scoring strategy has been used [24]. In particular, a combination of four docking methods and five scoring functions has been utilized to appropriately dock and rank all MMS-database candidates. The ellagic acid resulted in the top 5% of the ranked database independently from the nature of the used scoring function.

Molecular Modeling as a Valuable Tool to Understand PK Drug Resistance and Dynamics

Computational studies helped not only the design of new inhibitors but also to understand drug resistance and the protein dynamics. An important example was the structural analysis of the mutations in the ABL kinase domain resulting resistance to drug treatment in chronic myeloid leukaemia (CML) patients. Such studies contributed to the development of the second generation of ABL inhibitors able to overcome imatinib-resistant CML like nilotinib, bosutinib and dasatinib, recently approved by the US Food and Drug. Imatinib was the first kinase inhibitor drug approved and it still represents a highly effective therapy for CML. However more than 50 different ABL mutations have been identified leading to imatinib-resistant pathologies [25]. Initially it was thought that since this molecule is an ATP-competitive inhibitor mutations affecting its binding would have also affect ATP resulting in an inactive enzyme. However the docking of the inhibitor inside ABL is only partly superimposable over that of ATP and diverges of about an angle of 90 degree, stabilizing the protein in its inactive conformation. Therefore, mutations in the kinase domain were able to disrupt imatinib binding without interfering much with its enzymatic activity. Furthermore while several point mutations were in the active pocket impairing the binding others affected the activation loop, the P-loop and farer positions in the hydrophobic patch between helices in the C-terminal lobe of the enzyme. These in particular cannot be linked to a simple effect on the interaction with the inhibitor and it has been speculated that they could influence the kinase mechanism and its thermodynamic properties [26]. However this effect is very complex to be evaluated precisely *in silico*.

The protein kinase Hck [27] and CDK5 [28] have been studied by means of non-standard molecular dynamics (MD) to understand their flexibility and the conformational transition pathway between the inactive and active states. Hck plays a key role in the regulation of cell growth and proliferation and it is linked to several diseases as cancer. X-ray diffraction structures of active and inactive conformations of this enzyme are publically available in the protein structural data bank [29]. MD represented an important tool to obtain a complete mapping of the short lived conformational transitions between these two states that are very difficult to achieve with experimental methods. In this first study 78 all-atoms MD simulations (for a total of ~1 μS) in explicit solvent have been combined together to achieve a connectivity map of the transition event. This allowed identifying the collective dynamics of the conformational transition based on the clustering of the entire configuration space. Two intermediate states along the pathway and linking the active conformation to the inactive have been also identified. They were the most connected "hub"-like structural states and the authors speculate the possibility to exploit them to rational design new protein inhibitors. In particular their binding pockets would be able to accommodate a wider range of inhibitors than the inactive states. This study also supported the fact that the αC helix plays a more important role than the A-loop in terms of stabilizing the inactive conformation as result of its higher rigidity. In particular a switching of the interaction of a set of key residues involved in ion-pair proximal to the A-loop and the αC helix have been identify. Interesting Glu310 from the αC helix switched its interacting partner from Arg409 to Lys295, while Lys295 appeared to move away from Asp404 of the DFG motif, to Glu310.

In the second study a metadynamics approach has been used to study the conformational changes from the open to the close state of CDK5. This enzyme is implicated in neuronal development and architecture, while its deregulation has been associated with a number of neurodegenerative diseases, such as Alzheimer's disease and Parkinson's disease [30]. Starting from an open and close conformation a guess path has been built exploiting a morphing algorithm resulting in 5 interpolated intermediates. The final 7 conformations were used to generate the starting path that has been optimized by means of metadynamics using path collective variables [31]. This enhanced sampling algorithm within the framework of classical MD allows to efficiently explore the phase space of biological systems by adding a non-markovian (history-dependent) bias to the interaction potential, and estimating the relative free-energy of the relevant conformations [32]. It is based on a careful choice of collective variables able to describe the slow events that are relevant to the conformational change in study. Simulations showed that the process took place in two steps. In the first step, the αC-helix rotated by ~45°, and in the second step, the A-loop closed the catalytic cleft while the αC-helix completed its rotation. At the beginning the salt bridge between Lys33 and Glu51 anchoring the αC-helix in its open position was broken and a new salt bridge with Arg149 in the A-loop is formed resulting in the first partial rotation of the helix. In the second step, the αC-helix completed its rotation while the A-loop refolds to assume the final closed conformation. This switching of the electrostatic network together with hydration/dehydration of key residues resulted to play a fundamental role in the dynamics of the enzyme. A total of 150 ns MD simulation allowed also the reconstruction of the free energy surface corresponding to the conformational change and the estimation of the relative energy of the different states. In particular even in this study an intermediate was identified and it appeared to have enough elements of diversity to the open state of CDK5 to suggest the possibility to design new drugs targeting this peculiar state.

IN SILICO DESIGN OF ADENOSINE A$_3$ RECEPTOR ANTAGONIST

With at least 800 unique full-length members, human GPCRs comprise the largest family of cell-surface receptors [33]. They represent ubiquitous biological control points of the cell, translating the information encoded by diverse stimuli into readable signals [34]. Many distinct physiological responses are controlled by GPCRs: from cell growth and differentiation to cardiovascular function, metabolism, immune responses, and neurotransmission. They also represent the largest family of drug targets with about 50% of the existing drugs currently targeting GPCRs for their beneficial action [35], though their therapeutic potential might be discovered later to be considerably larger [36]. Structurally, GPCRs are cell receptors characterized by seven transmembrane (TMs) helices clustered in the form of a bundle and linked by three intracellular (IL) and three extracellular (EL) loops.

The adenosine receptors family of GPCRs includes a total of four distinct subtypes characterized pharmacologically, structurally and functionally into A$_1$, A$_{2A}$, A$_{2B}$ and A$_3$ (A3R). This last one is coupled to G$_{i2,3}$, which mediates inhibition of adenylyl cyclase, and to G$_{q/11}$, which stimulates phospholipase C [37]. Furthermore it is known to couple to phospholipase D and activate ERK1/2 [37]. The human A3R is involved in a variety of important physiological processes like inflammation, cell growth and immunosuppression [38]. Agonists or antagonists of A3Rs may be useful as therapeutic agents for the treatment of ischemic, inflammatory diseases and regulation of cell growth [39].

Ligand-based Homology Modeling

The mechanisms that control ligand-GPCR binding, receptor activation and signal transduction is highly complex and multifactorial. Understanding the structural determinants required for receptor function is crucial also for the success of computer-aided drug discovery. Indeed deciphering the structure-function relationships in GPCRs can help the prediction of the binding mode(s) of known ligands and the identification of the pharmacophores involved. In particular GPCR drug discovery requires a multidisciplinary approach, where molecular models represent a structural context to efficiently integrate experimental data derived from molecular biology, biophysics, bioinformatics, pharmacology and organic chemistry. In the last years these joined efforts have demonstrated that both active and inactive states can be described by several different receptor conformations depending on the nature of both ligand and receptor [40]. This structural flexibility of GPCRs and the ability of ligands to induce or to stabilize specific conformations can result in functional plasticity. The elaboration of successful methods to explore the multiconformational space of GPCRs is crucial to fully understand the intimate mechanism of ligand-receptor recognition and also to speed up the discovery of new potent and selective GPCR-ligands. The improving of the efficacy of virtual screening approach might be linked to the ability to include different conformations of the same pharmacological state of the receptor.

In an attempt to face this important challenge a new strategy called *ligand-based homology modeling* has been developed [41]. The final aim was the simulation of possible receptor structural reorganization induced by the antagonist binding. This modeling technique is an evolution of a conventional homology modeling algorithm based on a Boltzmann weighted randomized modeling procedure adapted from Levitt and combined with specialized logic for the proper handling of insertions and deletions, where any selected atom is included in the energy tests and in the minimization stages of the modeling procedure. The *ligand-based* option allows building a homology model in the presence of a small organic molecule docked to the primary template. In this specific case both model building and refinement take into account the presence of the ligand in terms of specific steric and chemical features.

The approach has been validated on more than 200 known human A3R antagonists in the corresponding putative ligand binding site [42-46]. In particular the receptor conformational change linked to the interaction with pyrazolotriazolopyrimidines derivatives has been analyzed. Docking studies on a starting homology model based on rhodopsin X-ray crystal structure showed that the binding of this class of antagonists occurred in the upper region of the TM helical bundle. In particular TMs 3, 5, 6 and 7 seemed to be crucial for the recognition of the antagonists. The quality of this first theoretical model has been confirmed experimentally by the effect on ligand binding of the mutation of several residues predicted to be involved in the antagonist recognition [47]. However the model could not explain the observed activity when bulky substituents were present. Independently from the used molecular docking algorithm, a strongly destabilizing van der Waals energy component avoided to sample reasonable antagonist-receptor complexes. Therefore this starting homology model was not appropriated to guarantee a good complementarity among the topology of the receptor's cleft and the shape of the bulkier derivatives. Starting from this conventional rhodopsin-based homology model and applying the ligand-based homology modeling implementation has been possible to generate other two antagonist-like conformational states of human A3 receptor in which the ligand recognition cavity has been expanded from 670 Å^3 to 770 and 1120 Å^3 respectively. These two new antagonist-like conformational states were able to rationalize the observed activities for all reported compounds.

A Successful Example of a Combined Target-based and Ligand-based Drug Design Approach

Using this multi-conformational states approach based on the ligand-based homology modeling, a consensus binding motif among all known antagonists has been found, and a novel "Y-shaped" 3D-pharmacophore model has been proposed. In particular five regions can be identified in the binding site that are critical for the recognition of antagonist structures: two hydrophilic regions, one in TM6 near the extracellular side and the other between TM3 and TM6 in the core of the binding pocket; three hydrophobic-aromatic area, the first near TM5 and EL2, the second surrounded by TM5, TM3 and TM6 and the latter among TM3, TM6 and TM7. This pharmacophore has been used with known experimental activity data to understand the key feature of inhibitors binding and create an effective protocol for lead optimization based on a combined target-based and ligand-based drug design approach (Fig. **4**) [48]. Using the pharmacophore model based on the docking results, a CoMFA approach was developed as a quantitative prediction tool for receptor binding affinity. CoMFA approach consists in the computation of steric and electrostatic field variables, calculated at grid points surrounding the whole molecule, followed by a PLS (Partial Least Squares) on these 3D structure descriptors and on the biological activity. The CoMFA results can be observed and analyzed as contour plot and regression plots. A remarkable correlation coefficient of 0.84 was obtained. Both steric and electrostatic contour plots, obtained from the CoMFA analysis, nicely fit on the hypothetical binding site achieved by molecular docking. Following this computational approach, 17 new derivatives have been designed, synthesized, and tested. Consistently, the predicted K_i values have been demonstrated to be very close to the experimental values [46]. It can be concluded that the integrated ligand-receptor based approach could be a very useful tool to speed up the design of new, potent and selective human A3R antagonists.

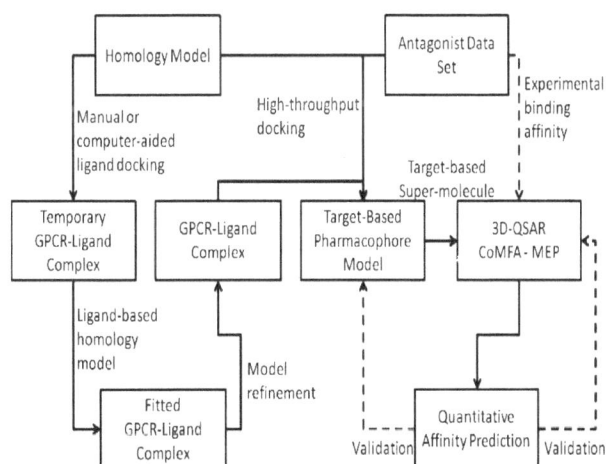

Figure 4: Flowchart that resumes the combined target-based and ligand-based drug design approach.

VIRTUAL SCREENING

Virtual screening is a powerful computational tool able to assess the similarity of a given probe against a database containing a large number of molecules. Virtual screening can be classified in two main classes: ligand based virtual screening and structure based virtual screening. The first uses as starting point only the structure of the active chemical whereas the second gains information also from the availability of the protein target of the compound under investigation.

Ligand based virtual screening can be performed in different ways and there are many commercially available software that claim to be able to screen collections of millions of molecules within few hours. Structure based virtual screening software can also perform the prediction of the placement and binding affinity of large number of molecules in the binding pocket of a protein (docking), however this is a much more demanding approach and accuracy is sometimes sacrificed for speed.

It is also been suggested that a combination of both, ligand based and structure based virtual screening could be a more accurate procedure in cases where the protein target or its homology model is also available: the ligand based virtual screening can allow to select a more restricted number of compounds from a large database, based on the chemical features of the ligand and/or its shape; docking can then refine this selection adding the information based on the interaction between the ligand under investigation and the putative active site of its protein target.

In this paragraph we are going to mainly focus on ligand based virtual screening, the theory underpinning it and an example of its application.

In cases when the mode of action of an active compound is unknown or its target protein structure is not available as crystal or as a sequence (making impossible the creation of the corresponding homology model), ligand based virtual screening has been proven to be a successful methodology to identify insightful SAR and/or suggest novel scaffolds. Ligand based virtual screening can be roughly classified as two-dimensional (2D) ligand based virtual screening [49, 50] and three-dimensional (3D) ligand based virtual screening [51, 52].

In 2D ligand based virtual screening methodology, molecules are described as vectors of numerical values of physical and chemical properties (such as molecular weight, clogP) or binary values (0 for absence, 1 for presence of a substructure feature for example) in a fingerprint. The advantages of the 2D ligand based virtual screening are the facility with which is possible to create the database to screen and the speed of the overall procedure. 2D ligand based approaches are usually very fast due to the simplistic approach to represent molecules and assess the similarity amongst them. The outcome of this exercise is a selection of compounds that are very similar to the one under investigation used as a probe: this can give a quick idea of the possible SAR available in the screened database. However, not taking into consideration any information on the geometry or the shape of the probe makes this approach lacking of any insight based on the possible bioactive conformation of the ligand and also makes this approach very unlikely to generate scaffold hopping and novel chemical classes.

In 3D based virtual screening methodology, molecules are describes as pharmacophores, define as arrangement in the space of chemical features present in the ligand under investigation (hydrogen bond donor, hydrogen bond acceptor, hydrophobic regions, negative charged and positively charged regions), or as a combination of volume to describe the ligand shape and chemical force fields to describe the ligand atom types. The creation of the database to query using a 3D approach can be computationally time and space (for storage) demanding. Each molecule contained into the database has to be investigated in terms its conformation and all the conformers so generated have to be stored and carried over into the full procedure to allow an exhaustive comparison between the probe and all the molecules contained in the database. The fact that the 3D dimensional methodologies are so dependent on the geometry of the compounds under study makes the analysis of the probe the most important step of this exercise. Conformational analysis of the probe can be carried out as a molecular or quantum mechanical level, to try to identify the most likely bioactive conformation: the minimum in energy is usually associated to the possible bioactive conformation of a ligand, however this can be a gamble since the conformation of a ligand can dramatically change from the one at its minimum in energy when bound to the putative active site of its protein target. A useful source of information to identify the bioactive conformation of a given ligand is the CCDC (Cambridge

Crystallographic Data Centre [53]) that contains crystal structure information, chemical, and bibliographic data for about 400,000 organic and metal-organic compounds. A search in the CCDC allows having insights on the possible bioactive conformation of the ligand under investigation or in most cases it allows to match parts of this ligand to fragments deposited in the CCDC to build then the complete hypothesis on its most likely bioactive conformation. 3D ligand based virtual screening can take from few hours to few days, it depend on the algorithm used in the software and the possibility to parallelised the computational procedure in more than one processor. The calculation can be demanding, but the outcome can be a very advantageous scaffold hopping, adding novelty in the chemical program of the compound under study.

A successful software that allows to perform 3D ligand bases virtual screening is ROCS (Rapid Overlay of Chemical Structures). ROCS is a fast shape comparison application, based on the idea that molecules have similar shape if their volumes overlay well and any volume mismatch is a measure of dissimilarity. It uses a smooth Gaussian function to represent the molecular volume [54], so it is possible to routinely minimize to the best global match. ROCS' high speed makes it possible, for virtual screening, to search multi-conformer representations of corporate collections in a day to find active compounds with similar shape to a lead compound (lead-hopping [55]). Recent work suggests that ROCS is competitive with, and often superior to, structure-based approaches in virtual screening [56, 57], both in terms of overall performance and consistency [58].

Application of Ligand Based Virtual Screening to Identify NTS1R Agonist

A successful example of this approach is the work reported by Fan *et al.* [59] where small molecule NTS1R agonist compounds have been identified through virtual screening of the corporate database using a ROCS approach.

Neurotensin (NT, pGlu-Leu-Tyr-Glu-Asn-Lys-Pro-Arg-Arg-Pro-Tyr-Ile-Leu) is a tridecapeptide that acts as a neuromodulator in the CNS [60]. Due to the evidences that have shown that NT modulates dopamine (DA) neurotransmission and its activity on behavioural animal models, NT has been implicated in the pathophysiology of schizophrenia and has been proposed as an endogenous neuroleptic [61]. Despite the research efforts in this field, the most advanced compounds remain peptidic in nature [61]. It is then easy to understand that small molecule, non-peptide NTS1 agonists would be of very interesting leads for development of a novel class of anti-psychotic compounds. To date, direct structural information about the agonist binding site in NTS1R is not available. With little information available for the receptor structure, this is the perfect scenario to apply a ligand based virtual screening approach. As stated before, the software chosen in this study to perform the ligand based virtual screening calculation is ROCS.

As a starting point for the ROCS search, SR-48527, the NTS1R selective antagonist, has been chosen as a probe (relying on the experimental observations that NT agonist and antagonist might share similar binding regions [62-64]). The structure of SR-48527 is shown in Fig. **5**.

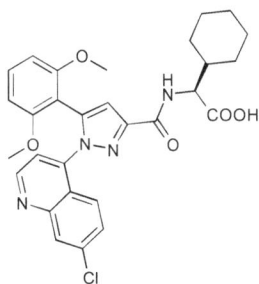

Figure 5: Chemical structure of SR-48527.

The probe structure was first minimized using MMFF94 force field [65] and then further analysed using OMEGA [66] with the energy cut-off of 5 kcal/mol. OMEGA allows a reliable conformational analysis and it is a robust tools in terms of speed and performance. Five representative conformers were retained as search queries. As mentioned before the selection of the starting conformation for the compound used as a probe in a three-dimensional virtual screening exercise is of great importance: it is very common to carry over a multi-query approach where more than one conformation of the same probe can be taken into account, where no definitive information is available on the most likely bioactive conformation of the compound under investigation. To identify a novel NTS1 small molecule agonist, ROCS shape-based searches were performed on Wyeth corporate collections (200,000 compounds with molecular weights between 180 and 400, less than 9 rotatable bonds, no more than one chiral center, a $logD_{7.4}$ of less than 4.0, fewer than 4 ionizable groups, and applied filter of non-reactive functionalities.). After finding the best alignment based on the shape, the program calculates the chemical complementarities refines shape-based superimpositions based on chemical similarity. The 'lead-like' molecules were also expanded into a set of 3D conformations using the program OMEGA. Virtual screening hits were selected based on the minimum combo score of 1.2 in the ROCS searches. After removing the unavailable compounds, 170 were finally submitted for testing with the consideration of chemical diversity. In the primary screening the effects of agonists were measured using CHOK1 cells stably expressing the human NTS1R and were monitored with intracellular Ca^{2+} measurements (fluorometric imaging plate reader, FLIPR). Two initial hits have been identified as NTS1R partial agonist with potency in the moderate micromolar range. To follow-up the most potent compound **1** (Fig. **5**), a 2D similarity searches has also been performed using topological torsion and atom pair based descriptors [67, 68]. The subsequent similarity searches led two additional potent compounds with compound **2** as a representative shown in Fig. **6**.

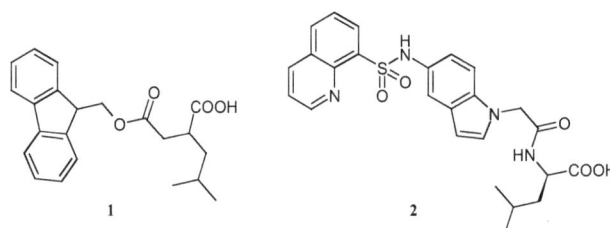

Figure 6: Two confirmed hits identified from virtual screening.

The EC_{50} values of compounds **1** and **2** resulting from the NTS1R FLIPR functional assay were of 215.50±84.15 nM and 178.01±77.58 nM respectively and their agonist efficacy expressed as a percentage of the maximal value produced by NT resulted to be of 60% and 17% respectively. The results indicated the production of the agonist activity by both compounds.

FUTURE DIRECTIONS

Structure-based approaches to drug discovery in the presence or in the absence of the real target structures requires a multi-disciplinary approach, where computational hypothesis represent a structural context to efficiently integrate experimental data and inferences derived from molecular biology, biophysics, bioinformatics, pharmacology and organic chemistry methods. Although not always achievable, the success of a synergistic effect among these disciplines is highly dependent on the experimental design. Synergy is best achieved when mutations are structurally interpretable, structural hypotheses are experimentally testable, ligands are well characterized pharmacologically, and the necessary chemical modifications of the ligands are available. The extent to which these conditions are met define the quality of the information derived to guide the lead compound optimization process.

REFERENCES

[1]   Dobson, C.M. Chemical space and biology. *Nature* **2004**, *432*, 824-8.

[2]   Lipinski, C.; Hopkins, A. Navigating chemical space for biology and medicine. *Nature* **2004**, *432*(7019), 855-861.

[3].   Couzin, J. Cancer drugs. Smart weapons prove tough to design. *Science* **2002**, *298*(5593), 522-525.

[4]   Fishman, M.C.; Porter, J.A. Pharmaceuticals: a new grammar for drug discovery. *Nature* **2005**, *437*(7058), 491-493.

[5]   Flower, R.J. The development of COX2 inhibitors. *Nat. Rev. Drug Discov.* **2003**, *2*(3), 179-191.

[6]   Picot, D.; Loll, P.J.; Garavito, R.M. The X-ray crystal structure of the membrane protein prostaglandin H2 synthase-1. *Nature* **1994**, *367*(6460), 243-249.

[7]   Kurumbail, R.G.; Stevens, A.M.; Gierse, J.K.; McDonald, J.J.; Stegeman, R.A.; Pak, J.Y.; Gildehaus, D.; Miyashiro, J.M.; Penning, T.D.; Seibert, K.; Isakson, P.C.; Stallings, W.C. Structural basis for selective inhibition of cyclooxygenase-2 by anti-inflammatory agents. *Nature* **1996**, *384*(6610), 644-648.

[8]   Chavatte, P.; Farce, A. A computational view of COX-2 inhibition. *Anticancer Agents Med. Chem.* **2006,** *6*(3), 239-249.

[9]   Mitchell, J.A.; Warner, T.D. COX isoforms in the cardiovascular system: understanding the activities of non-steroidal anti-inflammatory drugs. *Nat. Rev. Drug Discov.* **2006,** *5*(1), 75-86.

[10]   Topol, E.J. Failing the public health--rofecoxib, Merck, and the FDA. *N. Engl. J. Med.* **2004,** *351*(17), 1707-1709.

[11]   Okie, S. Raising the safety bar--the FDA's coxib meeting. *N. Engl. J. Med.* **2005,** *352* (13), 1283-1285.

[12]   Cowan-Jacob, S.W.; Mobitz, H.; Fabbro, D. Structural biology contributions to tyrosine kinase drug discovery. *Curr. Opin. Cell Biol.* **2009,** *21*(2), 280-287.

[13]   Zhang, J.; Yang, P.L.; Gray, N.S. Targeting cancer with small molecule kinase inhibitors. *Nat Rev Cancer* **2009,** *9*(1), 28-39.

[14]   Bortolato, A.; Cozza, G.; Moro, S. Protein kinase CK2 inhibitors: emerging anticancer therapeutic agents? *Anticancer Agents Med. Chem.* **2008,** 8(7), 798-806.

[15]   Cozza, G.; Bortolato, A.; Moro, S. How druggable is protein kinase CK2? *Med. Res. Rev.* **2010,** *30*(3), 419-462.

[16]   Battistutta, R.; De Moliner, E.; Sarno, S.; Zanotti, G.; Pinna, L.A. Structural features underlying selective inhibition of protein kinase CK2 by ATP site-directed tetrabromo-2-benzotriazole. *Protein Sci.* **2001,** 10(11), 2200-2206.

[17]   Pagano, M.A.; Bain, J.; Kazimierczuk, Z.; Sarno, S.; Ruzzene, M.; Di Maira, G.; Elliott, M.; Orzeszko, A.; Cozza, G.; Meggio, F.; Pinna, L.A. The selectivity of inhibitors of protein kinase CK2: an update. *Biochem. J.* **2008,** 415, (3), 353-365.

[18]   Battistutta, R.; Mazzorana, M.; Cendron, L.; Bortolato, A.; Sarno, S.; Kazimierczuk, Z.; Zanotti, G.; Moro, S.; Pinna, L.A. The ATP-binding site of protein kinase CK2 holds a positive electrostatic area and conserved water molecules. *Chembiochem* **2007,** 8(15), 1804-1809.

[19]   Richter, S.; Wenzel, A.; Stein, M.; Gabdoulline, R.R.; Wade, R.C. webPIPSA: a web server for the comparison of protein interaction properties. *Nucleic Acids Res.* **2008,** 36(Web Server issue), W276-80.

[20]   Bortolato, A.; Moro, S. In silico binding free energy predictability by using the linear interaction energy (LIE) method: bromobenzimidazole CK2 inhibitors as a case study. *J. Chem. Inf. Model.* **2007,** 47(2), 572-582.

[21]   Aqvist, J.; Medina, C.; Samuelsson, J.E. A new method for predicting binding affinity in computer-aided drug design. *Protein Eng.* **1994,** *7,* (3), 385-391.

[22]   Chilin, A.; Battistutta, R.; Bortolato, A.; Cozza, G.; Zanatta, S.; Poletto, G.; Mazzorana, M.; Zagotto, G.; Uriarte, E.; Guiotto, A.; Pinna, L. A.; Meggio, F.; Moro, S. Coumarin as attractive casein kinase 2 (CK2) inhibitor scaffold: an integrate approach to elucidate the putative binding motif and explain structure-activity relationships. *J. Med. Chem.* **2008,** *51*(4), 752-759.

[23]   Cozza, G.; Bonvini, P.; Zorzi, E.; Poletto, G.; Pagano, M. A.; Sarno, S.; Donella-Deana, A.; Zagotto, G.; Rosolen, A.; Pinna, L. A.; Meggio, F.; Moro, S. Identification of ellagic acid as potent inhibitor of protein kinase CK2: a successful example of a virtual screening application. *J. Med. Chem.* **2006,** *49*(8), 2363-2366.

[24]   Miteva, M.A.; Lee, W.H.; Montes, M.O.; Villoutreix, B.O. Fast structure-based virtual ligand screening combining FRED, DOCK, and Surflex. *J. Med. Chem.* **2005,** *48*(19), 6012-6022.

[25]   Weisberg, E.; Manley, P.W.; Cowan-Jacob, S.W.; Hochhaus, A.; Griffin, J.D. Second generation inhibitors of BCR-ABL for the treatment of imatinib-resistant *chronic* myeloid leukaemia. *Nat. Rev. Cancer.* **2007,** *7*(5), 345-356.

[26]   Gambacorti-Passerini, C.B.; Gunby, R.H.; Piazza, R.; Galietta, A.; Rostagno, R.; Scapozza, L. Molecular mechanisms of resistance to imatinib in Philadelphia-chromosome-positive leukaemias. *Lancet Oncol.* **2003,** *4*(2), 75-85.

[27]   Yang, S.; Banavali, N.K.; Roux, B. Mapping *the* conformational transition in Src activation by cumulating the information from multiple molecular dynamics trajectories. *Proc. Natl. Acad. Sci U S A* **2009,** *106*(10), 3776-3781.

[28]   Berteotti, A.; Cavalli, A.; Branduardi, D.; Gervasio, F.L.; Recanatini, M.; Parrinello, M. Protein conformational transitions: the closure mechanism of a kinase explored by atomistic simulations. *J. Am. Chem. Soc.* **2009,** *131*(1), 244-250.

[29]   Berman, H.M.; Westbrook, J.; Feng, Z.; Gilliland, G.; Bhat, T.N.; Weissig, H.; Shindyalov, I.N.; Bourne, P.E. The Protein Data Bank. *Nucleic Acids Res.* **2000,** *28*(1), 235-242.

[30]   Angelo, M.; Plattner, F.; Giese, K.P. Cyclin-dependent kinase 5 in synaptic plasticity, learning and memory. *J. Neurochem.* **2006,** *99*(2), 353-370.

[31]   Branduardi, D.; Gervasio, F.L.; Parrinello, M. From A to B in free energy space. *J. Chem. Phys.* **2007,** *126*(5), 054103.

[32]   Laio, A.; Parrinello, M. Escaping free-energy minima. *Proc. Natl. Acad. Sci. USA* **2002,** *99*(20), 12562-12566.

[33]   Gloriam, D.E.; Fredriksson, R.; Schioth, H.B. The G protein-coupled receptor subset of the rat genome. *BMC Genomics* **2007,** *8,* 338.

[34]   Marinissen, M.J.; Gutkind, J.S. G-protein-coupled receptors and signaling networks: emerging paradigms. *Trends Pharmacol. Sci.* **2001,** *22*(7), 368-376.

[35]   Overington, J.P.; Al-Lazikani, B.; Hopkins, A.L. How many drug targets are there? *Nat. Rev. Drug Discov.* **2006,** *5(*12), 993-996.

[36]   Strange, P.G. Signaling mechanisms of GPCR ligands. *Curr. Opin. Drug Discov. Devel.* **2008**, *11*(2), 196-202.

[37]   Fishman, P.; Bar-Yehuda, S. Pharmacology and therapeutic applications of A3 receptor subtype. *Curr. Top. Med. Chem.* **2003**, *3*(4), 463-469.

[38]   Gessi, S.; Merighi, S.; Varani, K.; Leung, E.; Mac Lennan, S.; Borea, P.A. The A3 adenosine receptor: an enigmatic player in cell biology. *Pharmacol. Ther.* **2008**, *117*(1), 123-140.

[39]   Hasko, G.; Linden, J.; Cronstein, B.; Pacher, P. Adenosine receptors: therapeutic aspects for inflammatory and immune diseases. *Nat. Rev. Drug Discov.* **2008**, *7*(9), 759-770.

[40]   Kobilka, B.K.; Deupi, X. Conformational complexity of G-protein-coupled receptors. *Trends Pharmacol. Sci.* **2007**, *28*(8), 397-406.

[41]   Moro, S.; Deflorian, F.; Bacilieri, M.; Spalluto, G. Ligand-based homology modeling as attractive tool to inspect GPCR structural plasticity. *Curr. Pharm. Des.* **2006**, *12*(17), 2175-2185.

[42]   Moro, S.; Deflorian, F.; Spalluto, G.; Pastorin, G.; Cacciari, B.; Kim, S.K.; Jacobson, K.A. Demystifying the three dimensional structure of G protein-coupled receptors (GPCRs) with the aid of molecular modeling. *Chem. Commun. (Camb)* **2003**, (24), 2949-2956.

[43]   Maconi, A.; Pastorin, G.; Da Ros, T.; Spalluto, G.; Gao, Z. G.; Jacobson, K.A.; Baraldi, P.G.; Cacciari, B.; Varani, K.; Moro, S.; Borea, P.A. Synthesis, biological properties, and molecular modeling investigation of the first potent, selective, and water-soluble human A(3) adenosine receptor antagonist. *J. Med. Chem.* **2002**, *45*(17), 3579-3582.

[44]   Pastorin, G.; Da Ros, T.; Spalluto, G.; Deflorian, F.; Moro, S.; Cacciari, B.; Baraldi, P.G.; Gessi, S.; Varani, K.; Borea, P.A. Pyrazolo[4,3-e]-1,2,4-triazolo[1,5-c]pyrimidine derivatives as adenosine receptor antagonists. Influence of the N5 substituent on the affinity at the human A 3 and A 2B adenosine receptor subtypes: a molecular modeling investigation. *J. Med. Chem.* **2003**, *46*(20), 4287-4296.

[45]   Colotta, V.; Catarzi, D.; Varano, F.; Calabri, F.R.; Lenzi, O.; Filacchioni, G.; Martini, C.; Trincavelli, L.; Deflorian, F.; Moro, S. 1,2,4-triazolo[4,3-a]quinoxalin-1-one moiety as an attractive scaffold to develop new potent and selective human A3 adenosine receptor antagonists: synthesis, pharmacological, and ligand-receptor modeling studies. *J. Med. Chem.* **2004**, *47*(14), 3580-3590.

[46]   Moro, S.; Braiuca, P.; Deflorian, F.; Ferrari, C.; Pastorin, G.; Cacciari, B.; Baraldi, P.G.; Varani, K.; Borea, P.A.; Spalluto, G. Combined target-based and ligand-based drug design approach as a tool to define a novel 3D-pharmacophore model of human A3 adenosine receptor antagonists: pyrazolo[4,3-e]1,2,4-triazolo[1,5-c]pyrimidine derivatives as a key study. *J. Med. Chem.* **2005**, *48*(1), 152-162.

[47]   Gao, Z.G.; Chen, A.; Barak, D.; Kim, S.K.; Muller, C.E.; Jacobson, K.A. Identification by site-directed mutagenesis of residues involved in ligand recognition and activation of the human A3 adenosine receptor. *J. Biol. Chem.* **2002**, *277*(21), 19056-19063.

[48]   Moro, S.; Deflorian, F.; Bacilieri, M.; Spalluto, G. Novel strategies for the design of new potent and selective human A3 receptor antagonists: an update. *Curr. Med. Chem.* **2006**, *13*(6), 639-645.

[49]   Casanola-Martin, G.M.; Marrero-Ponce, Y.; Tareq Hassan Khan, M.; Torrens, F.; Perez-Gimenez, F.; Rescigno, A. Atom- and bond-based 2D TOMOCOMD-CARDD approach and ligand-based virtual screening for the drug discovery of new tyrosinase inhibitors. *J. Biomol. Screen.* **2008**, *13*(10), 1014-1024.

[50]   Ewing, T.; Baber, J.C.; Feher, M. Novel 2D fingerprints for ligand-based virtual screening. *J Chem Inf Model* **2006**, *46*(6), 2423-2431.

[51]   Kinnings, S.L.; Jackson, R.M. LigMatch: a multiple structure-based ligand matching method for 3D virtual screening. *J. Chem. Inf. Model.* **2009**, *49*(9), 2056-2066.

[52]   Perez-Nueno, V.I.; Ritchie, D.W.; Rabal, O.; Pascual, R.; Borrell, J.I.; Teixido, J. Comparison of ligand-based and receptor-based virtual screening of HIV entry inhibitors for the CXCR4 and CCR5 receptors using 3D ligand shape matching and ligand-receptor docking. *J. Chem. Inf. Model.* **2008**, *48*(3), 509-533.

[53]   Allen, F.H. The Cambridge Structural Database: a quarter of a million crystal structures and rising. *Acta Crystallogr. B* **2002**, *58*(Pt 3 Pt 1), 380-388.

[54]   Grant, J.A.; Gallardo, M.A.; Pickup, B.T. A fast method of molecular shape comparison: A simple application of a Gaussian description of molecular shape. *J. Comp. Chem.* **1996**, *17*(14), 1653-1666.

[55]   Rush, T.S., 3rd; Grant, J.A.; Mosyak, L.; Nicholls, A. A shape-based 3-D scaffold hopping method and its application to a bacterial protein-protein interaction. *J. Med. Chem.* **2005**, *48*(5), 1489-1495.

[56]   Hawkins, P.C.; Skillman, A.G.; Nicholls, A. Comparison of shape-matching and docking as virtual screening tools. *J. Med. Chem.* **2007**, *50*(1), 74-82.

[57]   Venhorst, J.; Nunez, S.; Terpstra, J.W.; Kruse, C.G. Assessment of scaffold hopping efficiency by use of molecular interaction fingerprints. *J. Med. Chem.* **2008**, *51*(11), 3222-3229.

[58] Sheridan, R.P.; McGaughey, G.B.; Cornell, W.D. Multiple protein structures and multiple ligands: effects on the apparent goodness of virtual screening results. *J. Comput. Aided Mol. Des.* **2008**, *22*(3-4), 257-265.

[59] Fan, Y.; Lai, M.H.; Sullivan, K.; Popiolek, M.; Andree, T.H.; Dollings, P.; Pausch, M.H. The identification of neurotensin NTS1 receptor partial agonists through a ligand-based virtual screening approach. *Bio. Med. Chem. Lett.* **2008**, *18*(21), 5789-5791.

[60] Carraway, R.; Leeman, S.E. The isolation of a new hypotensive peptide, neurotensin, from bovine hypothalami. *J. Biol. Chem.* **1973**, *248*(19), 6854-6861.

[61] Boules, M.; Shaw, A.; Fredrickson, P.; Richelson, E. Neurotensin agonists: potential in the treatment of schizophrenia. *CNS Drugs* **2007**, *21*(1), 13-23.

[62] Barroso, S.; Richard, F.; Nicolas-Etheve, D.; Reversat, J.L.; Bernassau, J.M.; Kitabgi, P.; Labbe-Jullie, C. Identification of residues involved in neurotensin binding and modeling of the agonist binding site in neurotensin receptor 1. *J. Biol. Chem.* **2000**, *275*(1), 328-336.

[63] Labbe-Jullie, C.; Barroso, S.; Nicolas-Eteve, D.; Reversat, J. L.; Botto, J.M.; Mazella, J.; Bernassau, J.M.; Kitabgi, P. Mutagenesis and modeling of the neurotensin receptor NTR1. Identification of residues that are critical for binding SR 48692, a nonpeptide neurotensin antagonist. *J. Biol. Chem.* **1998**, *273*(26), 16351-16357.

[64] Pang, Y.P.; Cusack, B.; Groshan, K.; Richelson, E. Proposed ligand binding site of the transmembrane receptor for neurotensin(8-13). *J. Biol. Chem.* **1996**, *271*(25), 15060-15068.

[65] Halgren, T.A. Merck molecular force field. I. Basis, form, scope, parameterization, and performance of MMFF94. *J. Comp. Chem.* **1996**, *17*.

[66] Bostrom, J.; Greenwood, J.R.; Gottfries, J. Assessing the performance of OMEGA with respect to retrieving bioactive conformations. *J. Mol. Graph. Model.* **2003**, *21*(5), 449-462.

[67] Nilakantan, R.; Bauman, N.; Dixon, J.S.; Venkataraghavan, R. Topological torsion: a new molecular descriptor for SAR applications. Comparison with other descriptors. *J. Chem. Inf. Comp. Sci.* **1987**, *27*(2), 82-85.

[68] Carhart, R.E.; Smith, D.H.; Venkataraghavan, R. Atom pairs as molecular features in structure-activity studies: definition and applications. *J. Chem. Inf. Comp. Sci.* **1985**, *25*(2), 64-73.

CHAPTER 10

Visualisation and Efficient Communication in Structure-Based Lead Discovery

Brian D. Marsden[1,2], Ruben Abagyan[3,4] and Wen Hwa Lee[1,*]

[1] *SGC, University of Oxford, Headington, Oxford, United Kingdom,* [2] *Nuffield Department of Clinical Medicine University of Oxford, Headington, Oxford, United Kingdom,* [3] *Skaggs School of Pharmacy and Pharmaceutical Sciences, University of California San Diego, La Jolla, California, United States of America,* [4] *Molsoft LLC, La Jolla, San Diego, California, USA; E-mail: wenhwa.lee@sgc.ox.ac.uk*

Abstract: A key aspect of efficient scientific research is the efficient accumulation, exchange and presentation of relevant information. Multi-disciplinary fields of research, such as structure-based lead discovery (SBLD), depend upon many different data types and require the concurrent capture and display of such heterogeneous types of data. The visualisation of such data in a more intuitive and accessible format than the underlying data capture method is often extended to provide the viewer context within a wider range of data types. In the field of SBLD, experimental protein structure determination presents a need to visualise three-dimensional data and to further annotate such visualisations with additional information. The use of high-throughput methods in both chemistry and biology has resulted in a rapid accumulation of relevant information to support and prioritise SBLD. The appropriate integration and visualisation of this data maximises the impact of the underlying information within the context of a project both not only for computational chemists and structural biologists but also for biologists and medicinal chemists. In this chapter we will discuss current approaches and outstanding issues associated with this particular challenge in the context of SBLD.

INTRODUCTION

The efficient accumulation, exchange and presentation of information are critical components of research in both academia and industry. In the last 20 years the development of lower cost high-throughput methods and automation has resulted in a continued increase in the rate at which raw data is generated.

As a multi-disciplinary field of research, structure-based lead discovery (SBLD) projects depend upon several different data types, which can be broadly classified as representing chemical or biological information. These data types include chemical compounds, chemical (quantitative) structure-activity relationships (QSAR), target DNA and protein sequences, biological and metabolic pathways, protein target purification methodologies, protein-protein interaction partners, mutations, genetic variability, single nucleotide polymorphisms (SNPs), and protein structures (Fig. **1**). This has necessitated the development of concurrent capture and display approaches for these heterogeneous types of data. However, this has often resulted in a lack of consistency in which data of a given type is captured and stored. For example a chemical compound can be represented using a wide range of data formats including SDF (structure data format, firstly developed by Molecular Design Limited, now part of Symyx [1]), SMILES strings and inCHI [2]; coordinates of macromolecular structures can be represented using the PDB format [3] or mmCIF [4]. In addition different sources of data might report the readout of an assay with different unit standards or slightly different methodology (e.g. IC_{50} values being reported with different units such as micromolar /nanomolar or even as a pIC_{50} (-log IC_{50})).

Notwithstanding these issues, the task of displaying diverse types of data either alone or in conjunction with other types, for the purposes of analysis, can be broadly defined as 'data visualisation'.

Visualisation attempts to display data in a more accessible format than the underlying data capture method and is often extended to provide the viewer context within a wider range of data types. In general visualisation can be categorised into three types:

- Capture of information: By appropriately visualising raw or partly-processed data prioritisation or annotation of that data can be efficiently performed. For instance, in a collection of structures of a

Maria A. Miteva (Ed)

protein, visual inspection can reveal which residues are ordered in the protein crystal lattice influencing the definition of an active site for SBLD purposes

- Information integration: The appropriate visualisation of diverse data types provides a platform for further processing (e.g. clustering, sorting etc.) resulting in the identification of otherwise obfuscated trends or results

- Information sharing and dissemination: The ability to communicate data and conclusions to both experts and non-experts

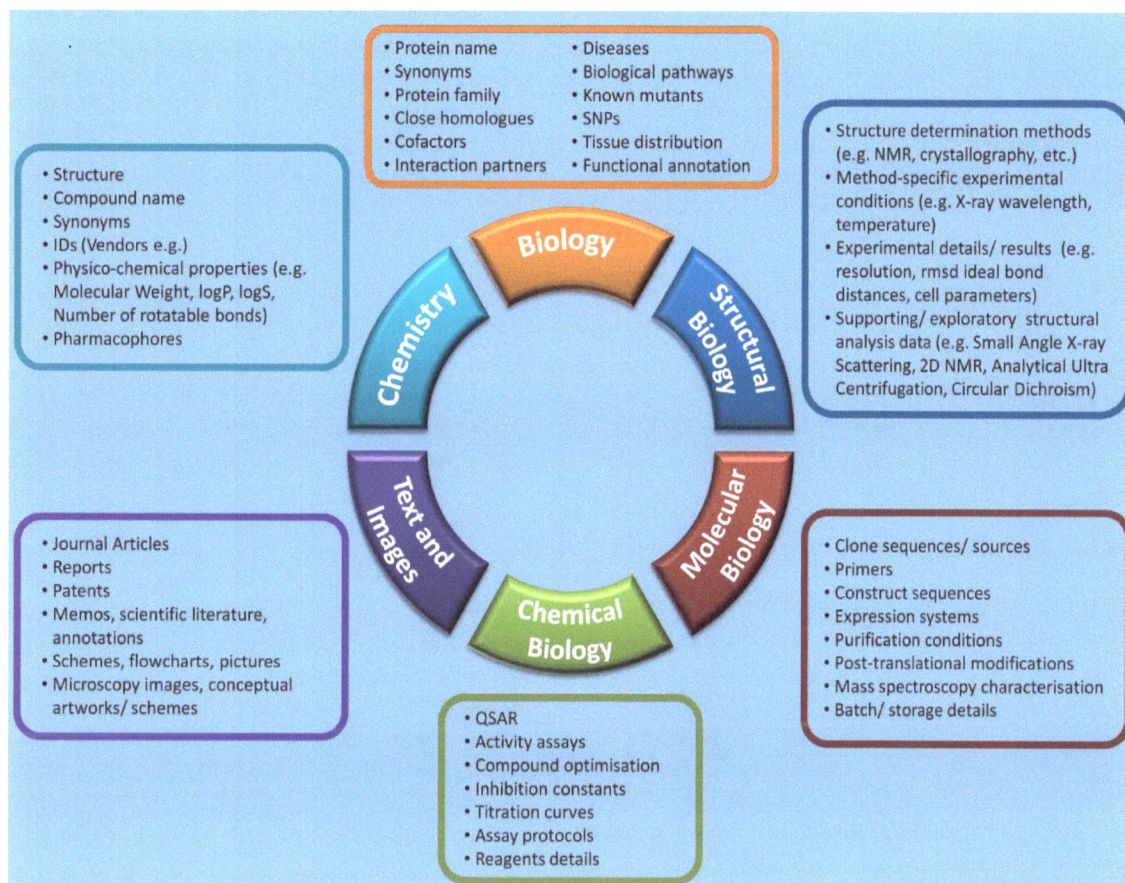

Figure 1: Main data types found in a drug development environment.

Data capture and visualisation solutions of chemical data are familiar concepts to scientists performing lead discovery. Chemistry-focussed data capture and visualisation software suites are often developed for use internally within pharmaceutical companies, for instance the ABCD platform used in Johnson & Johnson Pharmaceutical Research & Development [5] and Amgen's ADAAPT system [6]. Chemical data visualisation that allows users to manage different stages of chemistry design, development, synthesis etc. are available as off-the-shelf solutions (such as the Molsoft ICM [7, 8] or Schrödinger suites [9] or as a set of distinct solutions, often freely available (for an extensive list, please see Bruno Villoutreix's online Structural Bioinformatics resource [10]).

Biological data associated with lead discovery ranges from purely numeric data to high-content data such as that generated from cell-based imaging. Whilst numeric data is often captured, analysed and to an extent visualised in packages such as Microsoft Excel, GraphPad Prism [11] and Spotfire [12], high-content data is the domain of specialised software platforms and databases.

The establishment of protein structure determination presented the need to visualise three-dimensional data and further annotate this visualisation with additional information such as disease-causing mutations and single nucleotide polymorphisms (SNPs). The introduction of high-throughput methods, now applied to production of structural biology data (see below), has resulted in a rapid accumulation of relevant information to support and prioritise SBLD. Visualisation of this data is important to maximise the impact of the underlying data within the context of a project where the majority of scientists are not experts in protein structure. The remainder of this chapter focuses on the approaches and outstanding issues associated with this particular challenge.

HIGH-THROUGHPUT STRUCTURAL BIOLOGY – VISUALISING PROTEIN STRUCTURES TO ENABLE SBLD

The sequencing of the human genome enabled systematic technology-driven structural biology, or 'structural genomics' (SG), efforts to solve protein structures in a high throughput manner (for reviews see [13, 14]). First formalised in 1998 [15-17], several major SG efforts were subsequently started such as the NIH-funded Protein Structure Initiative (PSI) [18] European initiatives The Protein Structure Factory (PSF) [19] and Structural Proteomics in Europe (SPINE) [20] and Japanese Protein 3000 and RIKEN Structural Genomics/ Proteomics programmes [21]. The majority of these early programmes were tasked with exploring the breadth of the protein structural universe by focussing on prokaryotic targets whilst driving novel structural biology technology for a high-throughput environment [13, 22].

The availability of detailed structural descriptions of proteins and enzymes involved in the normal (or abnormal) functioning of the human body and associated pathogens provides a better understanding of the basis by which different proteins achieve selectivity and specific binding to their cognate ligands and drugs [13]. Indeed, several marketed drugs in recent years have been developed using structure-based methods, targeting a wide range of disease such as AIDS, leukaemia, cancers and venous thromboembolic events (for additional examples other than those explored in other chapters from this book see [13, 23-26]).

Despite these successes, determining the structures of human proteins in a high-throughput manner was seen to be technically challenging relative to those of less complex organisms (e.g. prokaryotes), where issues such as the lack of solubility of expressed proteins in simple bacterial systems pose major bottlenecks to scaling the structure determination process [27]. The SGC [28] was created in 2003 as a response to this challenge and operates as a not-for-profit organisation that aims to solve the three dimensional structures of proteins of medical relevance and place them into the public domain without encumbrance or restriction. The SGC has adopted a protein-family based approach whereby protein targets are chosen from discrete and medically-relevant human protein families. This approach enables comparative analysis and "blanket" methods which can be applied to members of the same family [27]. The SGC is driving the concept of "open-source science" to enable drug discovery by promoting pre-competitive structural biology. This approach, in conjunction with academic and industrial partners is being used to create open-access small molecule and biological probes [29-31] which might be used in a pre-competitive manner for target validation purposes, for example.

Since the production phase of the SGC began in 2004, over 1000 protein structures have been deposited and released into the public domain of which over 700 are novel and unique. The majority of these targets are human but around 10% are of proteins from parasitic organisms associated with neglected diseases, such as malaria and toxoplasmosis, and these make up 61% of publicly available apicomplexan structures to date.

HOW TO BEST DELIVER STRUCTURAL BIOLOGY DATA FOR LEAD DISCOVERY?

A clear mandate of SG efforts is to make results readily available to the scientific community. This data is necessarily heterogeneous in nature, ranging from materials and methods, construct and clone design, protein expression and purification to small molecule screening and protein structure determination information. The primary 'customers' for this information are not structural biologists but scientists with little structural biology experience such as medicinal chemists, assay developers, clinicians, pathologists and other professionals involved in the process of lead discovery. As such, the deposition of protein structures into specialist databases such as the Protein Data Bank [32] is not sufficient. This was emphasised by the Protein Structure Initiative (PSI) Assessment Panel in 2007: "The use of structural information is best driven by scientists who wish to understand biological mechanism, therefore deposition of structures in the PDB does not reach the key audience" [33]. As a consequence, the rich information contained in structural biology data and methods remains accessible only to those who understand protein structure and can fully interpret it.

There are four main challenges facing a non-structural biologist: Firstly, the need to be aware of where and how a protein structure can be retrieved. Secondly, it is necessary to visualise the structure (or the docking poses and results) using software that may not be intuitive. Thirdly, the interpretation of the structure/ docking results is dependent not only on an understanding of the biology and function of the protein, but also the structure and compound poses themselves. Finally, it is not trivial and often impossible to integrate other data sources into the structure such as mutation data.

The representation of a 3D structure using media, such as paper, that is in reality 2D poses an obstacle. The visualisation of protein structure on a computer screen as a movie is a clear improvement over a printed figure. Several movie-making tools (e.g. PMG [34], eMovie [35], VMD [36], MovieMaker [37], POLYVIEW-3D [38], YASARA movie [39])are an alternative to convey a sense of depth and perspective to structures. However, this approach provides a 'fixed' set of visualisations as defined by the movie authors resulting in the viewer being unable to interact directly with the structure displayed. In the specific case of sharing SBLD data, which may involve a significant amount of 3D information (e.g. docked compound poses in receptors), it is completely impractical to provide a movie for every pose.

This hurdle can be overcome using specialist visualisation software (e.g. PyMol [40], OpenEye VIDA [41], Schrodinger Maestro [9], Jmol [42], Kinemage/KiNG [43, 44], Swiss PDB Viewer [45], FirstGlance [46]. Some of these platforms were developed to cater to the needs of structural biologists/ computational chemists. As a result, these programs may be perceived to be overly feature-rich making them unintuitive to use by non-structural biologists. Other visualisation platforms are simple to use but at the cost of functionality and the ability to reproduce critical features seen in figures of peer-reviewed articles.

An optimal solution to these issues may be considered to include the following features:

1. **Annotation:** In order to make the underlying data accessible to both non-structural biologists and structural biologists, the visualisations of the data should be annotated and linked into a body of text, similar to a peer-review publication. The annotation text should be presented in close proximity to the graphical window. As such the associated visualisations are the equivalent to paper-printed figures, represented as a 3D object that can be rotated, translated and zoomed at will by the reader. This use of separate text and graphics areas means that there is no constraint upon the number to the number of these interactive 'figures' that may be incorporated into the associated text annotation.

2. **Pre-staged views:** Each of the pre-staged scenes is prepared by the author to illustrate the points being made in the explanatory text. Importantly, these pre-staged views should be able to be invoked in any order in a seamless fashion allowing the reader to go directly to the portion of the text that is most interesting.

3. **Other visualisations:** The pre-staged scenes should not be restricted to a 3D graphics window. By combining different delivery and visualisation methods the author can communicate data efficiently. For instance – displaying an SAR table of a scaffold with substituents highlighted side-by-side with a visualisation of the protein receptor within which compounds were docked. The docked poses can be used to help uncovering a rationale used to expand the series and how these correlate with structural information.

4. **Integrated authoring platform:** The analysis, interpretation and final visualisation should be made within the same software platform. This saves time by not having to duplicate the information derived from the analysis in another disconnected medium (e.g. PowerPoint slides to share 2D representations of chemical compounds), and also helps to deliver the data in its correct representation (e.g. 3D representations of active sites that can be manipulated *in silico*, instead of flat 2D images).

5. **Unified file:** All data, text annotations and visualisations should be integrated into one file. This obviates the need for the viewer to go to remote data repositories (such as the PDB) to obtain the underlying data.

THE iSee CONCEPT

The SGC has collaborated with MolSoft LLC to develop a platform known as iSee [47-49] which, as a dissemination tool for structural biology data, attempts to address the above requirements. iSee integrates key data from all stages of the high throughput structural biology pipeline, from cloning details and molecular biology/ protein chemistry protocols to the full coordinates of the solved structures. The structural information is interpreted, visualised and annotated by experts of the protein families within the SGC, focussing on making the results as accessible as possible to a wide range of readers [48].

The iSee concept has been implemented using Molsoft's ICM platform [7, 8, 50] which enables the bundling of a large number of heterogeneous data types together as a single, small computer file and provide high quality graphical visualisations of that data. Each file – which we dub an 'iSee datapack' – is usually smaller than 10 Mb and can be downloaded freely from the SGC website [47]. At the time of writing, there are more than 500 iSee datapacks freely available covering around 600 human and parasite protein structures that have been solved by the SGC.

Each datapack can be accessed online or offline. To view these datapacks online within a web browser, a free plugin is available for the most common internet browsers and platforms such as Windows, Linux and Mac OS/X with Internet Explorer, Mozilla Firefox, Google Chrome and Safari. To view the same datapacks offline, a free standalone reader program can be downloaded for all the three operating systems mentioned above.

A datapack is made up of two major interface components: an annotation text pane corresponding to the textual element of a paper, and a graphical pane where high-quality 3D visualisations are shown (Fig. **2**). Within the text, words or phrases are linked in the style of web hyperlinks to the triggering of an animation or transition to a pre-staged scene in the graphical pane depicting important structural features which are discussed in the text. Each of these scenes (or 3D slides) is prepared by the datapack's author in parallel with the writing of the text. Successive clicks on different links trigger animated transitions which are computed 'on-the-fly' from one scene to another, helping the reader to maintain context within the structure and establish the spatial relationship between features. Importantly, the ability to calculate transitions upon request means that the reader has direct control over the flow of reading as the links can be clicked on in any order desired.

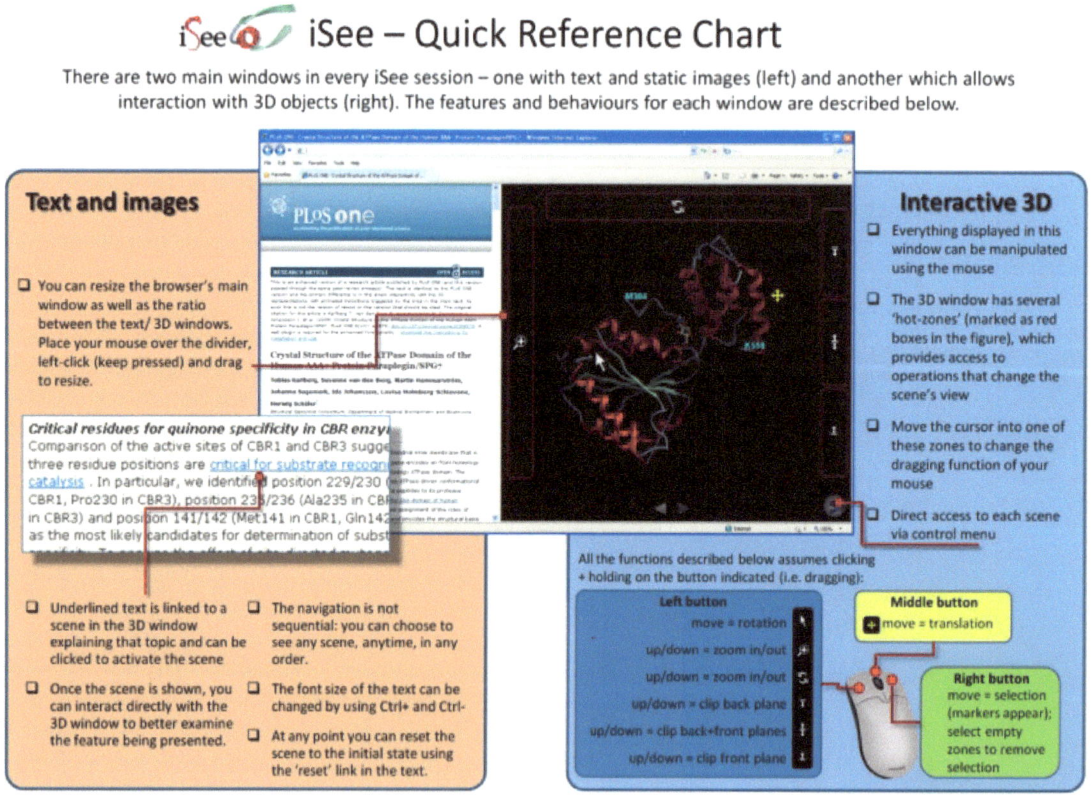

Figure 2: Main components of an iSee web-based session. On the left-hand side panel, text, static images, tables, etc. can be displayed. On the right-hand side panel, the interactive window allows the reader to directly manipulate 3D objects. The typical iSee session can be seen on-line (using a free web browser plugin) or offline (using the free standalone program ICM-Browser).

Every datapack contains all the data used in the analysis and annotation. By default, the datapacks are displayed in minimal mode for the non-specialist (as described above). For the specialist wishing to interact in more depth with the data contained in each datapack file, the free standalone reader software (ICM-Browser) can be used to work in 'expert mode' (Fig. **3**). This allows the reader to further explore the full body of data contained in the datapack in an offline fashion, using additional tools and methods such as interactive sequence and alignment viewers, alternative representations and mapping of properties like residue type and conservation, etc.

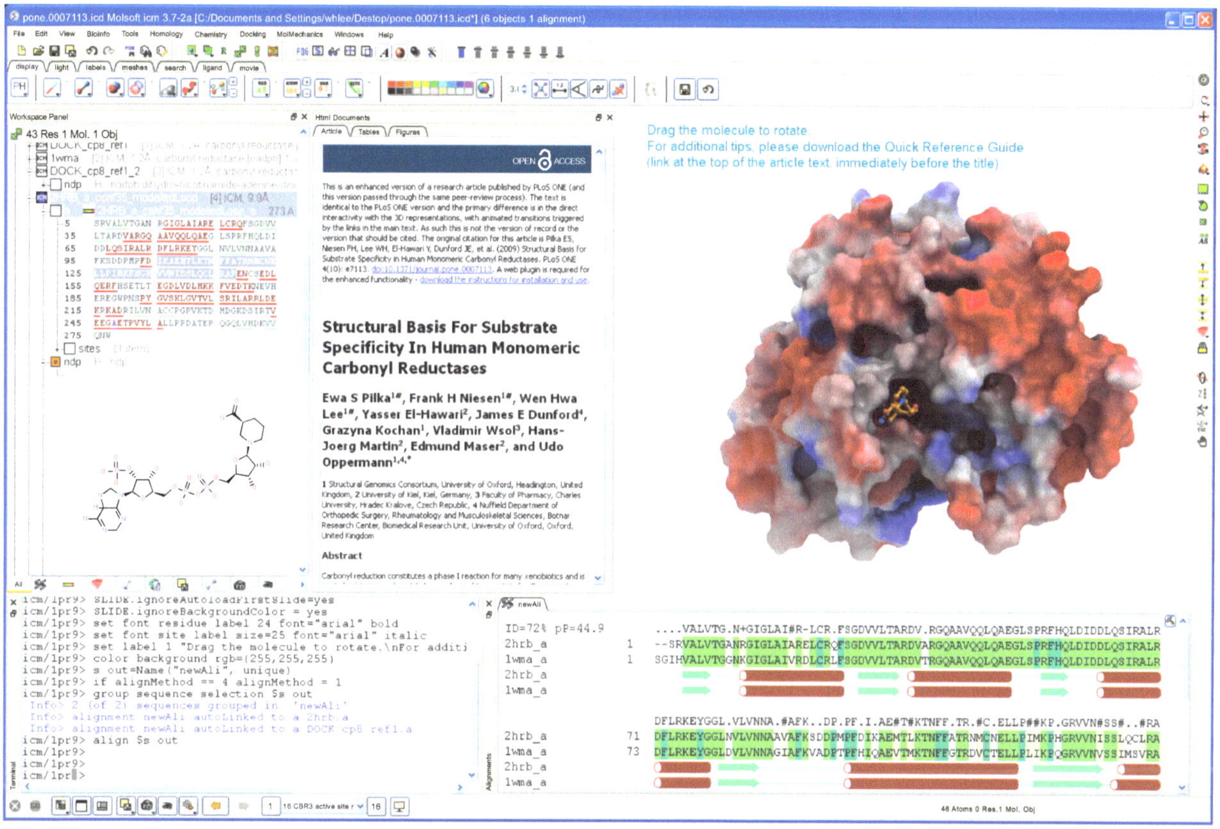

Figure 3: An iSee datapack being viewed in 'expert mode', using the standalone ICM-Browser program. In this mode, the specialist reader has access to the data used to the datapack (e.g. the sequences of the proteins being analysed, sequence alignments, chemical compounds, electron density maps, etc.). All relevant data types are linked – for instance selecting a region of a protein in the 3D graphics window will highlight the selected residues in the alignment window.

ELEMENTS OF iSee DATAPACK INTEGRATION AND VISUALISATIONS

An iSee datapack is generated and published in four main steps (Fig. **4**):

1. The capture of necessary data such as coordinates, sequences, compound structures etc.

2. The interpretation and annotation of the data. This includes the writing of the text component in parallel with the generation of visualisations in the form of slides

3. The saving of the datapack containing all the above elements as a single, compact, file

4. The publishing of the datapack as a 3D document either by sending the file via email to the recipient; by exporting to a web page which used the activeICM web plugin; by embedding the datapack within a Microsoft Office document such as PowerPoint using the activeICM activeX plugin

Currently, the activeICM plugin is directed towards 3D graphical visualisations and is wrapped within the text exported as HTML (for a web page, using Javascript to communicate the hyperlink clicks to the plugin) or in a PowerPoint slide (for example). The standalone ICM Browser platform can additionally visualise multiple sequence alignments, 2D chemical spreadsheets, tables and plots.

Figure 4: An example workflow of scientific data generation and publication.

The single 3D document and the possibility to export to web-delivered 3D content in this format provides a unique capability to easily distribute documents and electronic publications and to access them without direct web access in a platform-independent manner. Furthermore, write-access to the data contained in the datapack files can be controlled using user-defined passwords allowing sharing without compromising the integrity of the information. The datapacks can also be saved completely locked for reading (i.e. no content is shown until the reader enters the correct password).

Importantly in the context of SBLD, the ICM and activeICM software used by the current iSee implementation is able to visualise large and complex objects in high quality. The responsiveness of the visualisations is maximised by storing the data internally in a manner that does not require any further processing. The activeICM and ICM Browser software are written in C and C^{++} ensuring that optimal performance is achieved.

Multi-platform/Multi-browser Support

Supporting multiple platforms is essential as the lead discovery community is divided among the use of three main computer platforms, namely Windows, Mac OS/X and Linux. Despite the dominance of Windows, the market share of Mac OS/X and Linux in the research community is significant. In particular communities generating 3D content are significantly more dedicated to non-Windows platforms. For example the crystallographic community or researchers using electron microscopy tend to use Linux as their primary platform for data analysis and visualisation. Similarly, Mac OS/X has been steadily gaining popularity amongst scientists. However only a small minority of free 3D visualisation programs truly support multiple platforms, and only a fraction of those programs do so without any loss of functionality.

Taking different computing platforms into consideration, web-browser plug-ins can be divided into two main types. The first plug-in type is an ActiveX component for Internet Explorer and Microsoft Office integration; a second plug-in is written as Mozilla plug-in for Firefox and other compatible browsers such as Google Chrome and Safari. Finally, the plug-in code needs to be adjusted to a particular operating system to ensure uniformity in the final representation and operation. By supporting these two plug-in types, almost 100% percent of the "browsing community" can be covered.

3D Slides

A key benefit of 3D publication using the iSee method is the storage of a large number of different 3D visualisations (further referred to as 3D slides) along with the text and window layout. A 3D slide contains information about:

- a view point, e.g. rotation, translation, zoom, lighting and depth effects
- a set of atom-specific graphical representations such as surfaces, which can be represented as smooth, transparent or wireframe, ball-and-stick models, CPK (space-filling) models
- a set of atom, residue or distance labels on any of the atomic items
- a set of arbitrary 3D textual annotations using multiple styles and colours which is assigned to a point in space
- a set of arbitrary 2D annotations assigned to specific 2D coordinates on a screen
- parameters of the parametric animation (see below).
- (for the standalone method of publishing with ICM Browser) parameters that define the position and types of windows or panels that are shown

The metadata defining a slide is independent of screen resolution meaning that slides are always rendered at an optimal resolution and occupy minimal memory or file space.

Slides, when shown in the standalone ICM Browser platform, include information about the layout of the different elements/ tools/ data displaying panels in the application's window. In this mode, iSee represents a truly flexible communication tool, where the appropriate types of information can be visualised at the same time. For example a cellular assay image can be shown at the same time as the compound tested which itself is displayed within an SAR table and also in docked pose within the a receptor protein in the 3D graphics window.

Movie 1: example of animation/ transitions used to deliver 3D structural information. Double-click to launch. Different representations are shown here: the animation starts with a surface representation of an enzyme (a kinase) with an inhibitor and a peptide bound to it. It is followed by a rotation to show the overall shape and charge distribution. Then it proceeds to a zoomed view to show the ligand pockets being occupied, followed by a representation of the experimental electron density (yellow grid) seen for the ligands, allowing the researcher to validate the modelling. In a typical iSee session, the user can interact directly with the scene to explore additional views and features.

Animating Transitions Between and within Slides

The continuous and smooth transitioning of the 3D visualisations between slides is a key component of the iSee experience as it is performed in a manner that ensures that the viewer retains context of the spatial relationship between the two slides (Movie 1). The transitions are calculated on the fly, using a parametric method to calculate the time between frames, ensuring that the time taken between to transition between slides is platform-independent. The time taken to transition between slides is defined by the author and is commonly between half and one second unless gross movements are required whereupon a complex motion of more than one movement can be generated over a longer period of time. Conformational trajectories, illustrative or physical, with parametric interpolation such as displaying a transition from one loop conformation to another using a few snapshots of intermediate are also possible. Fading of elements between the slides is also possible and gives the viewer a clear dynamic appreciation of, for example, conformational changes within the visualised protein(s).

A further type of animation is a continuous movement around a current view, e.g. a rocking motion with different speeds and amplitudes and combinations of movements around X and Y axes, or rotations (with an infinite or fixed number of cycles).

Linking Text with 3D Slides

The linking of slides to relevant parts of the associated textual annotation is a trivial task during the authoring of a datapack and enables the viewer to click on the resulting hyperlink to cause a transition to the relevant slide.

Visualising Other Types of Structural Data

Visualisation of electron density derived from X-ray crystallographic experiments is not an intuitive operation especially when focussed around a discrete molecule such as a ligand. The ICM authoring platform is able to download maps from the Uppsala Electron Density Server [51], store and then visualize contours of density of important parts of the structure (e.g. ligand) with little storage or memory penalty.

For NMR experiments NOEs and other restraints used for model building can be dynamically shown upon user request. The NOEs are treated as a special type of three dimensional distance or torsion labels.

The ability to visualise experimentally or simulation-derived available conformational space (e.g. from NMR data) provides the viewer an intuitive understanding of the implied flexibility of a protein. Animations to show this, in contrast to simple discrete views, require more sophisticated techniques because intermediate conformations or frames must be calculated or interpolated for a smooth visualization. The approach employed in the ICM and activeICM platform is made up of a hybrid method. A small number of key intermediate frames are pre-generated and stored with the underlying structural data. Next, depending on the nature of those reference frames, the intermediate frames are generated to provide the smooth animation. The intermediate frames are stored in a compact manner and can be captured in internal coordinate space to provide a visual understanding of loop transitions in protein kinases or domain movements in which linkers undergo transitions and ligand:docking attempts.

Other Graphical Visualisations

The iSee implementation in the ICM and activeICM platform is able to store and display any arbitrary image, associated with any slide. For example, an image of a lipid bi-layer can be shown behind the structure of an integral membrane GPCR protein.

Any other three-dimensional object can be displayed by importing 3D files including OBJ (Wavefront), OFF (Object File Format), DAE (Collada) and 3DXML. An example where this is useful is to embed an integral membrane protein in the membrane of a cell, which is represented as a 3D shape with textures. The ICM browser also supports different types of stereo and can be directly used with a stereo-capable monitor.

Exporting Visualisations

The quality of the 3D visualisations can be altered within the ICM and activeICM platform. This is performed by enabling anti-aliasing where pixels are hinted depending on their intersection with drawn lines. In addition, by increasing the number and decreasing the size of triangles used to draw surfaces, complex objects are rendered in a smoother fashion. These enhancements and these of tiling on a virtual screen enable activeICM and ICM Browser to export high quality publication-ready images with larger dimensions than the screen. Images can be exported in PNG or JPEG format.

THE iSee CONCEPT ELSEWHERE

Since the release of the iSee concept, several implementations have been launched along similar lines. For instance Nature Structural Biology has adopted simple linking to an external 3D visualiser for articles with structures and associated PDB codes [52]; however this implementation is not intuitive to a non-structural biologist.

The Proteopedia initiative [53] employs a similar approach to the core of the iSee concept, whereby selected portions of an accompanying text are linked to interactive scenes. It is entirely web-based, employing Jmol [42] as the viewing engine. Its key innovation lies in the adoption of a community-authored Wiki model. This allows readers to find structures deposited at the PDB and author annotations to capture key scientific information about the protein. However, for the purposes of lead discovery, this solution might not appeal due to the nature of its unrestricted access on the web (both for reading and editing, although certain pages can be protected against unauthorised changes).

In September 2008 Kumar *et al.* [54] proposed and demonstrated the embedding and display of 3D objects within a PDF file. Their approach uses a pre-existing feature of the Adobe Acrobat package that has been little-used by the biomedical community (but commonly used by CAD/ engineering communities). Unfortunately, the process to produce embedded 3D objects within a PDF is non-trivial in the context of structural biology data. Moreover all the 3D objects using this format are merely described as surfaces and do not retain any of the underlying scientific information from the original structural data, making it suitable for simple viewing only. This is because the use of surfaces means that, apart from colour changes and viewpoints, every new interactive scene needs to be created as novel surfaces, at the expense of file size.

Recently publishers have indicated their awareness and desire to explore this field further. The initiative from Cell Press and Elsevier entitled 'Article of the Future' [55] explores the advantages of various modes in which a web-based publication of the future may be delivered. The Journal of Biological Chemistry also encourages the use of 3D PDFs to communicate structural information [56]. We welcome these initiatives and expect that more will come to light as the demands of authors and readerships change, aligning closely with the iSee concept.

PLoS ONE: iSee CONCEPT IN PEER-REVIEWED PUBLISHING

In October 2009 PLoS ONE, published by the Public Library of Science, launched a new collection of articles entitled 'Structural Biology and Human Health: Medically Relevant Proteins from the SGC' [57]. This collection contains a continually expanding series of articles documenting many of the novel protein structures determined by the SGC and work to further characterise their function and chemical biology. The most innovative aspect is the fact that the articles are published as enhanced interactive web-based papers using activeICM and the iSee concept (Fig. **5**), in addition to the standard HTML and PDF formats.

An important aspect of this PLoS ONE Collection is that, for the first time to our knowledge in an open-access journal, the peer-review process of these articles is carried out upon the enhanced non-paper based iSee datapack versions themselves. This allows reviewers to take advantage of the information integrated in a single file, in addition to the 3D data including live 2D data attachments such as chemical or standard spreadsheets and plots, alignments, etc. Additionally, this also saves the referees' time in that there is no immediate need to retrieve, analyse and validate the structural data, for these are all included in the iSee datapack itself.

The iSee versions of the articles in this collection are open-access, using the Creative Commons Attribution license [60] in common with all PLoS ONE articles, ensuring that the full content can be further extended if any reader wishes to do so. This allows other organisations to capture and even extend the information published in the articles and integrate it into novel platforms (e.g. electronic books).

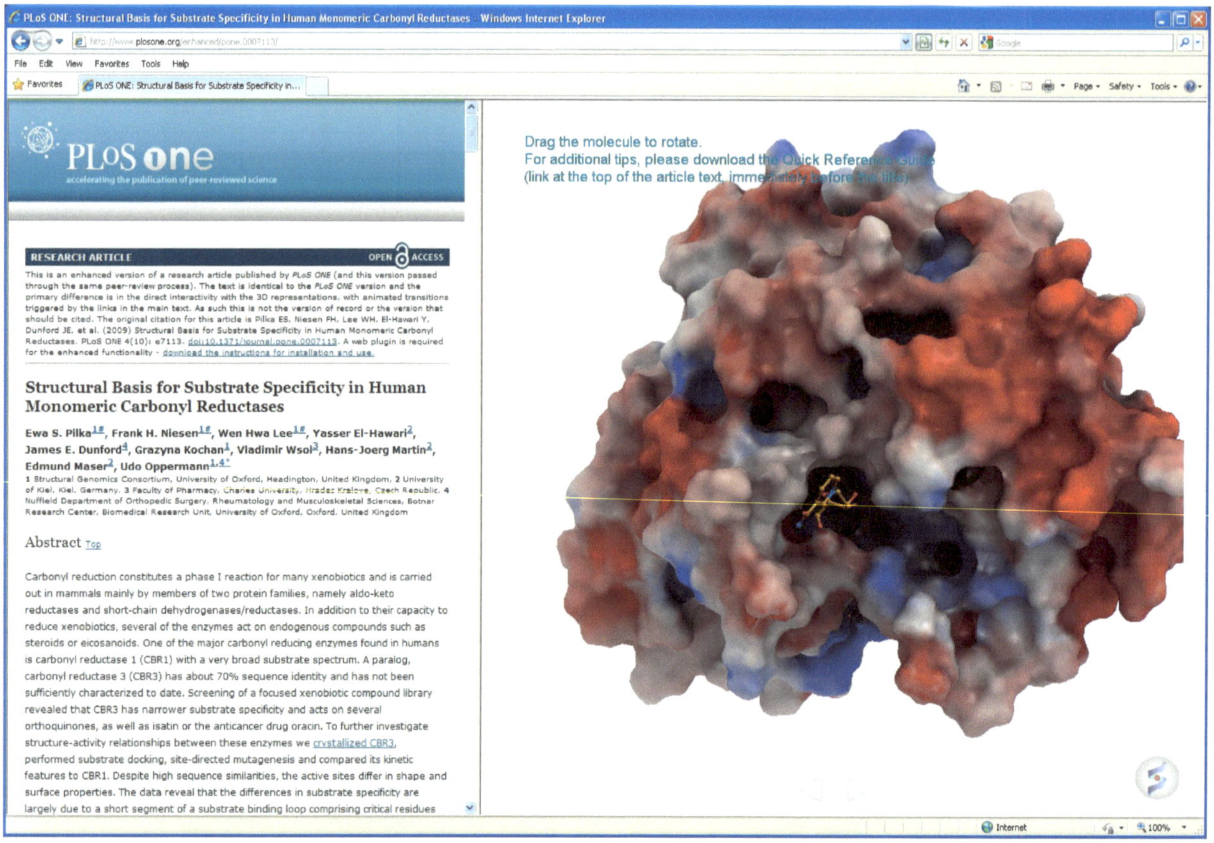

Figure 5: Example of the iSee concept built into an existing publishing platform (PLoS ONE Collection: Structural Biology and Human Health: Medically Relevant Proteins from the SGC [57]; example shown above is authored by Pilka *et al.*[58]). The interactive article can be accessed using the freely available activeICM plugin [59].

Both the web-based/PDF and the enhanced iSee versions will remain constant in terms of content and location. This is critically important for future reference and also archival purposes. For technical questions and feedback we have implemented and maintain an iSee wiki [61] to address the most common questions and to provide usage and authoring tips.

THREE-DIMENSIONAL DOCUMENTS - THE FUTURE

The iSee datapack files are currently provided in a binary file format to enable maximal compactness and fast download to the browser. A full text file format would increase the files' size resulting in relatively slow downloads into the browser/plugin and would ensure that the embedded data and visualisations are able to be suitably archived and curated in a platform-independent manner. We look forward to working with the visualisation communities to assist in defining such a file format.

With the recent success of publication of iSee datapacks in a peer-reviewed and fully electronic manner, we expect to see other journals and publishers showing increasing interest in the iSee approach as the 'paper of the future' is defined.

Finally, we are aware of a number of academic institutions who use the iSee approach as a teaching tool. Not only to allow students to gain a better understanding of particular proteins or molecular systems, but also as a platform where they may be able to write their own datapacks as assignments.

CONCLUSION

The appropriate delivery of data is as important as the generation of that data in the first place – unintepretable data is as useful as no data at all. Within the SBLD field the challenges of visualising and delivering heterogeneous types of data to an audience with a wide range of discrete skills and potentially located in diverse locations makes this task even more challenging. We believe that technology, such as described within this chapter, is now available for SBLD scientists to not only integrate data but also analyse, annotate and disseminate in an intuitive and integrated electronic medium.

ACKNOWLEDGEMENTS

We are grateful to Michael Sundström of CPR Copenhagen, Aled Edwards, Chas Bountra, Cheryl Arrowsmith, Johan Weigelt and Matthieu Schapira of SGC for their encouragement and support of the project. We wish to thank Eugene Raush and Maxim Totrov for their key efforts in the implementation of the iSee concept in activeICM and ICM. The SGC is a registered charity (Number 1097737) that receives funds from the Canadian Institutes for Health Research, the Canadian Foundation for Innovation, Genome Canada through the Ontario Genomics Institute, GlaxoSmithKline, Karolinska Institutet, the Knut and Alice Wallenberg Foundation, the Ontario Innovation Trust, the Ontario Ministry for Research and Innovation, Merck and Co., Inc., the Novartis Research Foundation, the Swedish Agency for Innovation Systems, the Swedish Foundation for Strategic Research, and the Wellcome Trust.

REFERENCES

[1] Symyx.

[2] Stein, S.E.; Heller, S.R.; Tchekhovskoi, D. In *An Open Standard for Chemical Structure Representation: The IUPAC Chemical Identifier*, International Chemical Information Conference, Nimes, France, 2003; Infonortics: Nimes, France, 2003; pp 131-143.

[3] Callaway, J.; Cummings, M.; Deroski, B.; Esposito, P.; Forman, A.; Langdon, P.; Libeson, M.; McCarthy, J.; Sikora, J.; Xue, D.; Abola, E.; Bernstein, F.; Manning, N.; Shea, R.; Stampf, D.; Sussman, J. Protein Data Bank Contents Guide: Atomic Coordinate Entry Format Description Version 2.1 (draft). In 1996.

[4] Bourne, P.E.; Berman, H.M.; McMahon, B.; Watenpaugh, K.D.; Westbrook, J.D.; Fitzgerald, P.M. Macromolecular Crystallographic Information File. *Methods Enzymol.* **1997,** 277, 571-90.

[5] Agrafiotis, D.K.; Alex, S.; Dai, H.; Derkinderen, A.; Farnum, M.; Gates, P.; Izrailev, S.; Jaeger, E.P.; Konstant, P.; Leung, A.; Lobanov, V.S.; Marichal, P.; Martin, D.; Rassokhin, D.N.; Shemanarev, M.; Skalkin, A.; Stong, J.; Tabruyn, T.; Vermeiren, M.; Wan, J.; Xu, X.Y.; Yao, X. Advanced biological and chemical discovery (ABCD): centralizing discovery knowledge in an inherently decentralized world. *J. Chem. Inf. Model.* **2007,** 47(6), 1999-2014.

[6] Cho, S.J.; Sun, Y.; Harte, W. ADAAPT: Amgen's data access, analysis, and prediction tools. *J. Comput. Aided Mol. Des.* **2006,** 20(4), 249-61.

[7] Abagyan, R.A.; Totrov, M.M.; Kuznetsov, D.N. ICM - a new method for protein modelling and design. Applications to docking and structure prediction fro the distorted native conformation. *J. Comp. Chem.* **1994,** 15, 488-506.

[8] *ICM - http://www.molsoft.com.*

[9] *Schrodinger - http://www.schrodinger.com.*

[10] Villoutreix, B.O. Structural Bioinformatics, Chemoinformatics & Virtual Ligand Screening.

[11] *GraphPad Prism - http://www.graphpad.com.*

[12] Spotfire - http://spotfire.tibco.com/.

[13] Edwards, A. Large-scale structural biology of the human proteome. *Annu. Rev. Biochem.* **2009,** 78, 541-68.

[14] Terwilliger, T.C.; Stuart, D.; Yokoyama, S. Lessons from structural genomics. *Annu. Rev. Biophys.* **2009,** 38, 371-83.

[15] Terwilliger, T.C.; Waldo, G.; Peat, T.S.; Newman, J.M.; Chu, K.; Berendzen, J. Class-directed structure determination: foundation for a protein structure initiative. *Protein Sci.* **1998,** 7(9), 1851-6.

[16] Sali, A. 100,000 protein structures for the biologist. *Nat. Struct. Biol.* **1998,** 5(12), 1029-32.

[17] Janin, J. Structural genomics: winning the second half of the game. *Structure* **2007,** 15(11), 1347-9.

[18] Protein Structure Initiative.

[19] The Protein Structure Factory.

[20] Structural Proteomics in Europe

[21] Protein 3000 and RIKEN Structural Genomics/ Proteomics programmes.

[22] Chandonia, J.M.; Brenner, S.E. The impact of structural genomics: expectations and outcomes. *Science* **2006,** 311(5759), 347-51.

[23] Congreve, M.; Murray, C.W.; Blundell, T.L. Structural biology and drug discovery. *Drug Discov. Today* **2005,** 10(13), 895-907.

[24] Thompson, W.J.; Fitzgerald, P.M.; Holloway, M.K.; Emini, E.A.; Darke, P.L.; McKeever, B.M.; Schleif, W.A.; Quintero, J.C.; Zugay, J.A.; Tucker, T.J.; *et al.* Synthesis and antiviral activity of a series of HIV-1 protease inhibitors with functionality tethered to the P1 or P1' phenyl substituents: X-ray crystal structure assisted design. *J. Med. Chem.* **1992,** 35(10), 1685-701.

[25] von Itzstein, M.; Wu, W.Y.; Kok, G.B.; Pegg, M.S.; Dyason, J.C.; Jin, B.; Van Phan, T.; Smythe, M.L.; White, H.F.; Oliver, S.W.; *et al.* Rational design of potent sialidase-based inhibitors of influenza virus replication. *Nature* **1993,** 363(6428), 418-23.

[26] Wlodawer, A. Rational approach to AIDS drug design through structural biology. *Annu. Rev. Med.* **2002,** 53, 595-614.

[27] Gileadi, O.; Knapp, S.; Lee, W.H.; Marsden, B.D.; Muller, S.; Niesen, F.H.; Kavanagh, K.L.; Ball, L.J.; von Delft, F.; Doyle, D.A.; Oppermann, U.C.; Sundstrom, M. The scientific impact of the Structural Genomics Consortium: a protein family and ligand-centered approach to medically-relevant human proteins. *J. Struct. Funct. Genomics* **2007,** 8(2-3), 107-19.

[28] The SGC.

[29] Edwards, A. Open-source science to enable drug discovery. *Drug Discov. Today* **2008,** 13(17-18), 731-3.

[30] Edwards, A.M.; Bountra, C.; Kerr, D.J.; Willson, T.M. Open access chemical and clinical probes to support drug discovery. *Nat. Chem. Biol.* **2009,** 5(7), 436-40.

[31] Weigelt, J. The case for open-access chemical biology. A strategy for pre-competitive medicinal chemistry to promote drug discovery. *EMBO Rep..* **2009,** 10(9), 941-5.

[32] Protein Data Bank.

[33] PSI Assessment Panel. http://www.nigms.nih.gov/News/Reports/PSIAssessmentPanel2007.htm

[34] Autin, L.; Tuffery, P. PMG: online generation of high-quality molecular pictures and storyboarded animations. *Nucleic Acids Res* **2007,** 35(Web Server issue), W483-8.

[35] Hodis, E.; Schreiber, G.; Rother, K.; Sussman, J.L. eMovie: a storyboard-based tool for making molecular movies. *Trends Biochem. Sci.* **2007,** 32(5), 199-204.

[36] Humphrey, W.; Dalke, A.; Schulten, K. VMD: visual molecular dynamics. *J. Mol. Graph.* **1996,** 14(1), 33-8, 27-8.

[37] Maiti, R.; Van Domselaar, G.H.; Wishart, D.S. MovieMaker: a web server for rapid rendering of protein motions and interactions. *Nucleic Acids Res.* **2005,** 33(Web Server issue), W358-62.

[38] Porollo, A.; Meller, J. Versatile annotation and publication quality visualization of protein complexes using POLYVIEW-3D. *BMC Bioinformatics* **2007,** 8, 316.

[39] *YASARA - http://www.yasara.org.*

[40] PyMol - http://www.pymol.org.

[41] OpenEye VIDA - http://www.eyesopen.com/products/applications/vida-vivant.html.

[42] Jmol - http://jmol.sourceforge.net.

[43] Chen, V.B.; Davis, I.W.; Richardson, D.C. KiNG (Kinemage, Next Generation): A versatile interactive molecular and scientific visualization program. *Protein Sci* **2009**.

[44] Richardson, D.C.; Richardson, J.S. The kinemage: a tool for scientific communication. *Protein Sci.* **1992,** 1(1), 3-9.

[45] Kaplan, W.; Littlejohn, T.G. Swiss-PDB Viewer (Deep View). *Brief Bioinform.* **2001,** 2(2), 195-7.

[46] FirstFlance - http://molvis.sdsc.edu/fgij.

[47] SGC iSee. http://www.thesgc.org/iSee

[48] Abagyan, R.; Lee, W.H.; Raush, E.; Budagyan, L.; Totrov, M.; Sundstrom, M.; Marsden, B.D. Disseminating structural genomics data to the public: from a data dump to an animated story. *Trends Biochem. Sci.* **2006,** 31(2), 76-8.

[49] Lee, W.H.; Atienza-Herrero, J.; Abagyan, R.; Marsden, B.D. SGC--structural biology and human health: a new approach to publishing structural biology results. *PLoS One* **2009,** 4(10), e7675.

[50] Raush, E.; Totrov, M.; Marsden, B.D.; Abagyan, R. A new method for publishing three-dimensional content. *PLoS One* **2009,** 4(10), e7394.

[51] Kleywegt, G.J.; Harris, M.R.; Zou, J.Y.; Taylor, T.C.; Wahlby, A.; Jones, T.A. The Uppsala Electron-Density Server. *Acta Crystallogr. D Biol. Crystallogr.* **2004,** 60(Pt 12 Pt 1), 2240-9.

[52] We're living in a 3D world. *Nat. Struct. Mol. Biol.* **2006,** 13(2), 93.

[53] Hodis, E.; Prilusky, J.; Martz, E.; Silman, I.; Moult, J.; Sussman, J.L. Proteopedia - a scientific 'wiki' bridging the rift between three-dimensional structure and function of biomacromolecules. *Genome Biol.* **2008,** 9(8), R121.

[54] Kumar, P.; Ziegler, A.; Ziegler, J.; Uchanska-Ziegler, B. Grasping molecular structures through publication-integrated 3D models. *Trends Biochem. Sci.* **2008,** 33(9), 408-12.

[55] Marcus, E. Cell launches a new format for the presentation of research articles online. http://beta.cell.com/index.php/2010/01/cell-launches-article-of-the-future-format

[56] 3D presentation of structural and image data. *J. Biol. Chem.* **2009,** 284(32), 21101.

[57] PLoS ONE Collection: Structural Biology and Human Health: Medically Relevant Proteins from the SGC. http://www.ploscollections.org/article/browseIssue.action?issue=info:doi/10.1371/issue.pcol.v02.i04

[58] Pilka, E.S.; Niesen, F.H.; Lee, W.H.; El-Hawari, Y.; Dunford, J.E.; Kochan, G.; Wsol, V.; Martin, H.J.; Maser, E.; Oppermann, U. Structural basis for substrate specificity in human monomeric carbonyl reductases. *PLoS One* **2009,** 4(10), e7113.

[59] activeICM plugin. http://www.molsoft.com/activeicm.html

[60] Creative Commons Attributon License. http://creativecommons.org/licenses/by/3.0/

[61] Lee, W.H.; Marsden, B.D.; Atienza-Herrero, J. iSee Wiki. http://www.whatisisee.org

Index

ABC family – transport proteins, 144-162

adenosine receptors, 168

adenosine triphosphate (ATP), 31-33

ADMET, 1, 6, 110

    absorption, 1, 5, 11, 110, 144

    ADMET Prediction, 1, 6

    BBB, See blood brain barrier

    bile, 10, 145

    bioavailability, 4, 144

    blood brain barrier, 11

    carcinogenicity, 6

    distribution, 1, 4, 5, 11, 110, 144

    efflux , 144

    excretion, 1, 2, 5, 10, 110, 144

    Lipinski rule of, 4, 6,11, 65, 66

    lipophilicity, 4, 5, 11, 147-150

    metabolism, 2, 4, 5, 10, 110, 152

    models, 5, 10, 11

    orally bioavailable, 4, 5, 8, 65, 127

    permeability, 4, 5, 10

    plasma proteins, 11

    plasma/serum, 11

    toxicity, 1, 4, 6, 10, 11, 110, 126, 131, 132

    transporters, 11, 144, 146

    urine, 10

administration, 2, 8, 65

Alchemical approaches, 106

Alzheimer's disease, 32, 165, 167

apoptosis, 11, 130

ATP, 31, 32, 33, 42, 144-146, 148, 152, 153, 155, 156, 164-167

BCRP, See breast cancer resistance protein

binding affinity, 10, 21, 23, 37, 61, 77

breast cancer resistance protein, 144

cancer, 2, 8, 11, 31, 32, 99, 100, 118, 127-130, 144, 145, 164, 167, 178

cardiovascular, 164, 168

cavities, 69, 88-91, 120, 124, 125

chemical libraries, 1, 2, 5, 8, 9, 33, 36, 48, 49, 53, 56

clinical trials, 1, 2, 11, 20, 60, 73, 127, 131, 132

CNS 171

coagulation, 2, 31, 76

CoMFA, 147, 169

www.ingramcontent.com/pod-product-compliance
Lightning Source LLC
Chambersburg PA
CBHW041701210326

41598CB00007B/491